Why Study Apollonius?

"[Apollonius's *Conics*] is one of the greatest scientific books of antiquity." "[Apollonius was a] giant, not simply as compared with men of antiquity, but even with men of all times. ... The ingenuity that enabled him to discover so much ... is truly admirable. Such achievements pass our imagination, they are almost weird."

George Sarton,
from *An Introduction to the History of Science*
and *A History of Science*

"If we want to read for ourselves authors like Kepler and Newton, or if we want to understand the significance of the Cartesian mathematics that has shaped the world we live in and shapes our minds as well—either way, whether to understand the past in its own terms or to understand the present as a deliberate transformation of the past—we need to study Apollonius."

Harvey Flaumenhaft
Dean, St. John's College, Annapolis

"Apollonius towers up as one of the greatest geometers of all time."

Eric Temple Bell,
Mathematics: Queen and Servant of Science

"His work is a *tour de force* of intellectual ingenuity in the Euclidean tradition, and continued to influence the work of mathematicians such as Newton."

Lancelot Hogben, *Mathematics in the Making*

"By pure geometry, Apollonius arrived at the properties of conics which we nowadays express by equations. … From our analytic geometry of conics he had clearly very little to learn except the notation… He worked with what is virtually an equation of the sixth degree in x and y, or its geometrical equivalent—in its day a wonderful feat."

Herbert Westren Turnbull, *The Great Mathematicians* in *The World of Mathematics*

"He who understands Archimedes and Apollonius will admire less the achievements of the foremost men of later times."

Liebniz

The reader of Apollonius "will see that there are some matters which no mind, however gifted, can present in such a way as to be understood in a cursory reading. There is need of meditation, and a close thinking through of what is said."

Kepler

Apollonius of Perga

Conics

Books I–IV

The Eight Books of Apollonius's *Conics*

Of the original eight books of Apollonius's *Conics*, the first four presented the elements of conic sections and the last four contained what he called "a fuller treatment." Eutocius (ca. 500 CE) edited and commented on Books I–IV, and these books (with the commentary) are available in the original Greek. The 1891–1893 edition of the Greek text edited by J. L. Heiberg is listed in the bibliography.

Books V–VII did not survive in Greek, but were preserved in an Arabic translation. In 1710 Edmond Halley published the *editio principes* of the Greek text, produced a Latin translation of Books V–VII, and attempted a reconstruction of the lost Book VIII. Books V–VII have also been published in Arabic with an English translation by Gerald Toomer (see the Bibliography).

In the late 1930s, in response to St. John's College's need of an English version, R. Catesby Taliaferro translated Books I–III. The other books were omitted from this translation because they did not contain any of the propositions regarded as fundamental to the study of mathematics and its historical development as it was (and is) approached in the St. John's College Great Books Program.

Sixty years later, Michael Fried filled the gap in the translations by bringing out Book IV in English. Originally published by Green Lion Press in a separate volume, Fried's translation is now included in the present publication, bringing English translations of all the surviving Greek texts of the *Conics* into a single volume.

More recently, Roshdi Rashid, Micheline Decorps-Foulquier, and Michel Federspiel have completed a massive project, publishing, in six volumes, a new Greek text of Books I–IV, the first complete text of the Arabic version, editions of a number of related Arabic texts, and French translations of all of the texts, with mathematical and philological commentaries. See the Bibliography for details.

—The Green Lion

Apollonius of Perga
Conics: Books I–IV

The entire surviving Greek text
in the only English translation

Books I–III translated by R. Catesby Taliaferro
Book IV translated by Michael N. Fried

Green Lion Press

Santa Fe, New Mexico

Manufactured in the United States of America.
Printed and bound by Sheridan Saline.

Published by Green Lion Press.
www.greenlion.com.

Cataloging-in-Publication Data:

Apollonius of Perga
Conics / by Apollonius of Perga
        Books I–III translated by R. Catesby Taliaferro,
        Book IV translated by Michael N. Fried,
        Introductions by Harvey Flaumenhaft and Michael N. Fried,
        Diagrams by William H. Donahue (Books I–III),
        and Michael N. Fried (Book IV).

Includes index, bibliography, introduction, and notes.

ISBN: 978-1-888009-41-5 (sewn binding with paper cover)
ISBN: 978-1-888009-40-8 (library binding)

1. Apollonius of Perga, Conics, English. 2. History of mathematics.
3. Conic sections. 4. Geometry. 5. History of Science.
I. Apollonius of Perga (born circa 262 BCE). II. Taliaferro, R. Catesby
(1907–1989). III. Donahue, William H. (1943–). IV. Fried, Michael N.
(1960–). V. Title.

Library of Congress Catalog Card Number 2013935366

Contents

The Green Lion's Preface   vii

Introductory Essay: *Approaching the Study of Apollonius*
  by Harvey Flaumenhaft   xiii

Translator's Note to Books I–III
  by R. Catesby Taliaferro   xxxi

Illustrator's Note to Books I–III
  by W. H. Donahue   xxxiii

Conics Book I   1

Conics Book II   117

Conics Book III   181

Translator's Introduction to Book IV
  by Michael N. Fried   267

Conics Book IV   289

Appendix A: *On Three- and Four-Line Loci*
  by R. Catesby Taliaferro   353

Appendix B: Corollary to I.38 and Revised Version   363

Bibliography   367

Index   371

About the Author   375

About the Translators   375

The Green Lion's Preface

The present volume unites, for the first time, English translations of the four books of Apollonius's *Conics* that survived in the original Greek (for the remaining books, see our note on the Eight Books of Apollonius's *Conics,* opposite the title page). The Taliaferro translation was originally done to serve the needs of the New Program at St. John's College in the 1930s, which required only the first three books. Green Lion Press undertook a careful revision of this translation and issued it in 2000. Book IV was translated more recently by Michael N. Fried, who offered it to Green Lion Press for publication. Largely because of a tight publication schedule for *Conics* I–III, it was decided to publish *Conics* IV as a separate volume, with the idea of eventually bringing them out together. That time has now come.

The original Green Lion editions, and the present combined edition, feature an index locating the first appearance, definition, and construction of key terms and geometrical entities, an introductory essay by Harvey Flaumenhaft, and a bibliography. The book is printed on high quality paper that will not show through or tear when notes are written, with a sewn binding so pages will not fall out when the book is opened flat.

Since the history and aims of the two translations were somewhat different, they will be described separately below.

Conics Books I–III

R. Catesby Taliaferro's translation of Apollonius's *Conics* was published in three previous editions, first by St. John's College, Annapolis, in their *The Classics of the St. John's Program* series and then by Encylopædia Britannica in the *Great Books of the Western World* series as part of Volume 11, and finally in a new revised Green Lion Press edition in 2000. All those editions are now out of print.

The Green Lion Press edition of the Taliaferro translation corrected a very large number of errors that appeared in the previous edition, both in text and diagrams. Improvements were made in the formatting, such as repeating diagrams so that they are present on every spread containing text of the associated proofs, enlarging the proposition numbers and including them in the running heads so that it is easier to locate particular propositions, and providing more space for notes between propositions, around diagrams, and in the form of wider margins. Diagrams were not only corrected but in many instances improved in significant ways.

Some translations, especially of the enunciations, were reworked to follow more conventional English style and to remedy often confusing statements where syntax and antecedents were formerly reconstructed only by recourse to the "setting out" and "specification" sections of the proof. This was always done in careful consultation with the Greek text.*

However, we did not undertake to retranslate or revise the translation generally, and what we have done has been done with caution and care. Even adding bracketed repetitions to supply antecedents has to be done with a thorough understanding of the work as a whole, a scrupulous attention to the Greek text, and an alertness to the ways Apollonius uses certain Greek phrases and constructions to refer to certain lines in previous and subsequent propositions. The reconstruction of antecedents is not child's play here. Furthermore, one might think one has reconstructed an antecedent by knowing the propositions and relationships between the lines in conic sections, but sometimes Apollonius has left things general or vague because he has not yet established which lines do what, or what limitations or relationships will emerge. In all such cases we have followed Taliaferro in his admirable literalness in leaving the statements as general as Apollonius wrote them.

In places the order of lines or points cited by Apollonius doesn't correspond to the order of correlate entities identified in constructions; some suggested "correcting" this. But here too the text was left as Apollonius wrote it. The great strength of Taliaferro's translation is its transparency as a window to the Greek text: what the translation says can be relied upon to be what the Greek says. It seemed better not to compromise this feature for minor pedagogical convenience.

The only change made to the Heiberg text is the moving of an apparently interpolated but unlabeled corollary, notorious for causing confusion, to an appendix, duly documented in the text.

Many citations of Euclidean or previous Apollonian propositions were corrected from those offered in the Britannica edition, and many others were added where it seemed they were needed. However, no attempt was made to check all the citations provided with the original translation, nor has there been any attempt to provide citations for every step.

---

* See *Apollonii Pergaei Quae Graece Exstant* in the Bibliography

New diagrams were drawn by William H. Donahue; his Illustrator's Note presents his account of the history of diagrams for this text and the changes he made from the previous edition. One comment is in order here about the general approach to providing diagrams for the *Conics*. We are dealing with three-dimensional geometric figures and representing them on a two-dimensional surface. Many of the propositions use constructions which do lie in one plane; these present no problem. Others require a third dimension; there are two ways to approach their representation.

They could be drawn with no attempt to indicate which lines are supposed to be out of the plane of the paper, while trying as much as possible to let all lines, in whatever plane they actually lie, look equal when supposed to be equal, and (again, when possible) letting lines making up sides of figures that are equal be measurably of the right length to do so *as drawn* in the plane of the paper. The drawback to this approach is that it can be difficult to see which lines are out of the plane of the paper; it may even misleadingly suggest that lines in different planes are in the same plane.

Alternatively, they could be handled using perspective drawing, which is clearer in indicating the different planes, bringing out the third dimension. As with even the simplest perspective drawing, such as the classic cube, to do this some equal lines must be drawn unequal; some lines measured along their projections on the plane of the paper will not have their true ratios. As long as one remembers that it is three dimensions being represented, and follows the proof using Apollonius's argument line rather than measuring lines in the diagram, this seems the less confusing and misleading method. Therefore in this edition we have used primarily the perspective view for the three-dimensional figures, with some diagrams flat where that was helpful and not misleading.

Many people provided us with suggestions for corrections and improvements to the pre-publication version of that revised edition, which was used at St. John's College during the 1997–98 school year and to the 2000 Green Lion edition. Corrections and other suggestions were sent to us by John Ross, John Michael MacDonald, Thomas Scally, Lynda Myers, Joe Sachs, Krishnan Venkatesh, Maximilian Gruner, James Wilkinson, Amanda Fuller, Cary Stickney, Claudia Honeywell, Stephen Houser, and Maren Cohn. Jim Wilkinson provided a particularly thorough list of corrections both of the old edition and of the interim edition. Our thanks to all of them. A second printing in 2002 included minor revisions; the Green Lion thanks David Bolotin, Eisso J. Atzema, and Michael N. Fried for their comments and suggestions.

A *corrigenda* sheet was compiled over the years by St. John's College tutors and students identifying a number of typos and mistranslations in the previous edition; corrections of these have been incorporated in the Green Lion editions as appropriate.

The Green Lion editors (Dana Densmore and William H. Donahue) evaluated each suggestion, often in consultation with the Heiberg Greek text. Only changes that seemed right after that consideration were implemented. In some cases the suggestion revealed a problem with a citation or footnote, but the implemented correction was different from the one suggested. Some suggested corrections which appeared plausible were not implemented either because the questioned word or construction was present in the Greek text and appeared not to be a textual error or because they were based on a misunderstanding of the translation. Where possible, in these latter cases, the translation was emended to make misunderstanding less likely, or, in the former, a footnote confirms that the Greek text has been followed.

Our thanks go also to others who helped in various ways. We gratefully offer special thanks to Nadine Shea for consultation and advice about type-setting software and book design. David Derbes sent us helpful information about the bibliographical history of *Conics*. Brother Robert Smith directed us to a microfilm of a 16th century edition that gave us good ideas for im-provements in the diagrams. Joe Sachs secured the microfilm and mailed it to us. Grant Franks provided a helpful consultation on copyright law.

Krishnan Venkatesh and Maximilian Gruner put many hours into proof-reading the text; their alert eyes caught much. Max Gruner also carefully examined all the new diagrams, working through both the interim and the final versions, and made many suggestions.

We thank the libraries on both campuses of St. John's College and especially librarians Kathryn Kinzer, Inga Waite, and Laura Cooley. The Greenfield Library in Annapolis loaned us their microfilm of the 1566 Commandino Latin edition of Apollonius and provided photocopied material from the 1939 St. John's College edition of the Taliaferro translation. The Meem Library in Santa Fe gave us two long-term loans of the 1891 Heiberg Greek text. We especially thank librarians Inga Waite, Laura Cooley, and Heather MacLean. Thanks go also to our Hodson Trust Internship Program summer intern Kathleen Kelly.

St. John's College Deans Eva Brann, Harvey Flaumenhaft and James Carey facilitated our efforts in a number of ways. The Deans' administra-tive assitants, Terry McGuire and Penny Russell, tracked down some elusive details and helped us coordinate with the College curriculum.

Finally, we thank Harvey Flaumenhalf for providing the useful and insightful essay inviting the modern reader into the study of Apollonius.

Conics Book IV

In 2002, the Green Lion presented, at long last, the hitherto untranslated Book IV of Apollonius's *Conics*, in Michael Fried's new translation, making all four of the books surviving in Greek available in English for the first time (three additional books survive only in Arabic). In his translation of Apollonius for St. John's College in the 1930s, R. Catesby Taliaferro, in the interest of brevity, and perhaps influenced by Sir Thomas Heath's low opinion of Book IV, chose to include only the first three of the four books that have survived in the original Greek. Books V–VII, which have survived only in an Arabic translation, have been retranslated into English by Gerald R. Toomer. Book VIII did not survive in any form.

In 1999, when we were preparing a new edition of Taliaferro's translation of *Conics* I–III, Joseph Cohen of St. John's College approached us with the idea that we might include a translation of Book IV, which his friend and former student Michael N. Fried had recently completed. The idea of course appealed to us. However, at that point we were too far along in production of Books I–III, with classes waiting for the book, to append the new Book IV translation in that edition. After some discussion, and with encouragement from Mr. Cohen, we decided to publish *Conics* IV as a separate volume, designed as a companion to *Conics* I–III.

Although an earlier version of the present translation was appended to Michael N. Fried and Sabetai Unguru, *Apollonius of Perga's Conica: Text, Context, Subtext* (for which see the bibliography), the translation has been thoroughly reviewed for this edition, and a number of significant changes have been made, improving both the accuracy and the clarity of presentation.

We retained the same layout as we provided for Books I–III, with generous margins and space around diagrams and between propositions. Although Apollonius's text was (as far as we know) written in narrative form with continuous paragraphs, we have followed Taliaferro in displaying statements of equality or proportionality on separate lines, centered. The translation departs from Taliaferro in eschewing modern mathematical symbolism, however, for reasons presented in the translator's introduction. As in all Green Lion books, notes are true footnotes rather than end notes, to make them easier to use.

Michael Fried's introduction argues convincingly that *Conics* IV has been misunderstood by most previous scholars, who were unable to shed the prejudices of their modern mathematical training. Once we appreciate the problems, and understand what Apollonius was trying to accomplish in this book, we can see it more nearly as the ancients saw it, as a fitting companion to the first three books, concluding the course in the "elements" of conic sections, as Apollonius put it.

The diagrams for Book IV were drawn by Michael Fried, and occasionally edited by William Donahue. They are based on the diagrams included in Heiberg's Greek text.

2013 Edition

When it became clear that the time had come for a consolidated edition of *Conics I–IV,* we were faced with a truly daunting task. There was no single page layout file for the entire book. Instead, we had the original 1998 edition, in an obsolete format, and we had correction pages that had been inserted over the years for several reprintings, also in old formats. In a few cases, we had the uneditable output files but not the underlying source files. It seemed at first a nearly impossible undertaking, given our other responsibilities.

It was Associate Editor Howard J. Fisher who saved the day. He painstakingly assembled all the surviving files, collated them, and determined which represented the latest version. He then converted them all to a current page layout program, and re-entered the portions that were missing. Next, he changed the fonts in Book IV to match those in Books I–III, and re-worked all the page breaks and diagram placements in that book. In the course of making the changes, he read over the entire combined book and made sure it formed a harmonious whole. Truly, this edition would not have come to be without his dedication.

The 2013 edition benefits from many suggestions and corrections to Books I–III made by alert reader Paul Kunkel. The Green Lion is most grateful for his thoughtful reading, and for his having taken the time to alert us to needed corrections and opportunities for improvement.

Dana Densmore and William H. Donahue
for the Green Lion Press

Approaching the Study of Apollonius
by Harvey Flaumenhaft

There's not much to be said about the life of Apollonius. Approximately half a century after the time that Euclid flourished, and a quarter-century after Archimedes was born—that is, about two-and-a-half centuries BCE—Apollonius was born in Perga, a small Greek city in southern Asia Minor. Later, he lived for a while in Alexandria, and visited Pergamum and Ephesus. In addition to the geometrical work which made him famous, his writings included works on optics and on astronomy, in which—according to Ptolemy—he made an essential contribution to what became the geometrization of the study of the stars. Reports about a number of his geometrical writings survive in the writings of other ancient authors, and there is a loose Arabic version of the two books of his *Cutting off of a Ratio*. Of the eight books of his *Conics*, the last is lost and the fifth through seventh exist only in an Arabic translation.

Thus, all that survives of Apollonius in the original Greek is the first half of the *Conics*—Books I through IV. About this surviving record of the working of his mind, there is much to say, though not much can be said here.*

In approaching the study of Apollonius, it's useful to know how his predecessors had studied the curves called conic sections. They obtained the curves upon a plane that cuts a right cone, and then they characterized the curves by using the unique diameter that makes right angles with the chords that it bisects. But that diameter, which Apollonius calls "the axis," is only one of the diameters he uses. Apollonius obtains the curves upon a plane that cuts any cone at all (whether it is a right cone, or is one that's oblique); and he characterizes them by using any diameter at all (whether it is the unique one that meets at right angles the chords that it bisects, or is one of those that meets them obliquely).

---

* For more, however, see my forthcoming *Insights and Manipulations: Classical Geometry and Its Transformation—A Guidebook. Volume I: Starting up with Apollonius; Volume II: From Apollonius to Descartes* .

That generalizing was not, however, the whole of his contribution to the study of conics. While many of the particulars that Apollonius presented in the first four books of his *Conics* (and others that he omitted) were known to his predecessors, he selected and arranged them in such a way as to provide so deep an insight and so generalized and systematized an overview that his work became the classic text on the curves which—following his innovative terminology—came to be called the "parabola," the "hyperbola," and the "ellipse." Other ancient treatises on those curves did not survive, and for some two thousand years no new treatise on the conics took the place of the one that had been written by Apollonius. It was Apollonius who was studied by, and who must be studied in order to understand, such writers as Kepler, Descartes, and Newton.

After Descartes, however, while the names of those curves persisted, and while they continued to be called collectively "the conic sections," they nonetheless eventually ceased to be studied as such. After Descartes, that is to say, they came to be studied algebraically—rather than as lines that are first obtained upon a plane by cutting a cone in various ways, and then are characterized by the relative sizes and shapes of certain boxes formed by associated straight lines standing in certain ratios. But although Descartes' *Geometry* may be in fact what it has been called—the greatest single step in the progress of the exact sciences—no one could see it *as* that without studying Apollonius.

So if we want to read for ourselves authors like Kepler and Newton, or if we want to understand the significance of the Cartesian mathematics that has shaped the world we live in and shapes our minds as well—either way, whether to understand the past in its own terms or to understand the present as a deliberate transformation of the past—we need to study Apollonius.

Apollonius tells his tale obliquely. He doesn't give us questions, but rather gives us only answers that are too hard to sort out and remember unless we ourselves figure out what questions to ask. About Apollonius as a teacher we must ask whether his work is informed by wisdom and benevolence.

Descartes, for one, didn't think so. In his own *Geometry* , and in his sketch of rules for giving direction to the native wit (especially in the 4th rule), Descartes found fault with the ancient mathematicians.

Descartes severely criticizes them—for being show-offs. He says that they made analyses in the course of figuring things out, but then, instead of being helpful teachers who show their students how to do what they themselves had done, they behaved like builders who get rid of the scaffolding that has made construction possible. Thus they sought to be admired for conjuring up one spectacular thing to look at after another, without a sign of how they might have found and put together what they present.

Descartes also suggests, however, that they did not fully know what they were doing. They didn't see that what they had could be a universal method. They operated differently for different sorts of materials because they treated materials for operation as simply objects to be viewed. Hence they learned haphazardly, rather than methodically, and therefore they didn't learn much. Mathematics for them was essentially a matter of wonderful spectacle rather than being material for methodical operation. The characteristic activity of the ancient mathematician was the presentation of theorems, not the transmission and application of the ability to solve problems. They had not discovered the first and most important thing to be discovered: the significance of discovery. They had not discovered the power that leads to discovery and the power that comes from discovery. They were not aware that the first tools to build are tools for making tools. They were too clever to be properly simple, and too simple to be truly clever. They were blinded by a petty ambition. Too overcome by their ambition, they couldn't be ambitious on the greatest scale.

Was Apollonius as a teacher liable to that criticism by Descartes? Was Apollonius guilty of the obtuseness of which Descartes, at the beginning of his *Geometry* accused classical mathematicians—was he guilty of the desultory fooling-around and disingenuous showing-off of which Descartes had accused them earlier, in his *Rules*? It remains to be seen.

In any case, readers of Apollonius gain access to the sources of the tremendous transformation in thought whose outcome has been the mathematicization of the world around us and the primacy of mathematical physics in the life of the mind. Scientific technology and technological science have depended upon a transformation in mathematics which made it possible for the sciences as such to be mathematicized, so that the exact sciences became knowledge par excellence. The modern project for mastering nature has relied upon the use of equations, often represented by graphs, to solve problems. When the equation replaced the proportion as the heart of mathematics, and geometric theorem-demonstration lost its primacy to algebraic problem-solving, an immense power was generated. It was because of this that Descartes' *Geometry* was called the greatest single step in the progress of the exact sciences. To determine whether it was indeed such a step, we need to know what it was a step from as well as what it was a step toward. We cannot understand what Descartes did to transform mathematics unless we understand what it was that underwent the transformation. By studying classical mathematics on its own terms, we prepare ourselves to consider Descartes' critique of classical mathematics and his transformation not only of mathematics but of the world of learning generally—and therewith his work in transforming the whole wide world.

Readers of Apollonius will gain a knowledge of the character of classical mathematics through an introductory treatment of the conic sections. It is in the classical study of the conic sections that the modern reader can most easily see both the achievement of classical mathematics and the difficulty that led Descartes and his followers to turn away from classical mathematics.

It has to do with ratio, and with notions of number and of magnitude. For Apollonius, as for Euclid before him, the handling of ratios is founded upon a certain view of the relation between numbers and magnitudes. When Descartes made his new beginning, almost two millenia later, he said that the ancients were handicapped by their having a scruple against using the terms of arithmetic in geometry. Descartes attributed this to their not seeing clearly enough the relation between the two mathematical sciences. Before modern readers can appreciate why Descartes wanted to overcome the scruple, and what he saw that enabled him to do it, they must be clear about what that scruple was.

Readers must, at least for a while, make themselves at home in a world where how-much and how-many are kept distinct—a world which gives an account of shapes in terms of geometric proportions rather than in terms of the equations of algebra. For a while, readers must stop saying "AB-squared," and must speak instead of "the square arising from the line AB"; they must learn to compound ratios instead of multiplying fractions; they must not speak of "the square root of 2."

The study of Euclid's definitions and theorems conveys an awareness of the foundation covered over by the later modern quantitative superstructure which was built upon it, and in which modern readers are so used to dwelling. With this foundation in Euclid's *Elements*, readers can move through the First Book of Apollonius's *Conics*. The road may be arduous, but reading Apollonius is essential for appreciating the character of the transformation of classical into modern mathematics. To fully understand the movement from Apollonian to Cartesian mathematics, however, the reader who has finished Book One must then go on to Books Two and Three of the *Conics,* and must consider in addition other authors ancient and modern: Pappus, Diophantus, Viète, and Descartes. Parts of Books Two and Three, along with passages from Pappus on loci, lead to the geometrical problem-solving that is the point of departure for Descartes' *Geometry*; and the classic numerical problem-solving of Diophantus and the innovative algebraic art of Viète present the problem-solving matrix of mathematical modernity.

Descartes' *Geometry* is not a book of theorems and their demonstration. By homogenizing what is studied, and by making the central activity the manipulative working of the mind, rather than its visualizing of form and its insight into what informs the act of vision, Descartes transformed mathematics into a tool with which physics can master nature. He went public with his project in a cunning discourse about the method of well conducting one's reason and seeking the truth in the sciences; and this discourse introduced a collection of scientific try-outs of this method, the third and last of which was his *Geometry*.

Among the questions that readers will consider, as they study Apollonius in a world transformed by Descartes, are the following: What is the relation between the demonstration of theorems and the solving of problems? What separates the notions of "how-much" and "how-many"? Why try to overcome that separation by the notion of quantity as represented by a number-line? What is the difference between a mathematics of proportions which arises to provide images for viewing being, and a mathematics of equations which arises to provide tools for mastering nature? How does mathematics get transformed into what can be taken as a system of signs referring to signs—as a symbolism which is meaningless until applied, when it becomes a source of immense power? What is mathematics, and why study it? What is learning, and what promotes it?

With minds shaped by the thinking of yesterday and of the day before it, we struggle to answer the questions of today, in a world transformed by the minds that did the thinking. We shall proceed more thoughtfully in the days ahead if we have thought through that thinking for ourselves. We can get some help in thinking for ourselves by reading Apollonius.

As you begin to study Apollonius—as you proceed through Book One of the *Conics* and look back on the steps that you're taking—the effort required will make it difficult for you to understand what he's doing. Apollonius himself seems to be stingy with his help. Perhaps his austerity was a favor to his readers at a time when books were few and life was more leisurely for serious students; nowadays, however, the result of such austerity is a scarcity of readers who'll spare the time to read him, and even those very few who do read him haven't the time to get nearly as much out of it as they would get if they had more help. With that in mind, I'll now supply some help. It won't be very useful before you read the text, but I hope that it will be of service to serious readers after they've made some effort to understand this very forbidding but very important writer. Here, then, to help you begin to study the *Conics*, are some suggestions for understanding the beginning of Book One.

First definitions

Cones had been defined in Euclid's *Elements* prior to the work of Apollonius. Euclid's definitions are, however, different from those of Apollonius. When Euclid gives definitions of solid figures at the beginning of the Eleventh Book of the *Elements*, he says that a cone is the figure comprehended when, taking one of the sides about the right angle in a right-angled triangle, you keep the side fixed and carry the triangle all the way around to the same position from which you began to move it. The cone's axis is the straight line which remains fixed, about which the triangle is turned; and the base is the circle swept out by the straight line which is carried round. Euclid, like Apollonius, defines a cone by generating it.

Apollonius, defining a conic surface by generating it, gives the conic surface before he gives the cone, and then he puts the cone together—out of a conic surface and a circular planar surface that serves as the base of the cone. Euclid gives the cone without giving the conic surface at all—thus leaving a conic surface to be merely what is left from the surface of a cone after you take away the part that is planar, the circular base. Euclid in his *Elements* was interested in a definite solid, whereas Apollonius here will be interested in cones and in conic surfaces only insofar as they are means for getting certain lines.

A *conic* surface is not the same thing as a *cone's* surface. A *cone's* surface is heterogeneous: a cone has two kinds of surface—one of them being conic, while the other one (the base) is planar. A *conic* surface has two parts, on opposite sides of the vertex, each one of which is itself a conic surface. One of the two surfaces is generated by the part of the moving straight line that extends above the fixed point; and the other one of the two surfaces, by the part of the moving straight line that extends below the generative circle. But although the movement of the straight line about the circumference of the circle generates a conic surface in which there are two surfaces, these surfaces are both of the same kind—conic.

Having shown the generation of cones, Apollonius considers their kinds. Not all cones here spoken of by Apollonius are *right* cones. Consider, by contrast, the cones with which Euclid is concerned in the definitions at the beginning of the Eleventh Book of his *Elements*. There, as we've seen, Euclid generates a cone by rotating a right triangle (as he generates a sphere by rotating a semicircle). He differentiates one kind of cone from another by the angle at the vertex, which depends upon the relative sizes of the legs of the generating right triangle. If the straight line that remains fixed be equal to the remaining side about the right angle that's carried round, the cone will be right-angled; if less, obtuse-angled; and if greater, acute-angled. Euclid's right-angled cone, and his obtuse-angled cone, and his acute-angled cone, are all right cones.

From Apollonius you can get right cones—when the axis is set up right. That is, the line drawn from the fixed point on the generating straight line, straight to the center of the generative circle, must be upright with respect to the plane of that circle. When the axis is upright, the cone is right. "Upright" and "right," with all their connotations, translate the same Greek term *orthos*, from which we get such words as "orthopedic" and "orthodox." The term is used in contrast to what is here translated as "oblique," *skalênos*, and could also be translated as "uneven." From Apollonius you can get not only cones that are right cones but also ones that are not—ones that stand unevenly rather than upright.

Before Apollonius, as has been said, right cones were all that mathematicians made use of to obtain curved lines that were conic sections. They obtained such lines by cutting into a right cone with a plane perpendicular to the hypotenuse of the generating right triangle in one of its positions. When the right cone was right-angled, then the plane would cut into the cone a line of one sort; but when the right cone was obtuse, then the line of section was of another sort; and it was of yet another sort when the right cone was acute .

It was Apollonius who introduced cuts at any angle (cuts made obliquely as well as at right angles) in any cone (oblique as well as right). It was also Apollonius who introduced the names "parabola" and "hyperbola" and "ellipse" for the sections. In the course of the *Conics* , he shows why he did so.

Note that a right cone is radially symmetrical, but an oblique cone is not. That is to say: a right cone, unlike an oblique cone, is the same all the way around. An oblique cone slants; it tilts to the side. Even an oblique cone, however, won't tilt to the side when it is viewed from a certain direction. When you view it from that direction, it tilts either directly toward you or directly away from you, rather than to the side. And so, we can say that—with respect to that direction—an oblique cone is bilaterally symmetrical, since the right-hand side of its appearance is the same as its left-hand side. A right cone, being radially symmetrical, could be said to be bilaterally symmetrical from *every* direction.

Whether or not a cone is a right cone, its nonplanar surface is the surface with which Apollonius will deal, and in dealing with that conic surface, he'll make use of the fact that even if its cone is oblique, and therefore is not symmetrical radially, that cone is at least symmetrical bilaterally.

But although he'll be dealing with the conic surface, he won't be studying it as such. He'll be dealing with the conic surface in order to study certain curved lines—as is suggested by the later items in the set of first definitions.

These deal with various sorts of diameters. Later, the 7th proposition, when read in the light of these first definitions, shows that if Apollonius had restricted himself to right cones as Euclid does, then the only diameters that his conic sections could have had to begin with would have been those that he'll soon call "axes."

The first definitions show that attention is going to be directed not to cones and conic surfaces as such, but to curved lines that they yield for study. A key to this study will be the relation between those curved lines called "conic sections" and certain other lines that are straight.

Cones may be interesting three-dimensional figures, and there may be various reasons for studying them, but the very title of the book indicates that the subject of the book is "conics"—"things conic"—rather than cones as such. The cone seems to be of interest to Apollonius here because its generation combines the straight line and the circular line, so that the cone, when cut by planes which are oriented in various ways, will yield as intersections various nonstraight lines which are also noncircular. These lines are the simplest of the nonstraight lines that are also noncircular; after the circle, they are the simplest of the curved lines.

The only lines whose study is presupposed by Apollonius in the First Book of his *Conics* are those that are either straight or circular. Starting from these, which are the only sorts of lines in Euclid's *Elements* , Apollonius engages the reader in a study of lines which—as we'll see—either are *not straight* but nonetheless are indefinitely prolongable (like lines that *are* straight), or else are *not circular* but nonetheless are closed (like lines that *are* circular).

In approaching the study of such curves obtained by intersection, readers should consider the cone in cross-section. They should do so not merely because two dimensions at a time are easier to visualize than three, but also because a planar view of the cone from above and a planar view of the cone from beside, taken together, will exhibit the two aspects of interest in exploring the sections that can be cut in a cone. From above, the cone in outline will come to view as the simplest figure that is bounded by a curved line—a circle. From beside, it will come to view as a triangle—the simplest figure bounded by straight lines. These two views will reveal the ratios that lurk within and can illuminate the character of the conic sections.

The definitions that precede the propositions begin with the conic surface and go on to deal with certain relations between straight lines and curves of an unspecified sort. The study of the propositions that follow them presupposes that we have completed an elementary Euclidean study of geometry—a study of figures involving lines that are all either straight or circular. With a knowledge of the straight line and the circle, we are prepared to study other lines—the simplest lines that are neither straight nor circular.

By putting *together* the straight line and the circle, we can generate a conic surface—a surface that is neither flat nor spherical; and if we cut that curvy surface with a plane surface, we shall get (as the intersection of the surfaces) some kind of a curvy line. By cutting the conic surface with a plane, we cut a conic section into the plane. We can begin to study this new kind of line in the plane by considering first the conic surface from which it was generated. And we can study this surface by considering the old kinds of lines which generated it. Both of the old kinds of lines (the straight lines and the circles) have an easy simplicity about them. There are no bumps or dips in either of them; neither of them has a ripple. But what about the conic surface generated by the two of them together: what is it like? Apollonius tells us in the propositions with which he begins.

The first ten propositions

First, Apollonius shows that a conic surface isn't wiggly in any direction—that [Prop. 1] it's like a straight line in being flat (if you go from a point on it toward its vertex); and that [Prop. 2] it's like a circle in being everywhere curvy bulging outward (if you go from a point on it toward another point on it, but without going straight toward its vertex).

Then (by cutting through the vertex, and also through the cone) he obtains a conic section that is straight-lined (being a triangle) [Prop. 3], and another conic section that is circular (by cutting so as to make an angle equal to one of the base angles: either parallel [Prop. 4] or subcontrariwise [Prop. 5]).

Then come propositions that act as a bridge from old lines (straight and circular) to new ones; the first of them [Prop. 6] is preparatory—it gives a bisecting surface; and the second [Prop. 7] gives the bisecting line that's prerequisite for presenting lines of section that are neither straight nor circular, and its porism tells us that there are in fact new ones, and identifies what it is that will tell us about them.

Then he obtains (by cutting only one side of a single surface, however much extended) [Prop.8] a conic section that is like a straight line in being indefinitely extendable, but is unlike it in having a diameter (being curved a certain way); and he obtains (by cutting both sides of a single surface—extended if necessary—but so as not to make an angle equal to one of the base angles) [Prop.9] another conic section that is like a circle in being closed but is unlike it in not having its perpendicular half-chord be a mean proportional between the segments into which it splits the diameter.

Finally [Prop. 10], he shows that a conic section is not wiggly.

In the next few propositions, Apollonius will present several varieties of the conic sections that are lines of a new sort. The first ten propositions thus constitute a prologue to the exhibition of these new lines.

The prologue begins by showing that the conic surface is nowhere wiggly, whether we view it from beside (1st proposition) or view it from above (2nd proposition); and immediately after that we are shown (3rd proposition) that the lines cut into a cone's surface by a plane that passes through the vertex, far from wiggling, will not even curve. The prologue ends by showing (10th proposition) that the lines cut into a cone's surface by a plane that doesn't pass through the vertex will, like the conic surface into which they are cut, curve without wiggling. Thus, the prologue's closing proposition shows for the conic section what the opening two propositions showed for the conic surface—that it's nowhere wiggly. Why is it important thus to emphasize the lack of wiggliness? Because from it we learn why it's a plane with which we cut the conic surface to get our first noncircular curves. Cutting the conic surface with a plane will guarantee that the noncircular curves that we get are simpler than the noncircular curves that we might get upon a conic surface in some other way. A line that's free of wiggles is simpler than a wiggly one. If a line that's drawn upon a conic surface is wiggly, then it cannot be cut into the conic surface with a plane—that is to say, we cannot slide it out for examination upon a flat surface.

In between the propositions about nonwiggliness that open and close the prologue, Apollonius places propositions from which we learn what lines of section the conic surface can yield when cut by a plane: not only lines of section that are perfectly straight or perfectly round, but also interesting hybrid lines that have properties like those that belong to the elemental lines—the straight and the circular. The conic surface gives us interesting hybrid lines because it is itself a hybrid surface—flat in one way and round in the other. Viewed from above, it's like a stack of larger and larger circles; viewed from beside, it is like a ladder of larger and larger similar triangles.

To say much more about the various new sorts of curved lines that can be obtained by cutting a cone, we must compare the sizes of straight lines in them. How so?

Well, the cone when viewed from *above* is like a stack of circles. If we slice ourselves a circle from the stack, we can (by using the perpendicular half-chord's mean-proportionality with respect to the segments into which it splits the diameter) get an equality of square and rectangle.

And the cone when viewed from *beside* is like a stack of similar triangles. If we slice ourselves a couple of triangles from the stack, we can (by using the proportionality of the straight lines that are the sides, and compounding the same ratios of different straight lines) get the equality of different rectangles.

By putting together the results of these two kinds of views—the straight-line aspect and the circular, it's possible to give an account of the lines that are formed in a plane when the plane intersects a cone so that the lines that are cut are neither straight nor circular.

It's possible to do so by making use of the equality of two rectangles to present in brief the proportionality of four straight lines, and by making use of the ratio of two rectangles to present in brief the ratio put together out of two ratios of straight lines.

The 11th, 12th, and 13th propositions

With the presentation of these propositions that follow the propositions of the prologue, the study of the conic sections has not yet left its beginnings behind, though it has defined the sections.

Apollonius gets hold of these curved shapes by articulating the relationship between the sizes of certain straight lines. By complicated
manipulation, he shows how the relationship of magnitudes informs the
look of figures.

Apollonius seems to have been the one who named these curved lines
according to whether the rectangles thrown out from their abscissas fall
alongside, or fall beyond, or fall short of the straight line he contrives and
calls the "upright" side. He is therefore also likely to have been the first to
derive this line for all three sections. What led him to do it, he does not say.

The lines that he contrives, and the *eidê* that they constitute for us to
look at, are displays of *logoi* . What informs the looks of the figures—what
he manipulates in order to show them—is the relationship in size of certain
straight lines that are attached to the lines of conic section. His work is done
with ratios, in two parts.

In the first part, what must be considered is what takes place in a circle.
The ordinate is the mean proportional between the segments into which the
line parallel to the axial triangle's base is split by the point where the ordinate meets the abscissa. In other words, the ratio of the one segment to the
ordinate is the same as the ratio of the ordinate to the other segment.

In the second part, what must be considered is what takes place in that
straight-line figure, the axial triangle. The one segment has a constant ratio
to the abscissa; and the other segment either is itself constant (in the case of
the parabola), or else, while not constant itself, has a constant ratio—either
to a whole line one part of which is the constant transverse side and the other
part of which is the abscissa (in the case of the hyperbola), or to a part of a
line the whole of which is the constant transverse side and the other part of
which is the abscissa (in the case of the ellipse).

The heart of the matter in defining the conic sections, then, is this: the
ordinate is the mean proportional between two lines—one of which is in
constant ratio with the abscissa, and the other of which is either constant
itself, or else is in constant ratio with the line resulting from the constant
transverse side when the abscissa is added on to it or taken away from it.

From the *top*-view we get the mean proportionality of the *ordinate*, and
from the *side*-view we get the constancy of the relationships that involve
the *abscissa*.

In all three of the propositions, when we speak of the ordinate's being
a mean proportional, it's easy to miss the centrality of the *sameness of two
ratios* in the *top*-view equality of square and rectangle (the former being
the square whose side is the ordinate, and the other being the mediating

rectangle whose sides are the segments into which the ordinate splits the line that runs across the axial triangle parallel to the base).

And in all three of the propositions, in the *side*-view equality of rectangles (one of them being the mediating rectangle, and the other one being the rectangle whose side is the abscissa), it's easy to miss the fact that the heart of the matter is the *constancy of two ratios*. Each of the two ratios involves one of the segments into which the ordinate splits the line that runs across the axial triangle parallel to the base (these segments being the sides of the mediating rectangle). The central ratios are buried deep within the body of the demonstration, where they mediate between the upright side's defining ratio (which is what we start with) and the ratio that contains as one of its terms the mediating rectangle that is equal to the abscissa-sided rectangle (which is therefore equal, in the end, to the ordinate-sided square).

There is something else that is not put in a prominent place but is central to the *side*-view equality. At the heart of the matter in distinguishing the sections is the constancy of another ratio. In the side-view part of the proposition for the parabola, the middle part flows without any intermediate step, directly from the upright side's defining ratio to the ratio of rectangles that's required for the concluding part. But in the side-view part of the propositions for the hyperbola and the ellipse, there is an intermediate step in the flow: another constant ratio mediates the movement from the upright side's defining ratio to the ratio of rectangles.

We have to labor much to glimpse what lies behind the complicated presentation in these last few propositions where Apollonius defines the conic sections. At the center of it all, as we have seen, is the sameness of certain ratios of straight lines, and the constancy of certain other ratios of straight lines. The straight lines are not of constant size, but their ratios are expressed in terms of certain constant other lines which are brought in for a purpose that is not made clear in the definitional propositions. That's what is responsible for much of the complexity which makes it hard to see what Apollonius is doing in these propositions.

Next will come the last of the definitional propositions, the 14th. Something more than definition will be glimpsed in it. And after that, Apollonius will present forty-six more propositions in this first book on conics. While it would take too long to consider here the way that they're put together, it is possible here to state briefly the chief reason why the rest of Book One is needed.

The ordinates of the sections are related to their abscissas by the equality and inequality of certain rectangles which have them as sides (and which involve certain features of other lines as well—their sizes, their ratios, and the shapes of certain rectangles that they make). It can be shown (and later will be shown) that these relations of ordinate and abscissa are peculiar to the sections. These are properties, that is to say, which characterize the curves—characteristics which they do not share with other lines.

But something can have several distinctive characteristics. In the circle, for example, both of the following are characteristic properties: (1) the equality of the distances to a given point (called the circle's center) from every one of the points of the containing line; and (2) the equality of the following two figures which can be made from straight lines in the circle: the square arising from a half-chord perpendicular to the circle's diameter, and the rectangle contained by the segments into which the half-chord splits the diameter.

As a definition of the circle, the first property is more like the inarticulate primary apprehension of perfect roundness—but it is a step toward the second property. What if we started with the second property? We might define the circle as a closed line such that: the perpendicular from each point of the closed line to a given straight line can give rise to a square which is equal to a rectangle contained by the segments into which the perpendicular splits the given straight line. From that definition, we could demonstrate as a theorem that the distance from any point on the curved line to the midpoint of the given straight line is equal to the distance from any other point on the curved line to that same midpoint. We could proceed in this way with the circle, but there doesn't seem to be any good reason to do so. What about the other conics?

In the conics, which of the following is essential—which of them says what the conic is: the particular way that the plane cuts the cone, or the particular square-rectangle property that relates the ordinate and abscissa?

Perhaps we should say that the cutting merely gives us the line—and that once the line has been given already, then the square-rectangle property will characterize it. We can determine whether a given line is a conic section of a determinate sort, by testing whether the square-rectangle property holds for it; or, if we are given a conic section of a determinate sort, we can infer as a consequence that the square-rectangle property holds for it. This would be like the situation in the *Elements*: Euclid's circle is given by the postulate; and his definition is applied as a test or consequence to a line that has already been given.

But the names of the sections (of the new lines, that is to say—the sections other than the ones that are circular or straight-lined), the very names of the sections refer, not to the looks of the curved line itself, but rather to the *eidos* formed by straight lines which are contrived in it. We may have the "wheel" (which in Greek is *kuklos*, usually translated "circle"), but we do not have the "egg"; we have, rather, the "ellipse." We want to name the curves, not by how we generated them, but by how they look—that is to say, by an *eidos* of some kind—and the "ellipse," together with the "parabola" and the "hyperbola," might be even harder to name without the square-rectangle property.

Apollonius names the sections neither by saying how they are generated, nor by saying what they look like. He names them, rather, by the looks of figures that he erects alongside them.

Apollonius has shown us the curves that we get by making conic sections. Do the looks that belong to those conic sections belong only to conic sections? Apollonius hasn't told us. For all we know, such looks may belong also to some curves which are not conic sections.

In other words, Apollonius has shown this: if a curve is obtainable by cutting some cone with a plane, then the curve will be a circle, or a parabola, or an hyperbola, or an ellipse—but he has not shown the converse, which is this: if a curve is a circle, or a parabola, or an hyperbola, or an ellipse, then it must be obtainable by cutting some cone with a plane.

Eventually, Apollonius will show this converse. His preparations for showing it will take up most of the rest of the First Book. What he has exhibited in these propositions defining the conic sections—namely, the figures that he has erected alongside the curves that he has cut into cones—these turn out to provide perhaps the best point of departure for showing the converse of the defining propositions. There is in fact a less difficult way to show the "if" by itself, but there isn't a more elegant way to show the "if" than by setting us on the way to showing the "*only* if" also. Nothing shows the elegance better than the set-up that Apollonius prepares in the last of the defining propositions, the one in which he presents the opposite sections: the 14th.

The 14th proposition

By the end of the 13th proposition, Apollonius has exhausted the possibilities for a plane's cutting of lines into a cone's nonplanar surface.

We can see this by using dichotomy successively, as follows: The plane either (I) goes through the vertex or (II) doesn't. If (I), then either (A) the plane does not cut through the surface or (B) it does. Possibility (I-A) yields a straight line, and (I-B) yields two intersecting straight lines. If (II), then either (C) the plane is parallel to the base or (D) it isn't (but its common section with the base is perpendicular to the axial triangle's common section with the base). Possibility (II-C) yields a circle. If (II-D), then either (i) the plane cuts subcontrariwise or (ii) it doesn't. Possibility (II-D-i) yields a circle. If (II-D-ii), then either (a) the plane is parallel to the axial triangle's other side or (b) it isn't (but meets the axial triangle's other side). Possibility (II-D-ii-a) yields a parabola. If (II-D-ii-b), then the plane either (1) meets the other side's extension beyond the vertex or (2) meets the other side below the vertex. Possibility (II-D-ii-b-1) yields an hyperbola, and possibility (II-D-ii-b-2) yields an ellipse.

Why, then, does Apollonius continue cutting? Well, into what does he cut now in the 14th proposition? Not into a cone. Instead, he cuts into a conic surface—or, rather, into vertically opposite conic surfaces, since the generation of a conic surface generates a surface with two parts, these two parts being two surfaces with the same shape, sharing a point.

Only once before did he cut into a conic surface rather than a cone. That was in the 4th proposition, when he cut into a conic surface parallel to its generating circle, and the new circle which he thus obtained became the base of a cone which the newly-cut circle made with the part of the conic surface above the cut. Why does he now again cut into a conic surface and not into a cone—or, rather, why does he cut into vertically opposite conic surfaces, rather than cutting into two cones put together at their vertices?

Suppose that he put two cones together at their vertices. Will he then, by cutting into them, get two sections that have the same size and shape, or are even of the same kind? Not necessarily. When is it that he can be sure that the two sections will be the same? When the surfaces of the two cones are vertically opposite surfaces.

When the surfaces are vertically opposite surfaces, then the axial triangles will be similar (since their bases will be parallel, and the angles at their vertices will be vertical angles). For this reason, the one plane in this case can cut into both surfaces only by cutting a hyperbola into both of them.

Moreover, in this case not only must the two sections be hyperbolas—they must also have their upright sides equal. Why? Because the single plane which cuts a hyperbola into each of the two surfaces will give them a transverse side which is common, as well as giving them axial triangles which are similar—and so, the same magnitude (the common transverse side of both sections) has the same ratio to the upright side of each section; and hence the upright sides are equal.

And why is that ratio (of transverse side to upright side) the same for both sections? Because in each section, the ratio of the transverse side to the upright side is compounded of ratios of lines in that section's own axial triangle; but the axial triangles here are similar, and in similar triangles corresponding lines have to each other the same ratio.

But why bother to do what is done in this 14th proposition? After all, the two lines of section are both of the same kind, and this is a kind of section which we have seen before; moreover, each of the curves has exactly the same size and shape as the other curve (the two of them being hyperbolas with equal upright sides), and their diameters are located in a straight line with each other. Is this case a perfect bore—or are we missing something that might make it interesting?

What we have here is not just a situation in which there are two hyperbolas of the same size and shape. Two hyperbolas of the same size and shape are not as such "opposite sections." Their position matters. But even two hyperbolas of the same size and shape which have their diameters in a straight line are not as such "opposite sections." For two hyperbolas to be "opposite sections," not only must they have equal upright sides, but they must also have the very same transverse side. The slicing of a single cut into vertically opposite conic surfaces guarantees that there will be vertically opposite sections—hyperbolas with equal upright sides and the very same transverse side. With the opposite sections we do not have a mere throwing together of two figures. We have, rather, a single definite configuration.

The way in which this proposition begins should have put us on the alert. The order of presentation differs from the order in those propositions where we got sections previously. Here, unlike before, Apollonius does the section-cutting prior to getting the axial triangle; indeed, he even obtains ordinates prior to getting the axial triangle. That is to say: here he proceeds so as to cut both surfaces with one slice without regard to which axial triangle, or which set of ordinates, he will have. In this proposition, what is interesting is not the straight line which is the diameter of each section. What is interesting, rather, is the straight line of which the diameters of both curves are prolongations: the transverse side.

The opposite sections together have something that keeps them apart: their common transverse side. Not only does it keep them apart, however; it also holds them together. They are made one by the same thing that makes them two. Their common transverse side makes them *one* pair and makes them one *pair*. Linked and separated by their common transverse side, the opposite sections are of interest not for each one's shape by itself, but rather for the figure which they both make together.

What has this to do with the three preceding conic sections, each of which is one because each of them is a single line? Even if we have seen compelling reasons to place the parabola's defining proposition before the defining propositions of the other sections, we have seen no compelling reasons to place the hyperbola's defining proposition before the ellipse's defining proposition. Wouldn't it have been better to present the ellipse before the hyperbola—interchanging the content of the 12th and the 13th propositions—so that the hyperbola would be placed immediately before the special pair of hyperbolas of the 14th proposition? Or is there some way in which this hyperbolic pair is related to the ellipse—some way in which it relates the ellipse to the hyperbola? Does this have anything to do with the transverse side? "Transverse" side, by the way, is the usual translation of the Greek term used by Apollonius, but a more literal translation would speak of the "oblique" side. Apollonius speaks of the "upright" side and the "oblique" side; his Greek words have connotations that are similar to those of the English.

We shall be seeing more of this in what follows. Indeed, even when we do not at first see it in what follows, it will be lurking there.

At the end of the 14th proposition, we have finished with the propositions that define the various conic sections. There were definitions long before these defining propositions, however. In his First Definitions at the beginning of the book, Apollonius gave, after the definition of a diameter, definitions for which we still have not seen the reason. Apollonius arranges for us to see it in what follows. Right after the next pair of propositions, we'll see a set of Second Definitions. And then we'll see much more.

Harvey Flaumenhaft
Dean, St. John's College, Annapolis
July 28, 1998

Translator's Note to Books I–III

by R. Catesby Taliaferro

If on first appearance this treatise should seem to the reader a jumble of propositions, rigorous indeed, but without much rhyme or reason in their sequence, then he can be sure he has not read aright, and as with the planets, he must look further to save the appearances. There are one or two hypotheses at least that can order the apparent wanderings of parabolas, hyperbolas, and ellipses through the first four books. Such hypotheses are the analogies between the three sections, and especially the development of the analogy between the hyperbola and the ellipse reaching its culmination, in the first book, with the final theorem, the construction of conjugate opposite sections.

In Book I Definition 5, Apollonius innocently defines two kinds of diameters, the transverse and the upright. Each one, in a conic section, bisects all the straight lines parallel to the other. But the upright diameter, defined here only as to position, has, in the case of the ellipse, natural bounds fixed by the section itself, and in Proposition I. 15 we find it is the mean proportional between the corresponding transverse diameter (or conjugate diameter) and its parameter. The transverse diameter, in turn, is the mean proportional between the upright diameter (or conjugate) and its parameter, so "upright" and "transverse" become meaningless terms, in the case of the ellipse, for something better expressed by the symmetrical relation "conjugate" (Book I Definition 6). Immediately, in Proposition I. 16, as if arbitrarily, the upright diameter of the hyperbola is bounded in the same way, given a definite magnitude, and becomes "the second diameter."

But so far transverse and upright diameters, or transverse and second diameters, are distinct things in the case of the hyperbola, and there seems to be little reason for giving this second diameter in magnitude: νομός has not yet become φύσις. That the upright diameter should be given even in position for the hyperbola becomes only very significant with two pairs of propositions—Propositions 1. 37 and 38, and 1. 39 and 40—where it is shown that certain properties holding for ordinates to the transverse diameter of the hyperbola and ellipse hold also for the ordinates to their conjugates. But it is only with the final proposition of the first book (I. 60) that the magnitude of the hyperbola's second diameter is justified in magnitude as well as position. It is the corresponding diameter of the opposite sections conjugate to the first.

And this analogy between the hyperbola and ellipse now stands on the threshold of a vast development. For this theorem, coming as a climax to the first book, makes possible the main theme of the second book: the asymptotes, those strange lines all but touching each opposite section (II. 2, 13, 14) and forming a single bound between each adjacent pair (II. 15, 17), so making the hyperbola an all but closed section, a puckered ellipse, a mouth turned inside out. And in the third book, the fruits of this analogy are gathered as in the especially nice case of Proposition III. 15.

Although this translation is literal, we have not hesitated to use such symbols and abbreviations as, without prejudicing any Greek number theory or introducing any modern theory of symbols, would yet make the reading and the mechanic of study easier and at the same time preserve all the rigor of Greek mathematics.

As for the Greek text, we have used Heiberg, and have constantly referred to the *editio princeps* of Halley. In certain instances we have been glad to consult the very excellent French translation of Paul Ver Eecke (Desclée de Brouwer, Bruges, 1923). We have also deferred, at all relevant points, to the English usage of T. L. Heath's translation of Euclid's *Elements*.

R. Catesby Taliaferro
Deer Isle, Maine
August 1939

Illustrator's Note to Books I–III

by William H. Donahue

The diagrams for Books I–III have all been newly drawn for this edition, and differ considerably from those in the Great Books, and, indeed, from those in any other edition. It therefore seems appropriate to explain briefly the reasons why they are as they are.

Of course, one would like to follow Apollonius as nearly as possible. However, there is no assurance that the diagrams as they exist in any of the manuscripts accurately represent what Apollonius himself might have drawn. J. L. Heiberg, editor of the Greek text, based his diagrams on the ones in the margins of Codex Vaticanus Graecus 206, which dates from the 12th or 13th century, supplemented by a few from Codex Vaticanus Gr. 203 (13th C) where those from CVG 206 had been lost. He expressed no opinion as to their authenticity.

On the other hand the diagrams in the second printed version, Federico Commandino's Latin translation (Bologna 1566), differ remarkably from Heiberg's. Although they may have been based upon those in the manuscripts available to Commandino, their style clearly shows the influence of Renaissance art, particularly in the use of perspective and their symmetry. Here, too, we are not dealing with authentic originals.

This absence of authentic exemplars shifts the responsibility for the diagrams to the illustrator. There are a number of conflicting criteria that must be balanced. Above all, the diagrams must accurately represent the points and lines which the demonstration treats. Next, they should be as easy to follow and as unambiguous in their labeling as accuracy permits. Finally, they need to be fit into the text in the appropriate places, and must be sized accordingly.

The diagrams in the Encyclopædia Britannica edition have much to recommend them as models. They are the ones familiar to most readers and are usually clear and accurate. They are evidently based on the Heiberg diagrams, as shown by the odd location of point F in III. 40 and the replication of an error in line RS in III. 12. On the other hand, they introduce new errors (such as in III. 36 and 48), and are drawn in a more expansive style than Heiberg's. They also often include points and lines not referred to in the propositions, making them unnecessarily cluttered (see, for example, the diagrams to III. 6–9 in this edition as compared with the Britannica edition). On one occasion (III. 50), the configuration of the diagram required a lengthy footnote to untangle the confusion it created. Other errors have been noted in the *Corrigenda* sheet that has been distributed at St. John's College for many years.

It was, however, the size of the Britannica diagrams that ultimately led to the decision not to emulate them. One of Green Lion Press's standards is that diagrams always be present on the same spread as the text that refers to them. This requires frequent repetition of diagrams, which in turn displace text. If the diagrams are large, they need to be included even more often because more text is displaced to subsequent pages. The cost of the book is thus increased without an improvement in usability. Further, many readers asked us to provide generous margins, both on the pages as a whole and around the diagrams, to allow for the writing of notes.

I therefore turned to the diagrams in the Commandino edition, which Brother Robert of the Annapolis campus of St. John's College had recommended to me. A look at the microfilm which the St. John's library generously lent us showed right away that we would not want to use these unchanged, even if we could make clean copies. The letters are small and hard to read, and are not always the same as those in the Taliaferro translation. But more important, the illustrator in this edition, possibly led by a Renaissance fondness for symmetry, chose to show the axis of a curve where the proposition applies to all diameters. In my experience, students already have a strong tendency to use the axes and the ordinates perpendicular to them, and to fail to grasp the full generality of some of Apollonius's propositions. So the Commandino diagrams could not be adopted without some amendment.

The diagrams in Thomas Heath's retelling of the Conics were also considered, but were not used as models because their lettering, being different from Apollonius's, would have made their use very difficult and open to new errors. They were consulted in difficult cases or where there was doubt as to the correct placement of a line.

The solution adopted for most of the propositions was to use Commandino's configuration as a starting point, but to construct the diagrams step by step as Apollonius described them. The drawing was done with Adobe Illustrator, and the resulting drafts were skewed, stretched, and resized to accommodate the letters. These draft diagrams were then included in the interim edition, and received many useful comments. As a result, in the present edition, the letters have all been made smaller, and many of the diagrams have been enlarged. This affords greater clarity in the indication of points. Greater attention was also paid to making the lengths of the lines appropriate to the proposition being proved. Some of the diagrams were completely redrawn as a result of suggestions from readers.

It is hoped that these diagrams keep the compactness and simplicity of the Commandino engravings while representing more general instances of the propositions. Another feature adopted from Commandino is the inclusion of auxiliary circles in the demonstration of focal properties (III. 48-50). Although these circles are not strictly speaking part of the diagram, and were not included by Heiberg, they are mentioned by Apollonius, and their inclusion serves to make the proof easier to follow.

In a departure from all previous models, some of the diagrams that involve lines in different planes were done in perspective, following suggestions from those who had used the Britannica edition. The diagram for I. 15 is based on a submission by J. M. MacDonald, and closely follows his drawing. Some others are similar in style, with hidden parts indicated by broken lines.

Finally, a few remarks, some general and some specific, that may be helpful in the use of these diagrams.

Above all, it should be noted that no attempt has been made to present all possible configurations that might be covered by a proposition. The diagram is intended only to represent a typical configuration. Other configurations, possibly very different, exist to which the proposition applies equally well. In some cases, the line of the proof might have to change somewhat to accommodate these different cases (see, for example, the proof of I. 50). Readers should therefore beware of putting too much trust in the diagrams. They are only pointers; they are not the things themselves.

Next, many of the propositions apply to circles as well as other conic sections, and Apollonius explicitly mentions circles in the enunciation. In the interest of saving space, diagrams of circles are usually omitted, on the grounds that they can be considered as special cases of ellipses.

Finally, and more specifically, at the end of Book II, there is a series of "Problems," which are solved by the technique of analysis and synthesis, characteristic of Greek geometry. The figure for the analytical part is not constructed, but supposed as given, and is analyzed in order to find out how to construct a figure that solves the problem. The figure that is then constructed in the synthetic part is very much like, but sometimes not identical to, the figure analyzed. The diagrams in Heiberg and in the Britannica edition do not acknowledge this distinction, but make the same figures do for both parts. Commandino, in contrast, provides different engravings for the two parts, where they differ. We have followed Commandino in this respect, and have gone beyond him in presenting given angles in lines that are thicker than those in the rest of the diagram. It is hoped that these features will make the demonstrations, which are such valuable presentations of the method, easier to follow.

William H. Donahue
Santa Fe, July 1998

ΑΠΟΛΛΩΝΙΟΥ ΠΕΡΓΑΙΟΥ
ΚΩΝΙΚΩΝ

ΒΙΒΛΙΑ Δ΄. ΤΑ ΠΡΟΤΕΡΑ

ΜΕΤΑ

ΠΑΠΠΟΥ ΑΛΕΞΑΝΔΡΕΩΣ

ΛΗΜΜΑΤΩΝ

ΚΑΙ

ΕΥΤΟΚΙΟΥ ΑΣΚΑΛΩΝΙΤΟΥ

ΥΠΟΜΝΗΜΑΤΩΝ

APOLLONII PERGÆI
CONICORUM

LIBRI IV. PRIORES

CUM

PAPPI ALEXANDRINI

LEMMATIS

ET

EUTOCII ASCALONITÆ

COMMENTARIIS.

Ex Codd. MSS. Græcis edidit EDMUNDUS HALLEIUS apud
Oxonienfes Geometriæ Profeffor *Savilianus.*

Title Page of the First Edition of the Greek Text
1710

CONICS
BOOK ONE

**APOLLONIUS to EUDEMUS,
greetings.**

If you are restored in body, and other things go with you to your mind, well and good; and we too fare pretty well. At the time I was with you in Pergamum, I observed you were quite eager to be kept informed of the work I was doing in conics. And so I have sent you this first book revised, and we shall dispatch the others when we are satisfied with them. For I don't believe you have forgotten hearing from me how I worked out the plan for these conics at the request of Naucrates, the geometer, at the time he was with us in Alexandria lecturing, and how on arranging them in eight books we immediately communicated them in great haste because of his near departure, not revising them but putting down whatever came to us with the intention of a final going over. And so finding now the occasion of correcting them, one book after another, we publish them. And since it happened that some others among those frequenting us got acquainted with the first and second books before the revision, don't be surprised if you come upon them in a different form.

Of the eight books the first four belong to a course in the elements. The first book contains the generation of the three sections and of the opposite branches, and the principal properties ($\tau\grave{\alpha}$ $\dot{\alpha}\rho\chi\iota\kappa\grave{\alpha}$ $\sigma\upsilon\mu\pi\tau\acute{\omega}\mu\alpha\tau\alpha$) in them worked out more fully and universally than in the writings of others. The second book contains the properties ($\tau\grave{\alpha}$ $\sigma\upsilon\mu\beta\alpha\acute{\iota}\nu\upsilon\tau\alpha$) having to do with the diameters and axes and also the asymptotes, and other things of a general and necessary use for limits of possibility ($\pi\rho\grave{\varsigma}$ $\tau\upsilon\grave{\varsigma}$ $\delta\iota\upsilon\rho\iota\sigma\mu\upsilon\acute{\varsigma}$). And what I call diameters and what I call axes you will know from this book. The third book contains many incredible theorems of use for the construction of solid loci and for limits of possibility of which the greatest part and the most beautiful are new. And when we had grasped these, we knew that the three-line and four-line locus had not been constructed by Euclid, but only a chance part of it and that not very happily. For it was not possible for this construction to be completed without the additional things found by us. The fourth book shows in how many ways the sections of a cone intersect with each other and with the circumference of a circle, and contains other things in addition none of which has been written up by our predecessors, that is in how many points the section of a cone or the circumference of a circle and the opposite branches meet the opposite branches. The rest of the books are fuller in treatment. For there is one dealing more fully with maxima and minima, and one with equal and similar sections of a cone, and one with limiting theorems, and one with determinate conic problems. And so indeed, with all of them published, those happening upon them can judge them as they see fit. Good-bye.

FIRST DEFINITIONS

1. If from a point a straight line is joined to the circumference of a circle which is not in the same plane with the point, and the line is produced in both directions, and if, with the point remaining fixed, the straight line being rotated about the circumference of the circle returns to the same place from which it began, then the generated surface composed of the two surfaces lying vertically opposite one another, each of which increases indefinitely as the generating straight line is produced indefinitely, I call a conic surface, and I call the fixed point the vertex, and the straight line drawn from the vertex to the center of the circle I call the axis.

2. And the figure contained by the circle and by the conic surface between the vertex and the circumference of the circle I call a cone, and the point which is also the vertex of the surface I call the vertex of the cone, and the straight line drawn from the vertex to the center of the circle I call the axis, and the circle I call the base of the cone.

3. I call right cones those having axes perpendicular to their bases, and I call oblique those not having axes perpendicular to their bases.

4. Of any curved line which is in one plane, I call that straight line the diameter which, drawn from the curved line, bisects all straight lines drawn to this curved line parallel to some straight line; and I call the end of the diameter situated on the curved line the vertex of the curved line, and I say that each of these parallels is drawn ordinatewise to the diameter ($\tau\epsilon\tau\alpha\gamma\mu\acute{\epsilon}\nu\omega\varsigma$ $\dot{\epsilon}\pi\grave{\iota}$ $\tau\grave{\eta}\nu$ $\delta\iota\acute{\alpha}\mu\epsilon\tau\rho\sigma\nu$ $\kappa\alpha\tau\tilde{\eta}\chi\theta\alpha\iota$).[*]

5. Likewise, of any two curved lines lying in one plane, I call that straight line the transverse diameter ($\delta\iota\acute{\alpha}\mu\epsilon\tau\rho\sigma\varsigma$ $\pi\lambda\alpha\gamma\acute{\iota}\alpha$) which cuts the two curved lines and bisects all the straight lines drawn to either of the curved lines parallel to some straight line; and I call the ends of the [transverse] diameter situated on the curved lines the vertices of the curved lines; and I call that straight line the upright diameter ($\delta\iota\acute{\alpha}\mu\epsilon\tau\rho\sigma\varsigma$ $\dot{\sigma}\rho\theta\acute{\iota}\alpha$) which, lying between the two curved lines, bisects all the straight lines intercepted between the curved lines and drawn parallel to some straight line; and I say that each of the parallels is drawn ordinatewise to the [transverse or upright] diameter.

---

[*] We shall follow modern usage and generally call these parallels *ordinates*. (Tr.)

6. The two straight lines, each of which, being a diameter, bisects the straight lines parallel to the other, I call the conjugate diameters ($\sigma \upsilon \zeta \upsilon \gamma \epsilon \hat{\iota} \varsigma$ $\delta \iota \alpha \mu \acute{\epsilon} \tau \rho o \iota$) of a curved line and of two curved lines.

7. And I call that straight line the axis of a curved line and of two curved lines which being a diameter of the curved line or lines cuts the parallel straight lines at right angles.

8. And I call those straight lines the conjugate axes of a curved line and of two curved lines which, being conjugate diameters, cut the straight lines parallel to each other at right angles.

PROPOSITION 1

The straight lines drawn from the vertex of the conic surface to points on the surface are on that surface.

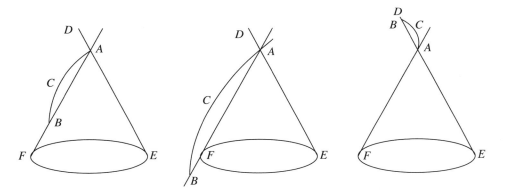

Let there be a conic surface whose vertex is the point *A,* and let there be taken some point *B* on the conic surface, and let a straight line *ACB* be joined.

I say that the straight line *ACB* is on the conic surface.

For if possible, let it not be, and let the straight line *DE* be the line generating the surface, and *EF* be the circle along which *ED* is moved. Then if, the point *A* remaining fixed, the straight line *DE* is moved along the circumference of the circle *EF,* it will also go through the point *B* (Def. 1), and two straight lines will have the same ends. And this is absurd.

Therefore the straight line joined from *A* to *B* cannot not be on the surface. Therefore it is on the surface.

PORISM

It is also evident that, if a straight line is joined from the vertex to some point among those within the surface, it will fall within the conic surface; and if it is joined to some point among those without, it will be outside the surface.

PROPOSITION 2

If on either one of the two vertically opposite surfaces two points are taken, and the straight line joining the points, when produced, does not pass through the vertex, then it will fall within the surface, and produced it will fall outside.

Let there be a conic surface whose vertex is the point *A,* and a circle *BC* along whose circumference the generating straight line is moved, and let two points *D* and *E* be taken on either one of the two vertically opposite surfaces, and let the joining straight line *DE,* when produced, not pass through the point *A.*

I say that the straight line *DE* will be within the surface, and produced will be without.

Let *AE* and *AD* be joined and produced. Then they will fall on the circumference of the circle (I. 1). Let them fall to the points *B* and *C,* and let *BC* be joined. Therefore the straight line *BC* will be within the circle, and so too within the conic surface.

Then let a point *F* be taken at random on *DE,* and let the straight line *AF* be joined and produced. Then it will fall on the straight line *BC;* for the triangle *BCA* is in one plane (Eucl. XI. 2). Let it fall to the point *G.* Since then the point *G* is within the conic surface, therefore the straight line *AG* is also within the conic surface (I. 1 porism), and so too the point *F* is within the conic surface. Then likewise it will be shown that all the points on the straight line *DE* are within the surface. Therefore the straight line *DE* is within the surface.

Then let *DE* be produced to *H*. I say then it will fall outside the conic surface.

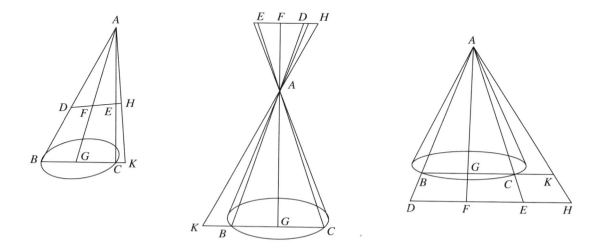

For if possible, let there be some point *H* of it not outside the conic surface, and let *AH* be joined and produced. Then it will fall either on the circumference of the circle or within (I. 1 and porism). And this is impossible, for it falls on *BC* produced, as for example to the point *K*. Therefore the straight line *EH* is outside the surface.

Therefore the straight line *DE* is within the conic surface, and produced is outside.

PROPOSITION 3

If a cone is cut by a plane through the vertex, the section is a triangle.

Let there be a cone whose vertex is the point *A* and whose base is the circle

BC; and let it be cut by some plane through the point A; and let it make, as sections, lines AB and AC on the surface, and the straight line BC in the base.

I say that ABC is a triangle.

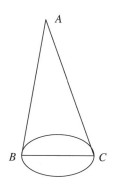

For since the line joined from A to B is the common section of the cutting plane and of the surface of the cone, therefore AB is a straight line. And likewise also AC. And BC is also a straight line. Therefore ABC is a triangle.

If then a cone is cut by some plane through the vertex, the section is a triangle.

PROPOSITION 4

If either one of the vertically opposite surfaces is cut by some plane parallel to the circle along which the straight line generating the surface is moved, the plane cut off within the surface will be a circle having its center on the axis, and the figure contained by the circle and the conic surface intercepted by the cutting plane on the side of the vertex will be a cone.

Let there be a conic surface whose vertex is the point A and whose circle along which the straight line generating the surface is moved is BC; and let it be cut by some plane parallel to the circle BC, and let it make on the surface as a section the line DE.

I say that the line DE is a circle having its center on the axis.

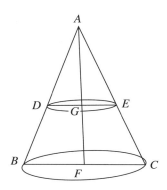

For let the point F be taken as the center of the circle BC, and let AF be joined. Therefore AF is the axis (Def. 1) and meets the cutting plane. Let it meet it at the point G, and let some plane be produced through AF. Then the section will be the triangle ABC (I. 3). And since the points D, G, E are points in the cutting plane, and are also in the plane of the triangle ABC, therefore DGE is a straight line (Eucl. XI. 3).

Then let some point H be taken on the line DE, and

let *AH* be joined and produced. Then it falls on the circumference *BC* (I. 1). Let it meet it at *K,* and let *GH* and *FK* be joined. And since two parallel planes, *DE* and *BC,* are cut by a plane *ABC,* their common sections are parallel (Eucl. XI. 16). Therefore the straight line *DE* is parallel to the straight line *BC.* Then for the same reason the straight line *GH* is also parallel to the straight line *KF.* Therefore

$$FA : AG :: FB : DG :: FC : GE :: FK : GH \text{ (Eucl. VI. 4).}$$

And

$$BF = KF = FC.$$

Therefore also

$$DG = GH = GE \text{ (Eucl. V. 9).}$$

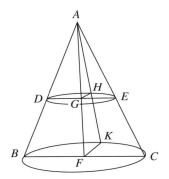 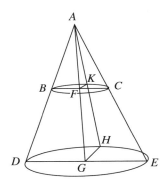 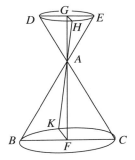

Then likewise we could show also that all the straight lines falling from the point *G* on the line *DE* are equal to each other.

Therefore the line *DE* is a circle having its center on the axis.

And it is evident that the figure contained by the circle *DE* and the conic surface cut off by it on the side of the point *A* is a cone.

And it is therewith proved that the common section of the cutting plane and of the axial triangle (triangle through the axis) is a diameter of the circle.

PROPOSITION 5

If an oblique cone is cut by a plane through the axis at right angles to the base, and is also cut by another plane on the one hand at right angles to the axial triangle, and on the other cutting off on the side of the vertex a triangle similar to the axial triangle and lying subcontrariwise (ὑπεναντίως), then the section is a circle, and let such a section be called subcontrary (ὑπεναντία).

Let there be an oblique cone whose vertex is the point *A* and whose base is the circle *BC*, and let it be cut through the axis by a plane perpendicular to the circle *BC*, and let it make as a section the triangle *ABC* (I. 3). Then let it also be cut by another plane perpendicular to the triangle *ABC* and cutting off on the side of the point *A* the triangle *AKG* similar to the triangle *ABC* and lying subcontrariwise, that is, so that the angle *AKG* is equal to the angle *ABC*. And let it make as a section on the surface, the line *GHK*.

I say that the line *GHK* is a circle.

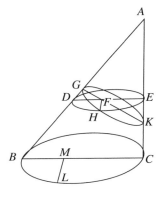

For let any points *H* and *L* be taken on the lines *GHK* and *BC*, and from the points *H* and *L* let perpendiculars be dropped to the plane through the triangle *ABC*. Then they will fall to the common sections of the planes (Eucl. XI. Def. 4). Let them fall as for example *FH* and *LM*. Therefore *FH* is parallel to *LM* (Eucl. XI. 6).

Then let the straight line *DFE* be drawn through *F* parallel to *BC*; and *FH* is also parallel to *LM*. Therefore the plane through *FH* and *DE* is parallel to the base of the cone (Eucl. XI. 15). Therefore it is a circle whose diameter is the straight line *DE* (I. 4).

Therefore
    rect. *DF, FE* = sq. *FH* (Eucl. III. 31, VI. 8 porism, and VI. 17).

And since *ED* is parallel to *BC*, angle *ADE* is equal to angle *ABC*. And angle *AKG* is supposed equal to angle *ABC*. And therefore angle *AKG* is equal to angle *ADE*. And the vertical angles at the point *F* are also equal. Therefore triangle *DFG* is similar to triangle *KFE*, and therefore

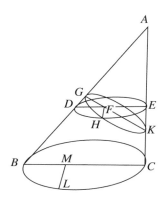

$EF : FK :: GF : FD$ (Eucl. VI. 4).

Therefore

rect. EF, FD = rect. KF, FG (Eucl. VI. 16).

But it has been shown that

sq. FH = rect. EF, FD;

and therefore

rect. KF, FG = sq. FH.

Likewise then all the perpendiculars drawn from the line GHK to the straight line GK could also be shown to be equal in square to the rectangle, in each case, contained by the segments of the straight line GK.

Therefore the section is a circle whose diameter is the straight line GK.

PROPOSITION 6

If a cone is cut by a plane through the axis, and if on the surface of the cone some point is taken which is not on a side of the axial triangle, and if from this point is drawn a straight line parallel to some straight line which is a perpendicular from the circumference of the circle to the base of the triangle, then that drawn straight line meets the axial triangle, and on being produced to the other side of the surface the drawn straight line will be bisected by the triangle.

Let there be a cone whose vertex is the point A and whose base is the circle BC, and let the cone be cut by a plane through the axis, and let it make a common section the triangle ABC (I. 3); and from some point M of those on the circumference, let the straight line MN be drawn perpendicular to the straight line BC. Then let some point D be taken on the surface of the cone, and through D let the straight line DE be drawn parallel to MN.

I say that the straight line DE produced will meet the plane of the triangle ABC, and, if further produced toward the other side of the cone until it meet its surface, will be bisected by the triangle ABC.

Let the straight line AD be joined and be produced. Therefore it will meet the circumference of the circle BC (I. 1). Let it meet it at K and from K let the straight line *KHL* be drawn perpendicular to the straight line *BC*. Therefore *KH* is parallel to *MN*, and therefore to *DE* (Eucl. XI. 9).

Let the straight line *AH* be joined from *A* to *H*. Since then in the triangle *AHK* the straight line *DE* is parallel to the straight line *HK*, therefore *DE* produced will meet *AH*. But *AH* is in the plane of *ABC*; therefore *DE* will meet the plane of the triangle *ABC*.

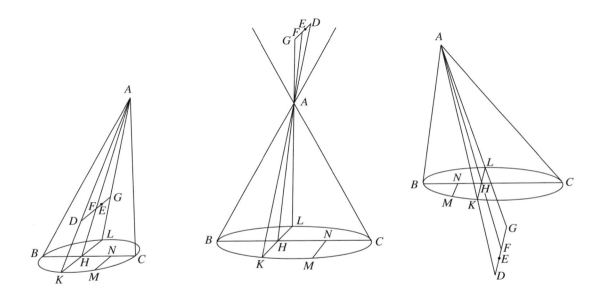

For the same reasons it also meets *AH*; let it meet it at *F*, and let *DF* be produced in a straight line until it meet the surface of the cone. Let it meet it at *G*.

I say that *DF* is equal to *FG*.

For since *A*, *G*, *L* are points on the surface of the cone, but also in the plane extended through the straight lines *AH*, *AK*, *DG*, *KL*, which is a triangle through the vertex of the cone (I. 3), therefore *A*, *G*, *L* are points on the common section of the cone's surface and of the triangle. Therefore the line through *A*, *G*, *L* is a straight line. Since then in the triangle *ALK* the straight line *DG* has been drawn parallel to the base *KHL* and some straight line *AFH* has been drawn across them from the point *A*, therefore

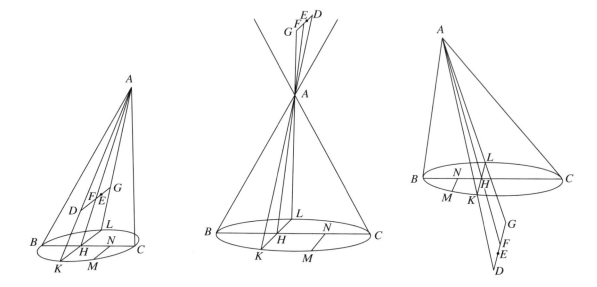

$KH : HL :: DF : FG$ (Eucl. VI. 4).
But *KH* is equal to *HL,* since *KL* is a chord in circle *BC* perpendicular to the diameter (Eucl. III. 3). Therefore *DF* is equal to *FG.*

PROPOSITION 7

If a cone is cut by a plane through the axis, and if the cone is also cut by another plane so that the plane of the base of the cone is cut in a straight line perpendicular either to the base of the axial triangle or to it produced, and if, from the cutting plane's resulting section on the cone's surface, straight lines are drawn parallel to the straight line perpendicular to the base of the triangle, then these straight lines will fall on the common section of the cutting plane and of the axial triangle, and, further produced to the other side of the section, these straight lines will be bisected by the common section; and if the cone is a right cone, then the straight line in the base will be perpendicular to the common section of the cutting plane and of the axial triangle, but if the cone is oblique, then the straight line in the base will be

*perpendicular to that common section only whenever the plane through the
axis is perpendicular to the base of the cone.*

Let there be a cone whose vertex is the point *A* and whose base is the circle
BC, and let it be cut by a plane through the axis and let it make as a section
the triangle *ABC* (I. 3). And let it also be cut by another plane cutting the
plane the circle *BC* is in, in the straight line *DE* perpendicular either to the
straight line *BC* or to it produced, and let it make as a section on the surface
of the cone the line *DFE.* Then the straight line *FG* is the common section
of the cutting plane and of the triangle *ABC.* And let any point *H* be taken
on the section *DFE,* and let the straight line *HK* be drawn through *H* parallel
to the straight line *DE.*

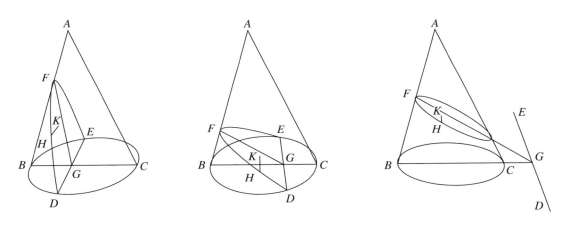

I say that the straight line *HK* meets the straight line *FG,* and, on being produced to the other side of the section *DFE,* will be bisected by *FG.*

For since a cone whose vertex is the point *A* and whose base is the circle *BC*
has been cut by a plane through its axis, and makes as a section the triangle
ABC, and, since some point *H* on the surface, not on a side of the triangle
ABC, has been taken, and since the straight line *DG* is perpendicular to the
straight line *BC,* therefore the straight line drawn through *H* parallel to *DG,*
that is *HK,* meets the triangle *ABC,* and if further produced to the other side
of the surface, will be bisected by the triangle (I. 6).

Then since the straight line drawn through *H* parallel to the straight line *DE*
meets the triangle *ABC* and is in the plane of the section *DFE,* therefore it

will fall on the common section of the cutting plane and of the triangle *ABC*. But the straight line *FG* is the common section of the planes. Therefore the straight line drawn through *H* parallel to *DE* will fall on *FG*, and, if further produced to the other side of the section *DFE*, will be bisected by the straight line *FG*.

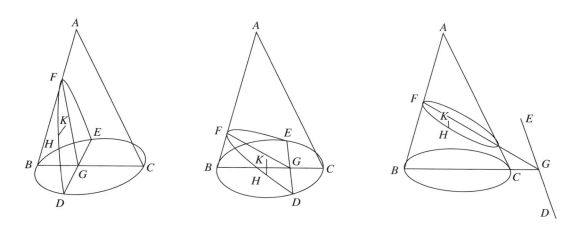

Then either the cone is a right cone, or the axial triangle *ABC* is perpendicular to the circle *BC*, or neither.

First let the cone be a right cone. Then the triangle *ABC* would be perpendicular to the circle *BC* (Def. 3; Eucl. XI. 18). Since then the plane *ABC* is perpendicular to the plane *BC*, and the straight line *DE* has been drawn in one of the planes, *BC*, perpendicular to their common section the straight line *BC*, therefore the straight line *DE* is perpendicular to the triangle *ABC* (Eucl. XI. Def. 4), and, therefore, to all the straight lines touching it and in the triangle *ABC* (Eucl. XI. Def. 3). And so *DE* is also perpendicular to the straight line *FG*.

Then let the cone not be a right cone. If, now, the axial triangle is perpendicular to the circle *BC*, we could likewise show that *DE* is perpendicular to *FG*.

Then let the axial triangle *ABC* not be perpendicular to the circle *BC*.— I say that *DE* is not perpendicular to *FG*. For if possible, let it be. And it is also perpendicular to the straight line *BC*. Therefore *DE* is perpendicular to both *BC* and *FG*, and therefore it will be perpendicular to the plane through *BC* and *FG*. But the plane through *BC* and *GF* is the triangle *ABC*, and therefore *DE* is perpendicular to the triangle *ABC*. And therefore all the planes through it are perpendicular to the triangle *ABC*. But one of the planes through *DE* is the circle *BC;* therefore the circle *BC* is perpendicular to the triangle *ABC*. And so the triangle *ABC* will also be perpendicular to the circle *BC*. And this is not supposed. Therefore the straight line *DE* is not perpendicular to the straight line *FG*.

PORISM

Then from this it is evident that the straight line *FG* is the diameter of the section *DFE,* since it bisects the straight lines drawn parallel to some straight line *DE,* and it is evident that it is possible for some parallels to be bisected by the diameter *FG* and not be perpendicular to *FG*.

PROPOSITION 8

If a cone is cut by a plane through its axis, and if the cone is cut by another plane cutting the base of the cone in a straight line perpendicular to the base of the axial triangle, and if the diameter of the resulting section on the surface is either parallel to one of the sides of the triangle or meets one of the sides extended beyond the vertex of the cone, and if both the surface of the cone and the cutting plane are produced indefinitely, then the section will also increase indefinitely, and some straight line drawn from the section of the cone parallel to the straight line in the base of the cone will cut off from the diameter on the side of the vertex a straight line equal to any given straight line.

Let there be a cone whose vertex is the point *A* and whose base is the circle *BC,* and let it be cut by a plane through its axis, and let it make as a section the triangle *ABC* (I. 3). And let it be cut also by another plane cutting the circle *BC* in a straight line *DE* perpendicular to the straight line *BC,* and let it make as a section on the surface the line *DFE.* And let the diameter *FG* of the section *DFE* be either parallel to the straight line *AC* or on being produced meet it beyond the point *A* (I. 7 and porism).

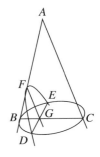

I say that, if both the surface of the cone and the cutting plane are produced indefinitely, the section *DFE* also will increase indefinitely.

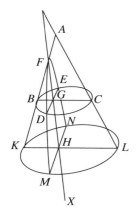

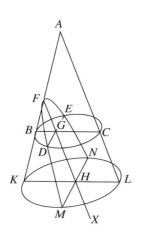

For let both the surface of the cone and the cutting plane be produced. Then it is evident that also the straight lines *AB*, *AC*, *FG* will be therewith produced. Since the straight line *FG* is either parallel to *AC* or produced meets it beyond the point *A*, therefore the straight lines *FG* and *AC* on being produced in the direction of *C* and *G* will never meet. Then let them be produced and let some point *H* be taken at random on the straight line *FG*, and let the straight line *KHL* be drawn through the point *H* parallel to the straight line *BC*, and *MHN* parallel to *DE*. Therefore the plane through *KL* and *MN* is parallel to the plane through *BC* and *DE* (Eucl. XI. 15). Therefore the plane *KLMN* is a circle (I. 4).

And since the points *D, E, M, N* are in the cutting plane and also on the surface of the cone, therefore they are on the common section. Therefore the section *DFE* has increased to the points *M* and *N*. Therefore, with the surface of the cone and the cutting plane increased to the circle *KLMN*, the section *DFE* has also increased to the points *M* and *N*. Then likewise we could show also, that if the surface of the cone and the cutting plane are extended indefinitely, the section *MDFEN* will also increase indefinitely.

And it is evident that some straight line will cut off on straight line *FH* on the side of point *F* a straight line equal to any given straight line. For if we lay down the straight line *FX* equal to the given straight line, and draw a parallel to *DE* through *X*, it will meet the section, just as the straight line through *H* was also proved to meet the section in the points *M* and *N*. And so some straight line is drawn meeting the section, parallel to *DE*, and cutting off on *FG* on the side of point *F* a straight line equal to the given straight line.

PROPOSITION 9

If a cone is cut by a plane which meets both sides of the axial triangle and is neither parallel to the base nor situated subcontrariwise, then the section will not be a circle.

Let there be a cone whose vertex is the point *A* and whose base is the circle *BC*, and let it be cut by some plane neither parallel to the base nor situated subcontrariwise, and let it make as a section on the surface the line *DKE*.

I say that the line *DKE* will not be a circle.

For if possible, let it be, and let the cutting plane meet the base, and let the straight line *FG* be the common section of the planes, and let the point *H* be the center of the circle *BC*, and from *H* let the straight line *HG* be drawn perpendicular to the straight line *FG*. And let a plane be extended through *GH* and the axis and let it make as sections on the conic surface the straight lines *BA* and *AC* (I.1). Since then *D*, *E*, *G* are points in the plane through the line *DKE*, and also in the plane through the points *A*, *B*, *C*, therefore *D*, *E*, *G* are points on the common section of the planes. Therefore *GED* is a straight line (Eucl. XI. 3).

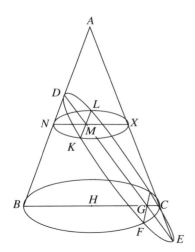 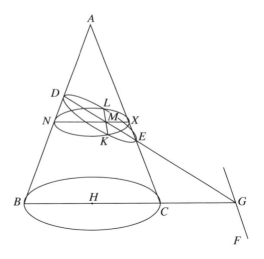

Then let some point *K* be taken on the line *DKE*, and through *K* let the straight line *KL* be drawn parallel to the straight line *FG;* then *KM* will be equal to *ML* (I. 7). Therefore the straight line *DE* is the diameter of the [supposed] circle *DKLE* (Def. 4). Then let the straight line *NMX* be drawn through *M* parallel to the straight line *BC*. But *KL* is also parallel to *FG*. And so the plane through the straight lines *NX* and *KM* is parallel to the plane through the straight lines *BC* and *FG*, that is to the base (Eucl. XI. 15), and the section will be a circle (I. 4). Let it be the circle *NKX*.

And since the straight line *FG* is perpendicular to the straight line *BG*, the straight line *KM* is also perpendicular to the straight line *NX* (Eucl. XI. 10).

And so

rect. *NM, MX* = sq. *KM* (Eucl. III. 31, VI. 8 porism, VI. 17).

But

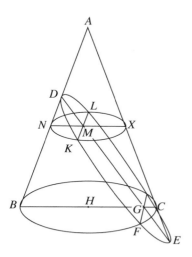

rect. *DM, ME* = sq. *KM,*

for the line *DKEL* is supposed a circle, and the straight line *DE* is its diameter.

Therefore

rect. *NM, MX* = rect. *DM, ME.*

Therefore

MN : MD :: EM : MX (Eucl. VI. 16).

Therefore triangle *DMN* is similar to triangle *XME* (Eucl. VI. 6; VI. Def. 1), and angle *DNM* is equal to angle *MEX.* But angle *DNM* is equal to angle *ABC,* for the straight line *NX* is parallel to the straight line *BC.* And therefore angle *ABC* is equal to angle *MEX.* Therefore the section is subcontrary (I. 5). And this is not supposed. Therefore the line *DKE* is not a circle.

PROPOSITION 10

If two points are taken on the section of a cone, the straight line joining the two points will fall within the section, and produced in a straight line it will fall outside.

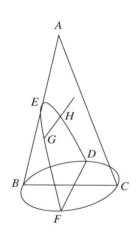

Let there be a cone whose vertex is the point *A,* and whose base is the circle *BC,* and let it be cut by a plane through the axis, and let it make as a section the triangle *ABC* (I. 3). Then let it also be cut [not through the vertex] by another plane, and let it make as a section on the surface of the cone the line *DEF,* and let two points *G* and *H* be taken on the line *DEF.*

I say that the straight line joining the two points *G* and *H* will fall within the line *DEF,* and produced in a straight line it will fall outside.

For since a cone, whose vertex is the point *A* and whose base is the circle *BC,* has been cut by a plane through the axis, and some points *G* and *H* have been taken on its surface which are not on a side of the axial triangle, and since the straight line joining *G* and *H* does not verge to the point *A,* therefore the straight line joining *G* and *H* will fall within the cone, and produced in a straight line it will fall outside (I. 2); consequently also outside the section *DFE.*

PROPOSITION 11

If a cone is cut by a plane through its axis, and also cut by another plane cutting the base of the cone in a straight line perpendicular to the base of the axial triangle, and if, further, the diameter of the section is parallel to one side of the axial triangle, and if any straight line is drawn from the section of the cone to its diameter such that this straight line is parallel to the common section of the cutting plane and of the cone's base, then this straight line to the diameter will equal in square the rectangle contained by (a) the straight line from the section's vertex to where the straight line to the diameter cuts it off and (b) another straight line which has the same ratio to the straight line between the angle of the cone and the vertex of the section as the square on the base of the axial triangle has to the rectangle contained by the remaining two sides of the triangle. And let such a section be called a parabola (παραβολή).

Let there be a cone whose vertex is the point *A,* and whose base is the circle *BC,* and let it be cut by a plane through its axis, and let it make as a section the triangle *ABC* (I. 3). And let it also be cut by another plane cutting the base of the cone in the straight line *DE* perpendicular to the straight line *BC,* and let it make as a section on the surface of the cone the line *DFE,* and let the diameter of the section *FG* (I. 7, and Def. 4) be parallel to one side *AC* of the axial triangle. And let the straight line *FH* be drawn from the point *F* perpendicular to the straight line *FG,*

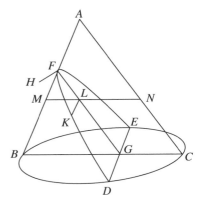

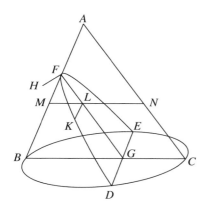

and let it be contrived that
$$\text{sq. } BC : \text{rect. } BA, AC :: FH : FA.$$
And let some point K be taken at random on the section, and through K let the straight line KL be drawn parallel to the straight line DE.

I say that sq. $KL = $ rect. HF, FL.

For let the straight line MN be drawn through L parallel to the straight line BC. And the straight line DE is also parallel to the straight line KL. Therefore the plane through KL and MN is parallel to the plane through BC and DE (Eucl. XI. 15), that is, to the base of the cone. Therefore the plane through KL and MN is a circle whose diameter is MN (I.4). And KL is perpendicular to MN since DE is also perpendicular to BC (Eucl. XI. 10). Therefore
$$\text{rect. } ML, LN = \text{sq. } KL \text{ (Eucl. III. 31, VI. 8 porism, VI. 17).}$$
And since
$$\text{sq. } BC : \text{rect. } BA, AC :: HF : FA,$$
and
$$\text{sq. } BC : \text{rect. } BA, AC :: BC : CA \text{ comp. } BC : BA \text{ (Eucl. VI. 23),}$$
therefore
$$HF : FA :: BC : CA \text{ comp. } BC : BA.$$
But
$$BC : CA :: MN : NA :: ML : LF \text{ (Eucl. VI. 4),}$$
and
$$BC : BA :: MN : MA :: LM : MF :: NL : FA \text{ (Eucl. VI. 4, VI. 2).}$$
20Therefore
$$HF : FA :: ML : LF \text{ comp. } NL : FA.$$
But
$$\text{rect. } ML, LN : \text{rect. } LF, FA :: ML : LF \text{ comp. } LN : FA \text{ (Eucl. VI. 23).}$$

Therefore
$$HF : FA :: \text{rect. } ML, LN : \text{rect. } LF, FA.$$

But, with the straight line FL taken as common height,
$$HF : FA :: \text{rect. } HF, FL : \text{rect. } LF, FA \text{ (Eucl. VI. 1),}$$
therefore
$$\text{rect. } ML, LN : \text{rect. } LF, FA :: \text{rect. } HF, FL : \text{rect. } LF, FA \text{ (Eucl. V. 11).}$$
Therefore
$$\text{rect. } ML, LN = \text{rect. } HF, FL \text{ (Eucl. V. 9).}$$

But
$$\text{rect. } ML,\ LN = \text{sq. } KL,$$
therefore also
$$\text{sq. } KL = \text{rect. } HF,\ FL.$$
And let such a section be called a parabola, and let *HF* be called the straight line to which the straight lines drawn ordinatewise to the diameter *FG* are applied in square (παρ' ἥν δύναται αἱ καταγόμεναι τεταγμένος ἐπὶ τὴν ZH διάμετρον), and let it also be called the upright side (ὀπθία).[*]

PROPOSITION 12

If a cone is cut by a plane through its axis, and also by another plane cutting the base of the cone in a straight line perpendicular to the base of the axial triangle, and if the diameter of the section produced meets one side of the axial triangle beyond the vertex of the cone, and if any straight line is drawn from the section to its diameter such that this straight line is parallel to the common section of the cutting plane and of the cone's base, then this straight line to the diameter will equal in square some area which is applied to a straight line [the parameter] (to which the straight line which is added along the diameter of the section—such that this added straight line subtends the exterior angle of the [vertex of the axial] triangle—has the same ratio as the square on the straight line drawn—parallel to the section's diameter—from the cone's vertex to the triangle's base has to the rectangle contained by the sections of the base which this straight line from the vertex makes when drawn), such that that applied area (which has as breadth the straight line on the diameter from the section's vertex to where the diameter is cut off by the straight line drawn from the section to the diameter) projects beyond (ὑπερβάλλον) by a figure (εἶδος), similar and similarly situated to

---

[*] The Greek of the phrase "the straight line to which the straight lines drawn ordinatewise to the diameter are applied in square," that is, ἡ παρ' ἥν δύνανται αἱ καταγόμεναι τεταγμένως ἐπὶ τὴν διάμετρον, soon becomes abbreviated to ἡ παρ' ἥν δύνανται αἱ καταγόμεναι and to ἡ παρ' ἥν δύνανται. We shall translate these abbreviations by the word "parameter." And we shall later on, after Proposition 14, shorten the long expression to "the parameter of the ordinates to the diameter." The Latin translation of ὀρθία (πλευρά) is *latus rectum*, which has become an English term too. (Tr.)

the rectangle contained by the straight line subtending the exterior angle of the [vertex of the axial] triangle and by the parameter. And let such a section be called an hyperbola (ὑπερβολή).

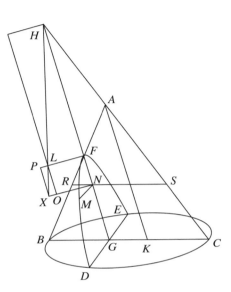

Let there be a cone whose vertex is the point *A* and whose base is the circle *BC*, and let it be cut by a plane through its axis, and let it make as a section the triangle *ABC* (I. 3). And let the cone also be cut by another plane cutting the base of the cone in the straight line *DE* perpendicular to *BC*, the base of the triangle *ABC*, and let this second cutting plane make as a section on the surface of the cone the line *DFE*, and let the diameter of the section *FG* (I. 7 and Def. 4) when produced meet *AC* one side of the triangle *ABC* beyond the vertex of the cone at the point *H*. And let the straight line *AK* be drawn through *A* parallel to the diameter of the section *FG*, and let it cut *BC*. And let the straight line *FL* be drawn from *F* perpendicular to *FG*, and let it be contrived that

sq. *KA* : rect. *BK, KC* :: *FH* : *FL*.

And let some point *M* be taken at random on the section, and through *M* let the straight line *MN* be drawn parallel to *DE*, and through *N* let the straight line *NOX* be drawn parallel to *FL*. And let the straight line *HL* be joined and produced to *X*, and let the straight lines *LO* and *XP* be drawn through *L* and *X* parallel to *FN*.

I say that *MN* is equal in square to the parallelogram *FX* which is applied to *FL*, having *FN* as breadth, and projecting beyond by a figure *LX* similar to the rectangle contained by *HF* and *FL*.

For let the straight line *RNS* be drawn through *N* parallel to *BC*; and *NM* is also parallel to *DE*. Therefore the plane through *MN* and *RS* is parallel to the plane through *BC* and *DE,* that is to the base of the cone (Eucl. XI. 15). Therefore if the plane is produced through *MN* and *RS,* the section will be a circle whose diameter is the straight line *RNS* (I.4). And *MN* is perpendicular to it. Therefore

$$\text{rect. } RN, NS = \text{sq. } MN.$$

And since

$$\text{sq. } AK : \text{rect. } BK, KC :: FH : FL,$$

and

$$\text{sq. } AK : \text{rect. } BK, KC :: AK : KC \text{ comp. } AK : KB \text{ (Eucl. VI. 23)},$$

therefore also

$$FH : FL :: AK : KC \text{ comp. } AK : KB.$$

But

$$AK : KC :: HG : GC :: HN : NS \text{ (Eucl. VI. 4)},$$

and

$$AK : KB :: FG : GB :: FN : NR.$$

Therefore

$$HF : FL :: HN : NS \text{ comp. } FN : NR.$$

And

rect. *HN, NF* : rect. *SN, NR* :: *HN* : *NS* comp. *FN* : *NR* (Eucl. VI. 23). Therefore also

rect. *HN, NF* : rect. *SN, NR* :: *HF* : *FL* :: *HN* : *NX* (Eucl. VI. 4). But, with the straight line *FN* taken as common height,

$$HN : NX :: \text{rect. } HN, NF : \text{rect. } FN, NX \text{ (Eucl. VI. 1)}.$$

Therefore also

rect. *HN, NF* : rect. *SN, NR* :: rect. *HN, NF* : rect. *XN, NF* (Eucl. V. 11). Therefore

$$\text{rect. } SN, NR = \text{rect. } XN, NF \text{ (Eucl. V. 9)}.$$

But it was shown

$$\text{sq. } MN = \text{rect. } SN, NR;$$

therefore also

$$\text{sq. } MN = \text{rect. } XN, NF.$$

But the rectangle contained by *XN* and *NF* is the parallelogram *XF*. Therefore the straight line *MN* is equal in square to *XF* which is applied to the straight line *FL,* having *FN* as breadth, and projecting beyond by the parallelogram *LX* similar to the rectangle contained by *HF* and *FL* (Eucl. VI. 24).

And let such a section be called an hyperbola, and let *LF* be called the straight line to which the straight lines drawn ordinatewise to *FG* are applied in square; and let the same straight line also be called the upright side, and the straight line *FH* the transverse side.

PROPOSITION 13

If a cone is cut by a plane through its axis and is also cut by another plane which on the one hand meets both sides of the axial triangle and which on the other hand, when extended, is neither parallel to the base nor subcontrariwise, and if the plane containing the base of the cone and the cutting plane meet in a straight line perpendicular either to the base of the axial triangle or to it produced, then any [straight] line which is drawn—parallel to the common section of the [base and cutting] planes—from the section of the cone to the diameter of the section will equal in square some area applied to a straight line [the parameter] (to which the diameter of the section has the same ratio as the square on the straight line drawn—parallel to the section's diameter—from the cone's vertex to the triangle's base has to the rectangle which is contained by the straight lines cut off [on the base] by this straight line in the direction of the sides of the [axial] triangle), an area which has as breadth the straight line on the diameter from the section's vertex to where the diameter is cut off by the straight line from the section to the diameter and which area is deficient (ἐλλεῖπον) by a figure similar and similarly situated to the rectangle contained by the diameter and parameter. And let such a section be called an ellipse (ἔλλειψις).

Let there be a cone whose vertex is the point *A* and whose base is the circle *BC,* and let it be cut by a plane through its axis, and let it make as a section the triangle *ABC*. And let it also be cut by another plane on the one hand

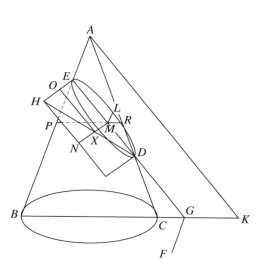

meeting both sides of the axial triangle and on the other extended neither parallel to the base of the cone nor subcontrariwise, and let it make as a section on the surface of the cone the line *DE*. And let the common section of the cutting plane and of the plane the base of the cone is in, be the straight line *FG* perpendicular to the straight line *BC,* and let the diameter of the section be the straight line *ED* (I. 7 and Def. 4). And let the straight line *EH* be drawn from *E* perpendicular to *ED*, and let the straight line *AK* be drawn through *A* parallel to *ED*, and let it be contrived that

$$\text{sq. } AK : \text{rect. } BK, KC :: DE : EH.$$

And let some point *L* be taken on the section, and let the straight line *LM* be drawn through *L* parallel to *FG*.

I say that the straight line *LM* is equal in square to some area which is applied to *EH*, having *EM* as breadth and deficient by a figure similar to the rectangle contained by *DE* and *EH*.

For let the straight line *DH* be joined, and on the one hand let the straight line *MXN* be drawn through *M* parallel to *HE*, and on the other let the straight lines *HN* and *XO* be drawn through *H* and *X* parallel to *EM*, and let the straight line *PMR* be drawn through *M* parallel to *BC*.

Since then *PR* is parallel to *BC*, and *LM* is also parallel to *FG*, therefore the plane through *LM* and *PR* is parallel to the plane through *FG* and *BC*, that is to the base of the cone (Eucl. XI. 15). If therefore a plane is extended through *LM* and *PR*, the section will be a circle whose diameter is *PR* (I. 4). And *LM* is perpendicular to it. Therefore

$$\text{rect. } PM, MR = \text{sq. } LM.$$

And since

$$\text{sq. } AK : \text{rect. } BK, KC :: ED : EH,$$

and

$$\text{sq. } AK : \text{rect. } BK, KC :: AK : KB \text{ comp. } AK : KC \text{ (Eucl. VI. 23),}$$

but

$$AK : KB :: EG : GB :: EM : MP \text{ (Eucl. VI. 4),}$$

and

$$AK : KC :: DG : GC :: DM : MR \text{ (Eucl. VI. 4),}$$

therefore

$$DE : EH :: EM : MP \text{ comp. } DM : MR.$$

But

rect. *EM, MD* : rect. *PM, MR* :: *EM* : *MP* comp. *DM* : *MR* (Eucl. VI. 23). Therefore

rect. *EM, MD* : rect. *PM, MR* :: *DE* : *EH* :: *DM* : *MX* (Eucl. VI. 4). And, with the straight line *ME* taken as common height,

$$DM : MX :: \text{rect. } DM, ME : \text{rect. } XM, ME \text{ (Eucl. VI. 1).}$$

Therefore also

rect. *DM, ME* : rect. *PM, MR* :: rect. *DM, ME* : rect. *XM, ME* (Eucl. V. 11). Therefore rect. *PM, MR* = rect. *XM, ME* (Eucl. V. 9).

But it was shown

rect. *PM, MR* = sq. *LM*;

therefore also

rect. *XM, ME* = sq. *LM*.

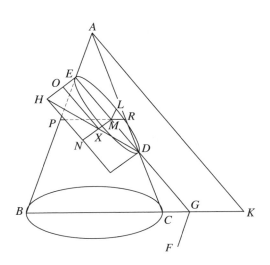

Therefore the straight line *LM* is equal in square to the parallelogram *MO* which is applied to the straight line *HE*, having *EM* as breadth and deficient by the figure *ON* similar to the rectangle contained by *DE* and *EH* (Eucl. VI. 24).

And let such a section be called an ellipse, and let *EH* be called the straight line to which the straight lines drawn ordinatewise to *DE* are applied in square, and let the same straight line also be called the upright side, and the straight line *ED* the transverse side.

PROPOSITION 14

If the vertically opposite surfaces are cut by a plane not through the vertex, the section on each of the two surfaces will be that which is called the hyperbola; and the diameter of the two sections will be the same straight line; and the straight lines, to which the straight lines drawn to the diameter parallel to the straight line in the cone's base are applied in square, are equal; and the transverse side of the figure, that between the vertices of the sections, is common. And let such sections be called opposite (ἀντικείμεναι).

Let there be the vertically opposite surfaces whose vertex is the point *A,* and let them be cut by a plane not through the vertex, and let it make as sections on the surface the lines *DEF* and *GHK*.

I say that each of the two sections *DEF* and *GHK* is the so-called hyperbola.

For let there be the circle *BDCF* along which the line generating the surface moves, and let the plane *XGOK* be extended parallel to it on the vertically opposite surface; and the straight lines *FD* and *GK* are common sections of the sections *GHK* and *FED*, and of the circles (I. 4). Then they will be parallel (Eucl. XI. 16). And let the straight line *LAU* be the axis of the conic surface, and the points *L* and U be the centers of the circles, and let a straight line drawn from *L* perpendicular to the straight line *FD* be produced to the points *B* and *C*, and let a plane be produced through the straight line *BC* and the axis. Then it will make as sections in the circles the parallel straight lines *XO* and *BC* (Eucl. XI. 16), and on the surface the straight lines *BAO* and *CAX* (I. 1 and Def. 1).

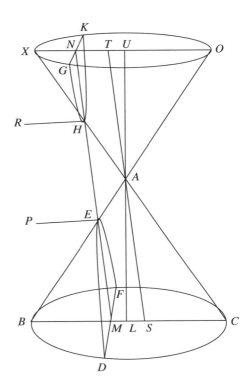

Then the straight line *XO* will be perpendicular to the straight line *GK*, since the straight line *BC* is also perpendicular to the straight line *FD*, and each of the two is parallel to the other (Eucl. XI. 10). And since the plane through the axis meets the sections in the points *M* and *N* within the [curved] lines [*FD* and *GK*], it is clear that the plane through the axis also cuts the [curved] lines. Let it cut them at *H* and *E*; therefore *M*, *E*, *H*, and *N* are points on the plane through the axis and in the plane the lines are in; therefore the line *MEHN* is a straight line (Eucl. XI. 3). It is also evident both that *X*, *H*, *A*, and *C* are in a straight line and *B*, *E*, *A*, and *O* also. For they are both on the conic surface and in the plane through the axis (I. 1).

Let then the straight lines *HR* and *EP* be drawn from *H* and *E* perpendicular to *HE*, and let the straight line *SAT* be drawn through *A* parallel to *MEHN*, and let it be contrived that

$$HE : EP :: \text{sq. } AS : \text{rect. } BS, SC,$$

and

$$EH : HR :: \text{sq. } AT : \text{rect. } OT, TX.$$

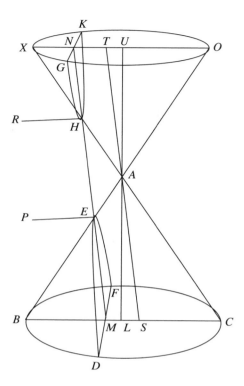

Since then a cone, whose vertex is the point *A* and whose base is the circle *BC,* has been cut by a plane through its axis, and it has made as a section the triangle *ABC*; and it has also been cut by another plane cutting the base of the cone in the straight line *DMF* perpendicular to the straight line *BC,* and it has made as a section on the surface the line *DEF*; and the diameter *ME* produced has met one side of the axial triangle beyond the vertex of the cone, and through the point *A* the straight line *AS* has been drawn parallel to the diameter of the section *EM,* and from *E* the straight line *EP* has been drawn perpendicular to the straight line *EM,* and

$$EH : EP :: \text{sq. } AS : \text{rect. } BS, SC,$$

therefore the section *DEF* is an hyperbola (I. 12), and *EP* is the straight line to which the straight lines drawn ordinatewise to *EM* are applied in square, and the straight line *HE* is the transverse side of the figure. And likewise *GHK* is also an hyperbola whose diameter is the straight line *HN* and whose straight line to which the straight lines drawn ordinatewise to *HN* are applied is *HR,* and the transverse side of whose figure is *HE.*

I say that the straight line *HR* is equal to the straight line *EP*.

For since *BC* is parallel to *XO*,
$$AS : SC :: AT : TX,$$
and
$$AS : SB :: AT : TO.$$
But
$$\text{sq. } AS : \text{rect. } BS, SC :: AS : SC \text{ comp. } AS : SB \text{ (Eucl. VI. 23)}$$
and
$$\text{sq. } AT : \text{rect. } XT, TO :: AT : TX \text{ comp. } AT : TO;$$
therefore
$$\text{sq. } AS : \text{rect. } BS, SC :: \text{sq. } AT : \text{rect. } XT, TO.$$
Also
$$\text{sq. } AS : \text{rect. } BS, SC :: HE : EP,$$
and
$$\text{sq. } AT : \text{rect. } XT, TO :: HE : HR.$$
Therefore also
$$HE : EP :: EH : HR \text{ (Eucl. V. 11).}$$
Therefore
$$EP = HR \text{ (Eucl. V. 9).}$$

PROPOSITION 15

If in an ellipse a straight line, drawn ordinatewise from the midpoint of the diameter, is produced both ways to the section, and if it is contrived that as the produced straight line is to the diameter so is the diameter to some straight line, then any straight line which is drawn—parallel to the diameter—from the section to the produced straight line will equal in square the area which is applied to this third proportional and which has as breadth the produced straight line from the section to where the straight line drawn parallel to the diameter cuts it off, but such that this area is deficient by a figure similar to the rectangle contained by the produced straight line to which the straight lines are drawn and by the parameter [i.e., the third proportional]; and if the straight line drawn parallel to the diameter is further produced to the other side of the section, this drawn straight line will be bisected by the produced straight line to which it has been drawn.

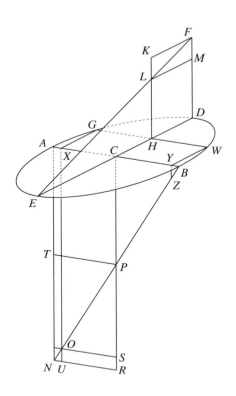

Let there be an ellipse whose diameter is the straight line *AB*, and let *AB* be bisected at the point *C*, and through *C* let the straight line *DCE* be drawn ordinatewise and produced both ways to the section, and from the point *D* let the straight line *DF* be drawn perpendicular to *DE*. And let it be contrived that

$$DE : AB :: AB : DF.$$

And let some point *G* be taken on the section, and through *G* let the straight line *GH* be drawn parallel to *AB*, and let *EF* be joined, and through *H* let the straight line *HL* be drawn parallel to *DF*, and through *F* and *L* let the straight lines *FK* and *LM* be drawn parallel to *HD*.

I say that the straight line *GH* is equal in square to the area *DL* which is applied to the straight line *DF*, having as breadth the straight line *DH* and deficient by a figure *LF* similar to the rectangle contained by *ED* and *DF*.

For let *AN* be the parameter of the ordinates to *AB*, and let *BN* be joined; and through *G* let the straight line *GX* be drawn parallel to *DE*, and through *X* and *C* let the straight lines *XO* and *CP* be drawn parallel to *AN*, and through *N*, *O*, and *P* let the straight lines *NU*, *OS*, and *TP* be drawn parallel to *AB*. Therefore

$$\text{sq. } DC = \text{ar. } AP,$$
$$\text{sq. } GX = \text{ar. } AO \text{ (I. 13)}.$$

And since

$$BA : AN :: BC : CP :: PT : TN \text{ (Eucl. VI. 4)},$$

and

$$BC = CA = TP,$$

and

$$CP = TA,$$

therefore

$$\text{ar. } AP = \text{ar. } TR,$$

and
$$\text{ar. } XT = \text{ar. } TU.$$
Since also
$$\text{ar. } OT = \text{ar. } OR \text{ (Eucl. I. 43)},$$
and area NO is common, therefore
$$\text{ar. } TU = \text{ar. } NS.$$
But
$$\text{ar. } TU = \text{ar. } TX,$$
and TS is common. Therefore
$$\text{ar. } NP = \text{ar. } PA = \text{ar. } AO + \text{ar. } PO;$$
and so
$$\text{ar. } PA - \text{ar. } AO = \text{ar. } PO.$$
Also
$$\text{ar. } AP = \text{sq. } CD, \ \text{ar. } AO = \text{sq. } XG,$$
and
$$\text{ar. } OP = \text{rect. } OS, \, SP;$$
therefore
$$\text{sq. } CD - \text{sq. } GX = \text{rect. } OS, \, SP.$$
Since also the straight line DE has been cut into equal parts at C, and into unequal parts at H, therefore
$$\text{rect. } EH, \, HD + \text{sq. } CH = \text{sq. } CD \text{ (Eucl. II. 5)},$$
or
$$\text{rect. } EH, \, HD + \text{sq. } XG = \text{sq. } CD.$$
Therefore
$$\text{sq. } CD - \text{sq. } XG = \text{rect. } EH, \, HD;$$
but
$$\text{sq. } CD - \text{sq. } XG = \text{rect. } OS, \, SP;$$
therefore
$$\text{rect. } EH, \, HD = \text{rect. } OS, \, SP.$$

And since
$$DE : AB :: AB : DF,$$
therefore
$$DE : DF :: \text{sq. } DE : \text{sq. } AB \text{ (Eucl. VI. 19 porism)},$$
that is
$$DE : DF :: \text{sq. } CD : \text{sq. } CB \text{ (Eucl. V. 15)}.$$
And
$$\text{rect. } PC, \, CA = \text{rect. } PC, \, CB = \text{sq. } CD \text{ (I. 13)};$$
and since
$$DE : DF :: EH : HL \text{ (Eucl. VI. 4)},$$
or

$$DE : DF :: \text{rect. } EH, HD : \text{rect. } DH, HL \text{ (Eucl. VI. 1, V. 11)},$$

and since

$$DE : DF :: \text{rect. } PC, CB : \text{sq. } CB,$$

and

$$\text{rect. } PC, CB : \text{sq. } CB :: \text{rect. } OS, SP : \text{sq. } OS,^*$$

therefore also

$$\text{rect. } EH, HD : \text{rect. } DH, HL :: \text{rect. } OS, SP : \text{sq. } OS.$$

And

$$\text{rect. } EH, HD = \text{rect. } OS, SP;$$

therefore

$$\text{rect. } DH, HL = \text{sq. } OS$$
$$= \text{sq. } GH.$$

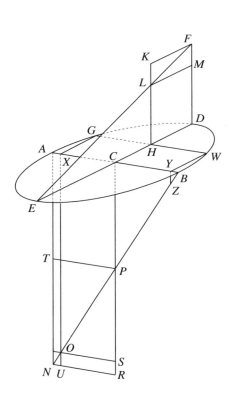

Therefore the straight line *GH* is equal in square to the area *DL* which is applied to the straight line *DF*, deficient by a figure *FL* similar to the rectangle contained by *ED* and *DF* (Eucl. VI. 24).

I say then that also, if produced to the other side of the section, the straight line *GH* will be bisected by the straight line *DE*.

For let it be produced and let it meet the section at *W*, and let the

---

* This follows from the proportions
$$PC : CB :: PS : OS \text{ (Eucl. VI. 4)},$$
and
$$PC : CB :: \text{rect. } PC, CB : \text{sq. } CB,$$
and
$$PS : OS :: \text{rect. } PS, OS : \text{sq. } OS \text{ (Eucl. VI. 1)}.$$

(Tr.)

straight line *WY* be drawn through *W* parallel to *GX*, and through *Y* let the straight line *YZ* be drawn parallel to *AN*. And since

$$GX = WY,$$

therefore also

$$\text{sq. } GX = \text{sq. } WY.$$

But

$$\text{sq. } GX = \text{rect. } AX, XO \text{ (I. 13)},$$

and

$$\text{sq. } WY = \text{rect. } AY, YZ \text{ (I. 13)}.$$

Therefore

$$OX : ZY :: YA : AX \text{ (Eucl. VI. 16)}.$$

And

$$OX : ZY :: XB : BY \text{ (Eucl. VI. 4)};$$

therefore also

$$YA : AX :: XB : BY.$$

And *separando*

$$YX : AX :: YX : BY \text{ (Eucl. V. 17)}.$$

Therefore

$$AX = YB.$$

And also

$$AC = CB;$$

therefore also the remainders

$$XC = CY;$$

and so also

$$GH = HW.$$

Therefore the straight line *HG*, produced to the other side of the section, is bisected by the straight line *DH*.

PROPOSITION 16

If through the midpoint of the transverse side of the opposite sections a straight line be drawn parallel to a straight line drawn ordinatewise, it will be a diameter of the opposite sections, conjugate to the diameter just mentioned.

Let there be the opposite sections whose diameter is the straight line *AB*, and let *AB* be bisected at *C*, and through *C* let the straight line *CD* be drawn parallel to a straight line drawn ordinatewise.

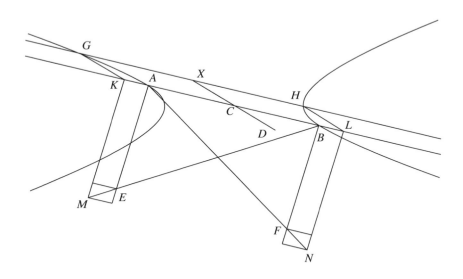

I say that the straight line *CD* is a diameter conjugate to *AB*.

For let the straight lines *AE* and *BF* be the parameters, and let the straight lines *AF* and *BE* be joined and produced, and let some point *G* be taken at random on either section, and through *G* let the straight line *GH* be drawn parallel to *AB*, and from *G* and *H* let the straight lines *GK* and *HL* be drawn ordinatewise, and through *K* and *L* let the straight lines *KM* and *LN* be drawn parallel to *AE* and *BF*. Since then

$$GK = HL \text{ (Eucl. I. 34)},$$

therefore also

$$\text{sq. } GK = \text{sq. } HL,$$

But

$$\text{sq. } GK = \text{rect. } AK, KM \text{ (I. 12)},$$

and

$$\text{sq. } HL = \text{rect. } BL, LN \text{ (I. 12)};$$

therefore

$$\text{rect. } AK, KM = \text{rect. } BL, LN.$$

And since

$$AE = BF \text{ (I. 14)},$$

therefore

$$AE : AB :: BF : BA \text{ (Eucl.V. 7)}.$$

But

$$AE : AB :: MK : KB \text{ (Eucl. VI. 4)},$$

and as

$$BF : BA :: NL : LA \text{ (Eucl. VI. 4)}.$$

And therefore

$$MK : KB :: NL : LA.$$

But, with KA taken as common height,

$$MK : KB :: \text{rect. } MK, KA : \text{rect. } BK, KA,$$

and, with BL taken as common height,

$$NL : LA :: \text{rect. } NL, LB : \text{rect. } AL, LB.$$

And therefore

$$\text{rect. } MK, KA : \text{rect. } BK, KA :: \text{rect. } NL, LB : \text{rect. } AL, LB.$$

And alternately

$$\text{rect. } MK, KA : \text{rect. } NL, LB :: \text{rect. } BK, KA : \text{rect. } AL, LB \text{ (Eucl. V. 16)}.$$

And

$$\text{rect. } AK, KM = \text{rect. } BL, LN \text{ (above)};$$

therefore

$$\text{rect. } BK, KA = \text{rect. } AL, LB;$$

therefore

$$AK = LB.^{*}$$

But also

$$AC = CB,$$

and therefore

$$KC = CL;$$

and so also

$$GX = XH.$$

Therefore the straight line GH has been bisected by the straight line XCD and is parallel to the straight line AB. Therefore the straight line XCD is a diameter and conjugate to the straight line AB (Defs. 5, 6).

---

* The intermediary steps to this conclusion are as follows. If

$$\text{rect. } BK, KA = \text{rect. } AL, LB,$$

then

$$BK : LB :: AL : KA \text{ (Eucl. VI. 16)}.$$

But

$$KB + BL : LB :: LA + AK : KA \text{ (Eucl. V. 18)};$$

that is,

$$KL : LB :: LK : KA.$$

Therefore

$$LB = KA \text{ (Eucl. V. 7 porism, V. 9)}.$$

(Ed.)

SECOND DEFINITIONS

9. Let the midpoint of the diameter of both the hyperbola and the ellipse be called the center of the section, and let the straight line drawn from the center to meet the section be called the radius of the section.

10. And likewise let the midpoint of the transverse side of the opposite sections be called the center.

11. And let the straight line drawn from the center parallel to an ordinate, being a mean proportional to the sides of the figure ($\epsilon \hat{\iota} \delta o \varsigma$) and bisected by the center, be called the second diameter.

PROPOSITION 17

If in a section of a cone a straight line is drawn from the vertex of the line, and parallel to an ordinate, it will fall outside the section (cf. Eucl. III. 16).

Let there be a section of a cone, whose diameter is the straight line *AB*.

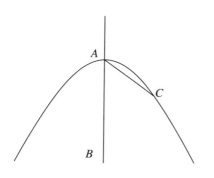

I say that the straight line drawn from the vertex, that is, from the point *A,* parallel to an ordinate, will fall outside the section.

For if possible, let it fall within as *AC*. Since then a point *C* has been taken at random on a section of a cone, therefore the straight line drawn from the point *C* within the section parallel to an ordinate will meet the diameter *AB* and will be bisected by it (I. 7). Therefore the straight line *AC* produced will be bisected by the straight line *AB*. And this is absurd. For the straight line *AC*, if produced, will fall outside the section (I. 10). Therefore the straight line drawn from the point *A* parallel to an ordinate will not fall within the line; therefore it will fall outside; and therefore it is tangent to the section.

PROPOSITION 18

If a straight line, meeting a section of a cone and produced both ways, falls outside the section, and some point is taken within the section, and through it a parallel to the straight line meeting the section is drawn, the parallel so drawn, if produced both ways, will meet the section.

Let there be a section of a cone and the straight line *AFB* meeting it, and let it fall, when produced both ways, outside the section. And let some point *C* be taken within the section, and through *C* let the straight line *CD* be drawn parallel to the straight line *AB*.

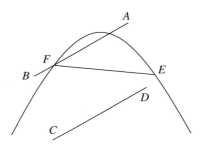

I say that the straight line *CD* produced both ways will meet the section.

For let some point *E* be taken on the section, and let the straight line *EF* be joined. And since the straight line *AB* is parallel to *CD*, and some straight line *EF* meets *AB*, therefore *CD* produced will also meet *EF*. And if it meets *EF* between the points *E* and *F*, it is evident that it also meets the section, but if beyond the point *E*, that it will first meet the section. Therefore *CD* produced to the side of points *D* and *E* meets the section. Then

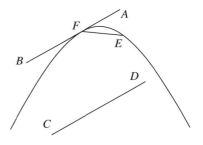

likewise we could show that, produced to the side of points *F* and *B*, it also meets it. Therefore the straight line *CD* produced both ways will meet the section.

PROPOSITION 19

In every section of a cone, any straight line drawn from the diameter parallel to an ordinate will meet the section.

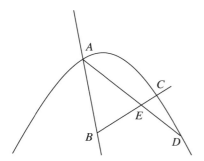 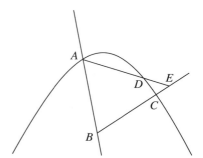

Let there be a section of a cone whose diameter is the straight line *AB,* and let some point *B* be taken on the diameter, and through *B* let the straight line *BC* be drawn parallel to an ordinate.

I say that the straight line *BC* produced will meet the section.

For let some point *D* be taken on the section. But *A* is also on the section; therefore the straight line joined from *A* to *D* will fall within the section (I. 10). And since the straight line drawn from *A* parallel to an ordinate falls outside the section (I. 17), and the straight line *AD* meets it, and the straight line *BC* is parallel to the ordinate, therefore *BC* will also meet *AD.* And if it meets *AD* between the points *A* and *D,* it is evident that it will also meet the section, but, if beyond point *D* as at *E,* that it will first meet the section. Therefore the straight line drawn from *B* parallel to an ordinate will meet the section.

PROPOSITION 20

If in a parabola two straight lines are dropped ordinatewise to the diameter, the squares on them will be to each other as the straight lines cut off by them on the diameter beginning from the vertex* are to each other.

---

* These are usually called "abscissas" from the Latin *abscindere,* to cut off. (Tr.)

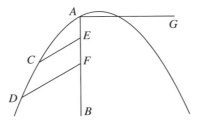

Let there be a parabola whose diameter is the straight line *AB*, and let some points *C* and *D* be taken on it, and from the points *C* and *D* let the straight lines *CE* and *DF* be dropped ordinatewise to *AB*.

I say that

$$\text{sq. } DF : \text{sq. } CE :: FA : AE.$$

For let *AG* be the parameter; therefore

$$\text{sq. } DF = \text{rect. } FA, AG,$$

and

$$\text{sq. } CE = \text{rect. } EA, AG \text{ (I. 11)}.$$

Therefore

$$\text{sq. } DF : \text{sq. } CE :: \text{rect. } FA, AG : \text{rect. } EA, AG.$$

But

$$\text{rect. } FA, AG : \text{rect. } EA, AG :: FA : AE. \text{ (Eucl. VI. 1)};$$

and therefore

$$\text{sq. } DF : \text{sq. } CE :: FA : AE.$$

PROPOSITION 21

If in an hyperbola or ellipse or in the circumference of a circle straight lines are dropped ordinatewise to the diameter, the squares on them will be to the areas contained by the straight lines cut off by them beginning from the ends of the transverse side of the figure, as the upright side of the figure is to the transverse, and to each other as the areas contained by the straight lines cut off (abscissas), as we have said.

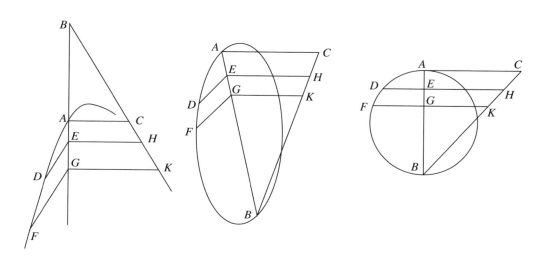

Let there be an hyperbola or ellipse or circumference of a circle whose diameter is *AB* and whose parameter is the straight line *AC*, and let the straight lines *DE* and *FG* be dropped ordinatewise to the diameter.

I say that

$$\text{sq. } FG : \text{rect. } AG, GB :: AC : AB$$

and

$$\text{sq. } FG : \text{sq. } DE :: \text{rect. } AG, GB : \text{rect. } AE, EB.$$

For let the straight line *BC* determining [διορίζουσα] the figure be joined, and through *E* and *G* let the straight lines *EH* and *GK* be drawn parallel to the straight line *AC*. Therefore

$$\text{sq. } FG = \text{rect. } KG, GA$$
$$\text{sq. } DE = \text{rect. } HE, EA \text{ (I. 12,13).}$$

And since

$$KG : GB :: CA : AB;$$

and, with *AG* taken as common height,

$$KG : GB :: \text{rect. } KG, GA : \text{rect. } BG, GA,$$

therefore

$$CA : AB :: \text{rect. } KG, GA : \text{rect. } BG, GA$$

or

$$CA : AB :: \text{sq. } FG : \text{rect. } BG, GA.$$

Then also for the same reasons

$$CA : AB :: \text{sq. } DE : \text{rect. } BE, EA.$$

And therefore
$$\text{sq. } FG : \text{rect. } BG,\, GA :: \text{sq. } DE : \text{rect. } BE,\, EA;$$
alternately
$$\text{sq. } FG : \text{sq. } DE :: \text{rect. } BG,\, GA : \text{rect. } BE,\, EA.^{*}$$

PROPOSITION 22

If a straight line cuts a parabola or hyperbola in two points, not meeting the diameter inside, it will, if produced, meet the diameter of the section outside the section.

Let there be a parabola or hyperbola whose diameter is the straight line *AB*, and let some straight line cut the section in two points *C* and *D*.

I say that the straight line *DC*, if produced, will meet the straight line *AB* outside the section.

For let the straight lines *CE* and *DB* be dropped ordinatewise from *C* and *D*; and first let the section be a parabola. Since then in the parabola
$$\text{sq. } CE : \text{sq. } DB :: EA : AB \text{ (I. 20)},$$
and
$$EA > AB,$$
therefore also
$$\text{sq. } CE > \text{sq. } DB.$$

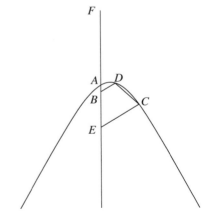

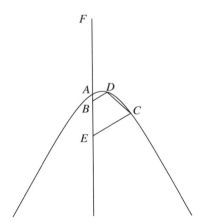

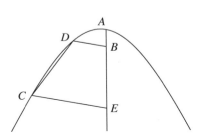

And so also
$$CE > DB.$$
And they are parallel; therefore *CD* produced will meet the diameter *AB* outside the section (I. 10).

But then let it be an hyperbola [with the transverse side *AF*]. Since then in the hyperbola
$$\text{sq. } CE : \text{sq. } DB :: \text{rect. } FE, EA : \text{rect. } FB, BA \text{ (I. 21)},$$
therefore also
$$\text{sq. } CE > \text{sq. } DB.$$
And they are parallel; therefore the straight line *CD* produced will meet the diameter of the section outside the section.

PROPOSITION 23

*If a straight line lying between the two (conjugate) diameters**cuts the ellipse, it will, when produced, meet each of the diameters outside the section.*

 So far Apollonius, by theorems I. 6, 13, 15, has shown, for every ellipse, the existence of at least one diameter and of one set of conjugate diameters, but of no more. He can therefore now speak of "the two diameters." Later on he will show the existence of an infinite number of such sets. The same is true of hyperbolas. (Tr.)

Let there be an ellipse whose diameters are the straight lines *AB* and *CD* (I. 15), and let some straight line *EF* lying between the diameters *AB* and *CD* cut the section.

I say that the straight line *EF*, when produced, will meet each of the straight lines *AB* and *CD* outside the section.

For let the straight lines *GE* and *FH* be dropped ordinatewise from *E* and *F* to *AB*; and the straight lines *EK* and *FL* ordinatewise to *CD*. Therefore

sq. *EG* : sq. *FH* :: rect. *BG, GA* : rect. *BH, HA* (I. 21)

and

sq. *FL* : sq. *EK* :: rect. *DL, LC* : rect. *DK, KC* (I. 21).

And

rect. *BG, GA* > rect. *BH, HA,*

for the point *G* is nearer the midpoint (Eucl. II. 5 [*]);
and

rect. *DL, LC* > rect. *DK, KC*;

therefore also

sq. *GE* > sq. *FH,*

and

sq. *FL* > sq. *EK*;

therefore also

GE > *FH,*

and

FL > *EK*.

And *GE* is parallel to *FH*, and *FL* to *EK*; therefore the straight line *EF* produced will meet each of the diameters *AB* and *CD* outside the section (I. 10; Eucl. I. 33).

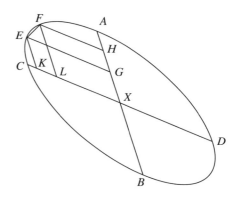

---

[*] By Euclid II.5,

sq. *XB* = rect. *BG, GA* + sq. *GX*
sq. *XB* = rect. *BH, HA* + sq. *HX.*

Therefore

rect. *GB, GA* + sq. *GX* = rect. *BH, HA* + sq. *HX.*

But

HX > *GX*

and so

sq. *HX* > sq. *GX.*

Therefore

rect. *BG, GA* > rect. *BH, HA.*

(Ed.)

PROPOSITION 24

If a straight line, meeting a parabola or hyperbola at a point, when produced both ways, falls outside the section, then it will meet the diameter.

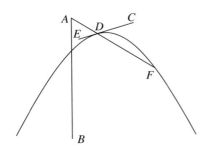

Let there be a parabola or hyperbola whose diameter is the straight line *AB,* and let the straight line *CDE* meet it at *D,* and, when produced both ways, let it fall outside the section.

I say that it will meet the diameter *AB.*

For let some point *F* be taken on the section, and let the straight line *DF* be joined; therefore *DF* produced will meet the diameter of the section (I. 22). Let it meet it at *A*; and the straight line *CDE* lies between the section and the straight line *FDA.* And therefore the line *CDE* produced will meet the diameter outside the section.

PROPOSITION 25

If a straight line, meeting an ellipse between the two (conjugate) diameters and produced both ways, falls outside the section, it will meet each of the diameters.

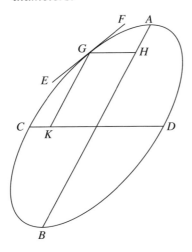

Let there be an ellipse whose diameters are the straight lines *AB* and *CD* (I. 15), and let *EF,* some straight line between the two diameters, meet it at *G,* and produced both ways fall outside the section.

I say that the straight line *EF* will meet each of the straight lines *AB* and *CD.*

Let the straight lines *GH* and *GK* be dropped ordinatewise to the straight lines *AB* and *CD* respectively. Since *GK* is parallel to *AB* (I. 15), and some straight line *GF* has met *GK*, therefore it will also meet *AB*. Then likewise *EF* will also meet *CD*.

PROPOSITION 26

If in a parabola or hyperbola a straight line is drawn parallel to the diameter of the section, it will meet the section in one point only.

Let there first be a parabola whose diameter is the straight line *ABC*, and whose upright side is the straight line *AD*, and let the straight line *EF* be drawn parallel to *AB*.

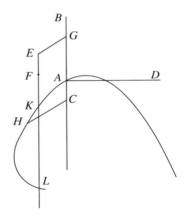

I say that the straight line *EF* produced will meet the section.

For let some point *E* be taken on *EF*, and from *E* let the straight line *EG* be drawn parallel to an ordinate, and let

rect. *DA, AC* > sq. *GE*,

and from *C* let *CH* be erected ordinatewise (I. 19). Therefore

sq. *HC* = rect. *DA, AC* (I. 11).

But

rect. *DA, AC* > sq. *EG*;

therefore

sq. *HC* > sq. *EG*;

therefore

HC > *EG*.

And they are parallel; therefore the straight line *EF* produced cuts the straight line *HC*; and so it will also meet the section.

Let it meet it at the point *K*.

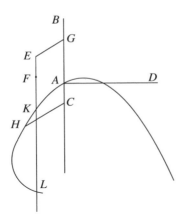

Then I say also that it will meet it in the one point *K* only.

For if possible, let it also meet it in the point *L*. Since then a straight line cuts a parabola in two points, if produced it will meet the diameter of the section (I. 22). And this is absurd, for it is supposed parallel. Therefore the straight line *EF* produced meets the section in only one point.

Next let the section be an hyperbola, and the straight line *AB* the transverse side of the figure, and the straight line *AD* the upright side, and let

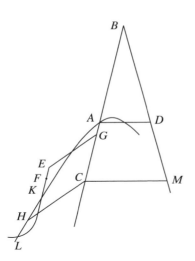

the straight line *DB* be joined and produced. Then with the same things being constructed, let the straight line *CM* be drawn from *C* parallel to *AD*. Since then

$$\text{rect. } MC,\ CA > \text{rect. } DA,\ AC,$$

and

$$\text{sq. } CH = \text{rect. } MC,\ CA,$$

and

$$\text{rect. } DA,\ AC > \text{sq. } GE,$$

therefore also

$$\text{sq. } CH > \text{sq. } GE.$$

And so also

$$CH > GE,$$

and the same things as in the first case will come to pass.

PROPOSITION 27

If a straight line [within the section] cuts the diameter of a parabola, then produced both ways it will meet the section.

Let there be a parabola whose diameter is the straight line *AB,* and let some straight line *CD* cut it within the section.

I say that the straight line *CD* produced both ways will meet the section.

For let some straight line *AE* be drawn from *A* parallel to an ordinate; therefore the straight line *AE* will fall outside the section (I. 17).

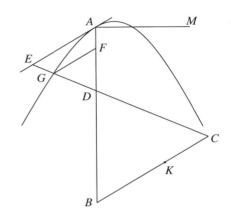

Then either the straight line *CD* is parallel to the straight line *AE* or not.

If now it is parallel to it, it has been dropped ordinatewise, so that produced both ways it will meet the section (I. 18).

Next let it not be parallel to *AE*, but produced let it meet *AE* at *E*. Then it is evident that it meets the section the side the point *E* is on; for if it meets *AE*, *a fortiori* it cuts the section.

I say that, produced the other way, it also meets the section. For let the straight line *MA* be the parameter and the straight line *GF* an ordinate, and let

$$\text{sq. } AD = \text{rect. } BA, AF \text{ (Eucl. VI. 11, VI. 17)},$$

and let the straight line *BK*, parallel to the ordinate, meet the straight line *DC* at *C*. Since

$$\text{rect. } BA, AF = \text{sq. } AD,$$

hence

$$AB : AD :: AD : AF;$$

and therefore,

$$BD : DF :: AB : AD \text{ (Eucl. V. 19)}.$$

Therefore also

$$\text{sq. } BD : \text{sq. } DF :: \text{sq. } AB : \text{sq. } AD.$$

But since

$$\text{sq. } AD = \text{rect. } BA, AF,$$

hence

$$AB : AF :: \text{sq. } AB : \text{sq. } AD :: \text{sq. } BD : \text{sq. } FD \text{ (Eucl. VI. 20, Porism)}.$$

But

$$\text{sq. } BD : \text{sq. } DF :: \text{sq. } BC : \text{sq. } FG,$$

and

$$AB : AF :: \text{rect. } BA, AM : \text{rect. } FA, AM.$$

Therefore

$$\text{sq. } BC : \text{sq. } FG :: \text{rect. } BA, AM : \text{rect. } FA, AM;$$

and alternately

sq. BC : rect. BA, AM :: sq. FG : rect. FA, AM.

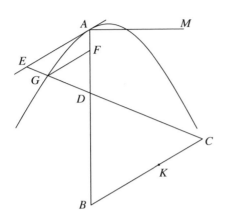

But

$$\text{sq. } FG = \text{rect. } FA,\ AM$$

because of the section (I. 11). Therefore also

$$\text{sq. } BC = \text{rect. } BA,\ AM.$$

But the straight line *AM* is the upright side,* and the straight line *BC* is parallel to an ordinate. Therefore the section passes through the point *C* (I. 11 converse), and the straight line *CD* meets the section at the point *C*.

PROPOSITION 28

If a straight line touches one of the opposite sections, and some point is taken within the other section, and through it a straight line is drawn parallel to the tangent, then produced both ways, it will meet the section.

Let there be opposite sections whose diameter is the straight line *AB*, and let some straight line *CD* touch the section *A*, and let some point *E* be taken within the other section, and through *E* let the straight line *EF* be drawn parallel to the straight line *CD*.

I say that the straight line *EF* produced both ways will meet the section.

Since then it has been proved that the straight line *CD* produced will meet the diameter *AB* (I. 24), and *EF* is parallel to it, therefore *EF* produced will meet the diameter. Let it meet it at *G*, and let *AH* be made equal to *GB*, and through *H* let *HK* (I. 18) be drawn parallel to *EF*, and let the straight line *KL* be dropped ordinatewise, and let *GM* be made equal to *LH*, and let the straight line *MN* be drawn parallel to an ordinate and let *GN* be further

---

* The text reads πλαγία, which is impossible. I have corrected to ὀρθία. (Tr.)

produced in the same straight line. And since *KL* is parallel to *MN*, and *KH* to *GN*, and *LM* is one straight line, triangle *KHL* is similar to triangle *GMN*. And

$$LH = GM;$$

therefore

$$KL = MN.$$

And so also

$$\text{sq. } KL = \text{sq. } MN.$$

And since

$$LH = GM,$$

and

$$AH = BG,$$

and *AB* is common, therefore

$$BL = AM;$$

therefore

$$\text{rect. } BL,\, LA = \text{rect. } AM,\, MB.$$

Therefore

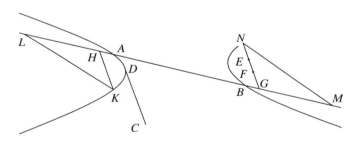

$$\text{rect. } BL,\, LA : \text{sq. } LK :: \text{rect. } AM,\, MB : \text{sq. } MN.$$

And

$$\text{rect. } BL,\, LA : \text{sq. } LK :: \text{the transverse : the upright (I. 21)};$$

therefore also

$$\text{rect. } AM,\, MB : \text{sq. } MN :: \text{the transverse : the upright.}$$

Therefore the point *N* is on the section. Therefore the straight line *EF* produced will meet the section at the point *N* (I. 21).

Likewise then it could be shown that produced to the other side it will meet the section.

PROPOSITION 29

If in opposite sections a straight line is drawn through the center to meet either of the sections, then produced it will cut the other section.

Let there be opposite sections whose diameter is the straight line *AB*, and whose center is the point *C*, and let the straight line *CD* cut the section *AD*.

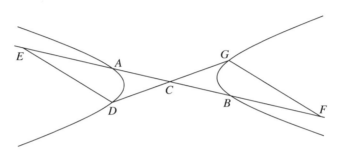

I say that it will also cut the other section.

For let the straight line *ED* be dropped ordinatewise, and let the straight line *BF* be made equal to the straight line *AE*, and let the straight line *FG* be drawn ordinatewise (I. 19). And since

$$EA = BF,$$

and *AB* is common, therefore

$$\text{rect. } BE, EA = \text{rect. } BF, FA.$$

And since

$$\text{rect. } BE, EA : \text{sq. } DE :: \text{the transverse} : \text{the upright (I. 21),}$$

but also

$$\text{rect. } BF, FA : \text{sq. } FG :: \text{the transverse} : \text{the upright (I. 21),}$$

therefore also

$$\text{rect. } BE, EA : \text{sq. } DE :: \text{rect. } BF, FA : \text{sq. } FG \text{ (I. 14).}$$

But

$$\text{rect. } BE, EA = \text{rect. } BF, FA;$$

therefore also

$$\text{sq. } DE = \text{sq. } FG.$$

Since then
$$EC = CF,$$
and
$$DE = FG,$$
and *EF* is a straight line, and *ED* is parallel to *FG*, therefore *DG* is also a straight line (Eucl. VI. 32). And therefore *CD* will also cut the other section.

PROPOSITION 30

If in an ellipse or in opposite sections a straight line is drawn in both directions from the center, meeting the section, it will be bisected at the center.

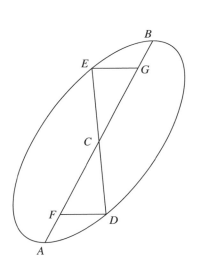
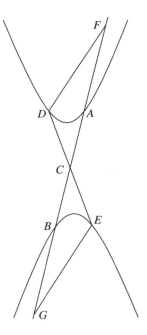

Let there be an ellipse or opposite sections, and their diameter the straight line *AB,* and their center *C,* and through *C* let some straight line *DCE* be drawn (I. 29).

I say that the straight line *CD* is equal to the straight line *CE.*

For let the straight lines *DF* and *EG* be drawn ordinatewise. And since

> rect. *BF, FA* : sq. *FD* :: the transverse : the upright (I. 21),

but also

> rect. *AG, GB* : sq. *GE* :: the transverse : the upright (I. 21),

therefore also

> rect. *BF,FA* : sq. *FD* :: rect. *AG, GB* : sq. *GE* (Eucl. V. 11).

And alternately

> rect. *BF, FA* : rect. *AG, GB* :: sq. *FD* : sq. *GE*.

But

> sq. *FD* : sq. *GE* :: sq. *FC* : sq. *CG* (Eucl. VI. 4, V. 16, VI. 22);

therefore alternately

> rect. *BF, FA* : sq. *FC* ::
> rect. *AG, GB* : sq. *CG*.

Therefore also, *componendo*[*] in the case of the ellipse, and inversely and *convertendo*[**] in the case of the opposite sections,

> sq. *AC* : sq. *CF* ::
> sq. *BC* : sq. *CG* (Eucl. II. 5,6);

and alternately. But

> sq. *CB* = sq. *AC*;

therefore also

> sq. *CG* = sq. *CF*.

Therefore

> *CG* = *CF*.

And the straight lines *DF* and *GE* are parallel; therefore also

> *DC* = *CE*.

---

[*]By Euclid V. 18 (*componendo*),

> rect *BF, FA* + sq. *FC* : sq. *FC* :: rect *AG, GB* + sq. *CG* : sq. *CG*.

By Euclid II.5,

> sq. *AC* = rect. *BF, FA* + sq. *CF*
> sq. *BC* = rect. *AG, GB* + sq. *CG*.

Therefore, substituting,

> sq. *AC* : sq. *FC* :: sq. *BC* : sq. *CG*.

(Ed.)

[**]By Euclid V. 19 porism (*convertendo*), and inverting,

> sq. *CF* − rect. *BF, FA* : sq. *CF* :: sq. *CG* − rect. *AG, GB* : sq. *CG*.

By Euclid II. 6,

> sq. *CF* − rect. *BF, FA* = sq. *AC*
> sq. *CG* − rect. *AG, GB* = sq. *CB*.

Substituting,

> sq. *AC* : sq. *FC* :: sq. *BC* : sq. *CG*.

(Ed.)

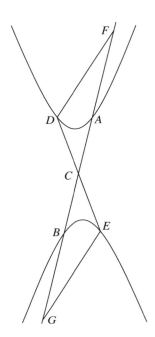

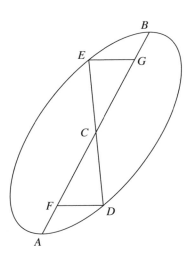

PROPOSITION 31

If on the transverse side of the figure of an hyperbola some point be taken cutting off from the vertex of the section not less than half of the transverse side of the figure, and a straight line be drawn from it to meet the section, then, when further produced, it will fall within the section on the near side of the section.

Let there be an hyperbola whose diameter is the straight line *AB,* and let *C* some point on the diameter be taken cutting off the straight line *CB* not less than half of *AB,* and let some straight line *CD* be drawn to meet the section.

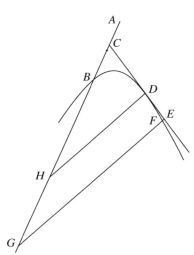

I say that the straight line *CD* produced will fall within the section.

For if possible, let it fall outside the section as the line *CDE* (I. 24), and from *E* a point at random let the straight line *EG* be dropped ordinatewise, also *DH*; and first let

$$AC = CB.$$

And since

sq. EG : sq. DH > sq. FG : sq. DH (Eucl. V. 8, VI. 22),

but

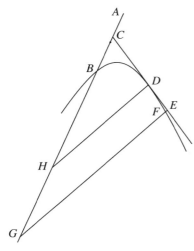

sq. EG : sq. DH :: sq. CG : sq. CH
because of EG's being parallel to DH, and
sq. FG : sq. DH :: rect. AG, GB : rect. AH, HB
because of the section (I. 21),
therefore
sq. CG : sq. CH > rect. AG, GB : rect. AH, HB.[*]
Alternately therefore
sq. CG : rect. AG, GB > sq. CH : rect. AH, HB.
Therefore *separando*
sq. CB : rect. AG, GB > sq. CB : rect. AH, HB
(Eucl. V. 17, II. 6);
and this is impossible (Eucl. V. 8). Therefore the straight line CDE will not fall outside the section; therefore inside.

[Porism]

And for this reason the straight line from some one of the points on the straight line AC will *a fortiori* fall inside, since it will also fall inside CD.

PROPOSITION 32

If a straight line is drawn through the vertex of a section of a cone, parallel to an ordinate, then it touches the section, and another straight line will not fall into the space between the conic section and this straight line.

Let there be a section of a cone, first the so-called parabola whose diameter is the straight line AB, and from A let the straight line AC be drawn parallel to an ordinate.

Now it has been shown that it falls outside the section (I. 17).

---

[*] The rules governing operations on inequalities in proportions are not developed by Euclid in Book V of the *Elements.* But they can be deduced on Euclid's principles. (Tr.)

Then I say that also another straight
line will not fall into the space between the straight line *AC* and the
section.

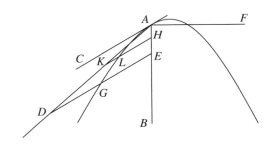

For if possible, let it fall in, as the
straight line *AD,* and let some point *D*
be taken on it at random, and let the
straight line *DE* be dropped ordinate-
wise, and let the straight line *AF* be the parameter of the ordinates. And since
$$\text{sq. } DE : \text{sq. } EA > \text{sq. } GE : \text{sq. } EA \text{ (Eucl. V. 8, VI. 22)},$$
and
$$\text{sq. } GE = \text{rect. } FA,\ AE \text{ (I. 11)},$$
therefore also
$$\text{sq. } DE : \text{sq. } EA > \text{rect. } FA,\ AE : \text{sq. } EA,$$
or
$$> FA : EA.$$
Let it be contrived then that
$$\text{sq. } DE : \text{sq. } EA :: FA : HA.^{*}$$
and through the point *H* let the straight line *HLK* be drawn parallel to *ED*.
Since then
$$\text{sq. } DE : \text{sq. } EA :: FA : AH :: \text{rect. } FA,\ AH : \text{sq. } AH,$$
and
$$\text{sq. } DE : \text{sq. } EA :: \text{sq. } KH : \text{sq. } HA \text{ (Eucl. VI. 4, VI. 22)},$$
and
$$\text{sq. } HL = \text{rect. } FA,\ AH \text{ (I. 11)},$$
therefore also
$$\text{sq. } KH : \text{sq. } HA :: \text{sq. } LH : \text{sq. } HA.$$
Therefore
$$KH = HL;$$
and this is absurd. Therefore another straight line will not fall into the space
between the straight line *AC* and the section.

---

* The intermediary steps to this construction are as follows:
Find length *M* such that
$$DE : EA :: EA : M \text{ (Eucl. VI. 11)}.$$
$$DE : M :: \text{sq. } DE : \text{sq. } EA \text{ (Eucl. VI. 19 porism)}.$$
Find *H* on *AB* such that
$$DE : M :: FA : HA \text{ (Eucl. VI. 12)}.$$
Therefore
$$\text{sq. } DE : \text{sq. } EA :: FA : HA \text{ (Eucl. V. 11)}.$$

(Ed.)

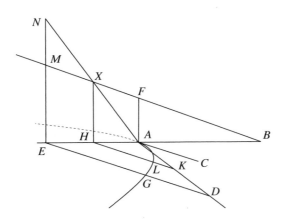
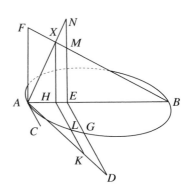

Next let the section be an hyperbola or ellipse or circumference of a circle whose diameter is the straight line *AB,* and whose upright side is the straight line *AF;* and let the straight line *BF* be joined and produced, and from the point *A* let the straight line *AC* be drawn parallel to an ordinate.

Now it has been shown that it falls outside the section (I. 17).

Then I say that also another straight line will not fall into the space between the straight line *AC* and the section.

For if possible, let it fall, as the straight line *AD,* and let some point *D* be taken at random on it, and from it let the straight line *DE* be dropped ordinatewise, and through *E* let the straight line *EM* be drawn parallel to the straight line *AF.*

And since
$$\text{sq. } GE = \text{rect. } AE, EM \text{ (I. 12,13)}$$
let it be contrived that
$$\text{rect. } AE, EN = \text{sq. } DE,$$
and let the straight line joining *AN* cut the straight line *FM* at *X,* and through *X* let the straight line *XH* be drawn parallel to *FA,* and through *H, HLK* parallel to *AC.* Since then
$$\text{sq. } DE = \text{rect. } AE, EN,$$
hence
$$NE : ED :: DE : EA;$$
and therefore
$$NE : EA :: \text{sq. } DE : \text{sq. } EA \text{ (Eucl. VI. 19 porism, VI. 22, V. 11).}$$

But
$$NE : EA :: XH : HA,$$
and
$$\text{sq. } DE : \text{sq. } EA :: \text{sq. } KH : \text{sq. } HA.$$
Therefore
$$XH : HA :: \text{sq. } KH : \text{sq. } HA;$$
therefore
$$XH : HK :: KH : HA \text{ (Eucl. VI. 19 porism converse).}$$
Therefore
$$\text{sq. } KH = \text{rect. } AH, HX;$$
but also
$$\text{sq. } LH = \text{rect. } AH, HX$$
because of the section (I. 12,13);
therefore
$$\text{sq. } KH = \text{sq. } HL;$$
and this is absurd. Therefore another straight line will not fall into the space between the straight line AC and the section.

PROPOSITION 33

If on a parabola some point is taken, and from it an ordinate is dropped to the diameter, and, to the straight line cut off by it on the diameter from the vertex, a straight line in the same straight line from its extremity is made equal, then the straight line joined from the point thus resulting to the point taken will touch the section.

Let there be a parabola whose diameter is the straight line AB, and let the straight line CD be dropped ordinatewise, and let the straight line AE be made equal to the straight line ED, and let the straight line AC be joined.

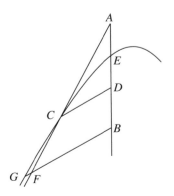

I say that the straight line AC produced will fall outside the section.

For if possible, let it fall within, as the straight line CF, and let the straight line

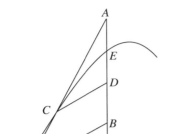

GB be dropped ordinatewise. And since

$$\text{sq. } BG : \text{sq. } CD > \text{sq. } FB : \text{sq. } CD,$$

but

$$\text{sq. } FB : \text{sq. } CD :: \text{sq. } BA : \text{sq. } AD,$$

and

$$\text{sq. } BG : \text{sq. } CD :: BE : DE \text{ (I. 20)},$$

therefore

$$BE : DE > \text{sq. } BA : \text{sq. } AD.$$

But

$$BE : DE :: 4 \text{ rect. } BE, EA : 4 \text{ rect. } DE, EA;$$

therefore also

$$4 \text{ rect. } BE, EA : 4 \text{ rect. } DE, EA > \text{sq. } AB : \text{sq. } AD.$$

Therefore alternately

$$4 \text{ rect. } BE, EA : \text{sq. } AB > 4 \text{ rect. } DE, EA : \text{sq. } AD;$$

and this is absurd; for since

$$AE = DE,$$

hence

$$4 \text{ rect. } DE, EA = \text{sq. } AD.$$

But

$$4 \text{ rect. } BE, EA < \text{sq. } AB;$$

for *E* is not the midpoint of *AB* (Eucl. II. 5). Therefore the straight line *AC* does not fall within the section; therefore it touches it.

PROPOSITION 34

If on an hyperbola or ellipse or circumference of a circle some point is taken, and if from it a straight line is dropped ordinatewise to the diameter, and if the straight lines which the ordinatewise line cuts off from the ends of the figure's transverse side have to each other a ratio which other segments of the transverse side have to each other, so that the segments from the vertex are corresponding, then the straight line joining the point taken on the transverse side and that taken on the section will touch the section.

Let there be an hyperbola or ellipse or circumference of a circle whose diameter is the straight line *AB,* and let some point *C* be taken on the section, and from *C* let the straight line *CD* be drawn ordinatewise, and let it be contrived that

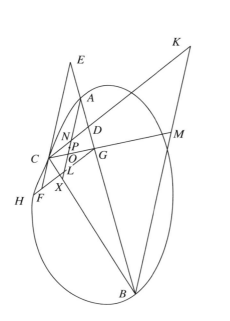
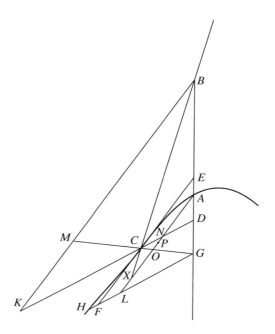

$$BD : DA :: BE : EA,^*$$

and let the straight line *EC* be joined.

I say that the straight line *CE* touches the section.

For if possible, let it cut it, as the straight line *ECF,* and let some point *F* be taken on it, and let the straight line *GFH* be dropped ordinatewise, and let the straight lines *AL* and *BK* be drawn through *A* and *B* parallel to the straight line *EC,* and let the straight lines *DC, BC,* and *GC* be joined and produced to the points *K, X,* and *M.* And since

$$BD : DA :: BE : EA,$$

but

$$BD : DA :: BK : AN,$$

and

$$BE : AE :: BC : CX :: BK : XN \text{ (Eucl. VI. 4)},$$

---

* This construction is easy. In the case of the hyperbola, *componendo,*
$$BD + DA : DA :: BA : EA;$$
and in the case of the ellipse, *separando,*
$$BD - DA : DA :: BA : EA.$$

(Tr.)

therefore
$$BK : AN :: BK : XN;$$

therefore
$$AN = NX.$$

Therefore
$$\text{rect. } AN,\ NX > \text{rect. } AO,\ OX \text{ (Eucl. VI. 27; II. 5).}$$

Therefore
$$NX : XO > OA : AN.^{*}$$

But
$$NX : XO :: KB : BM \text{ (Eucl. VI. 4);}$$

therefore
$$KB : BM > OA : AN.$$

Therefore
$$\text{rect. } KB,\ AN > \text{rect. } BM,\ OA.$$

And so
$$\text{rect. } KB,\ AN : \text{sq. } CE > \text{rect. } BM,\ OA : \text{sq. } CE \text{ (Eucl. V. 8).}$$

---

* Eutocius, commenting, says : "For since
$$\text{rect. } AN,\ NX > \text{rect. } AO,\ OX,$$

let
$$\text{rect. } AN,\ NX = \text{rect. } AO,\ XP$$
where XP is some line such that
$$XP > XO;$$

therefore
$$OA : AN :: NX : XP.$$

But
$$NX : XO > NX : XP \text{ (Eucl. V. 8)}$$
and therefore
$$NX : XO > OA : AN.$$

"Then the converse is also evident that, if
$$NX : XO > OA : AN,$$

then
$$\text{rect. } XN,\ NA > \text{rect. } AO,\ OX.$$

"For let it be that
$$OA : AN :: NX : XP,$$

where
$$XP > XO;$$

therefore
$$\text{rect. } XN,\ NA = \text{rect. } AO,\ XP;$$

and so
$$\text{rect. } XN,\ NA > \text{rect. } AO,\ OX."$$

(Tr.)

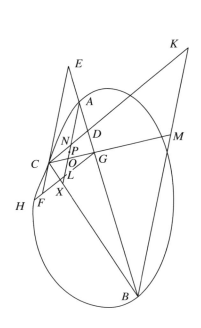
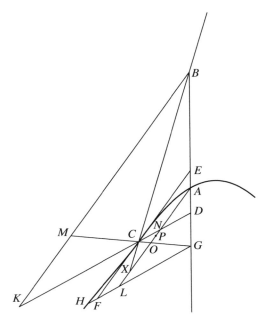

But

$$\text{rect. } KB, AN : \text{sq. } CE :: \text{rect. } BD, DA : \text{sq. } DE$$

through the similarity of the triangles *BKD*, *ECD*, and *NAD*,[*]
and

$$\text{rect. } BM, OA : \text{sq. } CE :: \text{rect. } BG, GA : \text{sq. } GE;$$

therefore

---

[*] Eutocius, commenting, says : "Since then, because *AN, EC,* and *KB* are parallel,

$$AN : EC :: AD : DE,$$

and

$$EC : KB :: ED : DB,$$

therefore *ex aequali*

$$AN : KB :: AD : DB;$$

therefore also

$$\text{sq. } AN : \text{rect. } AN, KB :: \text{sq. } AD : \text{rect. } AD, DB.$$

But

$$\text{sq. } EC : \text{sq. } AN :: \text{sq. } ED : \text{sq. } AD;$$

therefore *ex aequali*

$$\text{sq. } EC : \text{rect. } AN, KB :: \text{sq. } ED : \text{rect. } AD, DB;$$

and inversely

$$\text{rect. } KB, AN : \text{sq. } EC :: \text{rect. } AD, DB : \text{sq. } ED."$$

A similar proof holds for the proportion following. (Tr.)

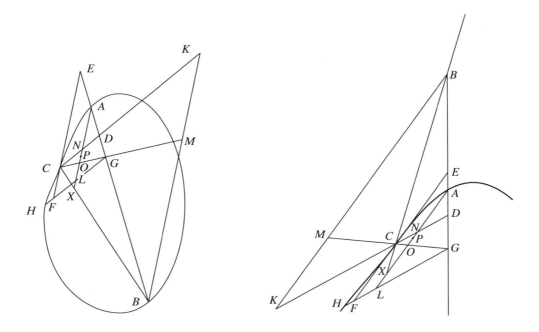

rect. *BD, DA* : sq. *DE* > rect. *BG, GA* : sq. *GE*.

Therefore alternately

rect. *BD, DA* : rect. *BG, GA* > sq. *DE* : sq. *GE*.

But

rect. *BD, DA* : rect. *AG, GB* :: sq. *CD* : sq. *GH* (I. 21),

and

sq. *DE* : sq. *EG* :: sq. *CD* : sq. *FG* (Eucl. VI. 4, VI. 22),

therefore also

sq. *CD* : sq. *HG* > sq. *CD* : sq. *FG*.

Therefore

HG < *FG* (Eucl. V. 10);

and this is impossible. Therefore the straight line *EC* does not cut the section; therefore it touches it.

PROPOSITION 35

If a straight line touches a parabola, meeting the diameter outside the section, the straight line drawn from the point of contact ordinatewise to the diameter will cut off on the diameter beginning from the vertex of the section a straight line equal to the straight line between the vertex and the [diameter's intersection with the] tangent, and no straight line will fall into the space between the tangent and the section.

Let there be a parabola whose diameter is the straight line *AB,* and let the straight line *BC* be erected ordinatewise, and let the straight line *AC* be tangent to the section.

I say that the straight line *AG* is equal to the straight line *GB.*

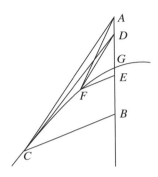

For if possible, let it be unequal to it, and let the straight line *GE* be made equal to *AG,* and let the straight line *EF* be erected ordinatewise, and let the straight line *AF* be joined. Therefore *AF* produced will meet the straight line *AC* (I. 33); and this is impossible. For two straight lines will have the same ends. Therefore the straight line *AG* is not unequal to the straight line *GB*; therefore it is equal.

Then I say that no straight line will fall into the space between the straight line *AC* and the section.

For if possible, let the straight line *CD* fall in between, and let *GE* be made equal to *GD,* and let the straight line *EF* be erected ordinatewise. Therefore the straight line joined from *D* to *F* touches the section (I. 33); therefore produced it will fall outside it. And so it will meet *DC,* and two straight lines will have the same ends; and this is impossible. Therefore a straight line will not fall into the space between the section and the straight line *AC.*

PROPOSITION 36

If some straight line, meeting the transverse side of the figure touches an hyperbola or ellipse or circumference of a circle, and if a straight line is dropped from the point of contact ordinatewise to the diameter, then as the straight line cut off by the tangent from the end of the transverse side is to the straight line cut off by the tangent from the other end of that side, so will the straight line cut off by the ordinate from the end of the side be to the straight line cut off by the ordinate from the other end of the side in such a way that the corresponding straight lines are continuous; and another straight line will not fall into the space between the tangent and the section of the cone.

Let there be an hyperbola or ellipse or circumference of a circle whose diameter is the straight line *AB,* and let the straight line *CD* be tangent, and let the straight line *CE* be dropped ordinatewise.

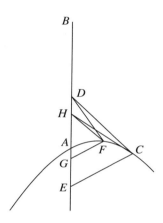

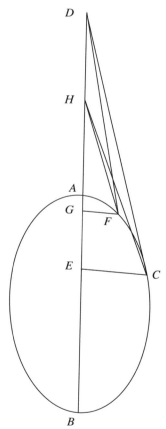

I say that
> *BE : EA :: BD : DA.*

For if it is not, let it be
> *BD : DA :: BG : GA,*

and let the straight line *GF* be erected ordinatewise; therefore the straight line joined from *D* to *F* will touch the section (I. 34); therefore produced it will meet *CD.* Therefore two straight lines will have the same ends; and this is impossible.

I say that no straight line will fall between the section and the straight line *CD.*

For if possible, let it fall between, as the straight line *CH,* and let it be contrived that

$$BH : HA :: BG : GA,$$

and let the straight line *GF* be erected ordinatewise; therefore the straight line joined from *H* to *F,* when produced, will meet *HC* (I. 34). Therefore two straight lines will have the same ends; and this is impossible. Therefore a straight line will not fall into the space between the section and the straight line *CD.*

PROPOSITION 37

If a straight line touching an hyperbola or ellipse or circumference of a circle meets the diameter, and from the point of contact to the diameter a straight line is dropped ordinatewise, then the straight line cut off by the ordinate from the center of the section with the straight line cut off by the tangent from the center of the section will contain an area equal to the square on the radius of the section, and with the straight line between the ordinate and the tangent will contain an area having the ratio to the square on the ordinate which the transverse has to the upright.

Let there be an hyperbola or ellipse or circumference of a circle whose diameter is the straight line *AB,* and let the straight line *CD* be drawn tangent, and let the straight line *CE* be dropped ordinatewise, and let the point *F* be the center.

I say that

$$\text{rect. } DF, FE = \text{sq. } FB,$$

and

rect. *DE, EF* : sq. *EC* :: the transverse : the upright.

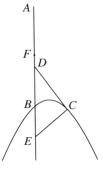

For since *CD* touches the section, and *CE* has been dropped ordinatewise, hence

$$AD : DB :: AE : EB \text{ (I. 36).}$$

Therefore *componendo*

$$AD + DB : DB :: AE + EB : EB.$$

And let the halves of the antecedents be taken (Eucl. V. 15); in the case of the hyperbola we shall say: but

$$\text{half } (AE + EB) = FE,$$

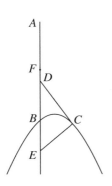

and
$$\text{half } AB = FB;$$
therefore
$$FE : EB :: FB : BD.$$
Therefore *convertendo*
$$FE : FB :: FB : FD,$$

therefore
$$\text{rect. } EF, FD = \text{sq. } FB.$$
And since
$$FE : EB :: FB : BD :: AF : BD,$$
alternately
$$AF : FE :: DB : BE;$$

componendo

$$AE : EF :: DE : EB;$$

and so

$$\text{rect. } AE, EB = \text{rect. } FE, ED.$$

But
$$\text{rect. } AE, EB : \text{sq. } CE :: \text{the transverse : the upright (I. 21);}$$
therefore also
$$\text{rect. } FE, ED : \text{sq. } CE :: \text{the transverse : the upright.}$$

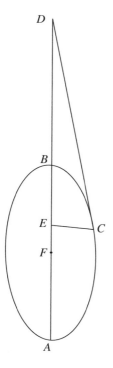

And in the case of the ellipse and of the circle we shall say: but
$$\text{half } (AD + DB) = DF,$$
and
$$\text{half } AB = FB;$$
therefore
$$FD : DB :: FB : BE.$$
Therefore *convertendo*
$$DF : FB :: BF : FE.$$
Therefore
$$\text{rect. } DF, FE = \text{sq. } BF.$$
But
rect. DF, FE = rect. DE, EF + sq. FE (Eucl. II. 3), and
    sq. BF = rect. AE, EB + sq. FE (Eucl. II. 5).
Let the common square on EF be subtracted; therefore
$$\text{rect. } DE, EF = \text{rect. } AE, EB.$$
Therefore

rect. *DE, EF* : sq. *CE* :: rect. *AE, EB* : sq. *CE.*

But

rect. *AE, EB* : sq. *CE* :: the transverse : the upright (I. 21).

Therefore

rect. *DE, EF* : sq. *CE* :: the transverse : the upright.

PROPOSITION 38

If a straight line touching an hyperbola or ellipse or circumference of a circle meets the second diameter, and if from the point of contact a straight line is dropped to the same [i.e., second] diameter parallel to the other diameter, then the straight line cut off from the center of the section by the dropped straight (κατηγμένη)* line, together with the straight line cut off [on the second diameter] by the tangent from the center of the section will contain an area equal to the square on the half of the second diameter, and, together with the straight line [on the second diameter] between the dropped straight line and the tangent, will contain an area having a ratio to the square on the dropped straight line which the upright side of the figure has to the transverse.

Let there be an hyperbola or ellipse or circumference of a circle whose diameter is the straight line *AGB*, and whose second diameter is the straight line *CGD*, and let the straight line *ELF*, meeting *CD* at *F*, be a tangent to the section, and let the straight line *HE* be parallel to *AB*.

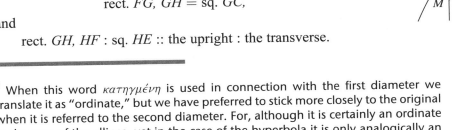

I say that

rect. *FG, GH* = sq. *GC,*

and

rect. *GH, HF* : sq. *HE* :: the upright : the transverse.

---

* When this word κατηγμένη is used in connection with the first diameter we translate it as "ordinate," but we have preferred to stick more closely to the original when it is referred to the second diameter. For, although it is certainly an ordinate in the case of the ellipse, yet in the case of the hyperbola it is only analogically an ordinate. This analogy, however, becomes stronger and stronger as the treatise moves on. It is, therefore, no accident that κατηγμένη is used in both cases. On the other hand, in Definition 5 Apollonius definitely calls both cases ordinates as if announcing the culmination of an analogy to be worked out in the course of the treatise. (Tr.)

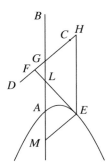 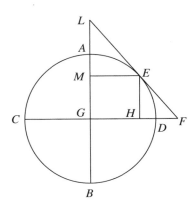

Let the straight line *ME* be drawn ordinatewise; therefore

rect. *GM, ML* : sq. *ME* :: the transverse : the upright (I. 37).

But

the transverse *BA* : *CD* :: *CD* : the upright (see Def. 11);

and therefore

the transverse : the upright :: sq. *BA* : sq. *CD* (Eucl. VI. 19 porism) ;

and as the quarters of them, that is

the transverse : the upright :: sq. *GA* : sq. *GC*;

therefore also

rect. *GM, ML* : sq. *ME* :: sq. *GA* : sq. *GC*.

But

rect. *GM, ML* : sq. *ME* :: *GM* : *ME* comp. *LM* : *ME*,

or

rect. *GM, ML* : sq. *ME* :: *GM* : *GH* comp. *LM* : *ME*.

Therefore inversely

sq. *CG* : sq. *GA* :: *EM* : *MG* or *HG* : *GM* comp. *EM* : *ML* or *FG* : *GL*.

Therefore

sq. *GC* : sq. *GA* :: *HG* : *GM* comp. *FG* : *GL*,

which is the same as

rect. *FG, GH* : rect. *MG, GL*.

Therefore

rect. *FG, GH* : rect. *MG, GL* :: sq. *CG* : sq. *GA*.

And alternately therefore

rect. *FG, GH* : sq. *CG* :: rect. *MG, GL* : sq. *GA*.

But

rect. *MG, GL* = sq. *GA* (I. 37),

therefore also

rect. *FG, GH* = sq. *CG*.

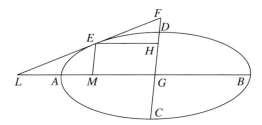

Again since

   the upright : the transverse :: sq. *EM* : rect. *GM, ML* (I. 37),

and

   sq. *EM* : rect. *GM, ML* :: *EM* : *GM* comp. *EM* : *ML*

or

   sq. *EM* : rect. *GM, ML* :: *HG* : *HE* comp. *FG* : *GL* or *FH* : *HE*

which is the same as

   rect. *FH, HG* : sq. *HE*;

therefore

   rect. *FH, HG* : sq. *HE* :: the upright : the transverse.[*]

PROPOSITION 39

If a straight line touching an hyperbola or ellipse or circumference of a circle meets the diameter, and if from the point of contact a straight line is dropped ordinatewise to the diameter, then whichever of the two straight lines is taken—of which one is the straight line between the [intersection of the] ordinate [with the diameter] and the center of the section, and the other is between [the intersections of] the ordinate and the tangent [with the diameter]—the ordinate will have to it the ratio compounded of the ratio of the

---

[*] This is the conclusion of the proposition which Apollonius enunciated; it is the proportion he said he would demonstrate. There follows in the Heiberg Greek text what seems to be an extrapolation, apparently by a later meddler, which is both incorrectly stated and incorrectly proved. For completeness, this extrapolation is included in Appendix B, along with a corrected statement and proof of the extrapolated theorem. (Ed.)

other of the two straight lines to the ordinate and of the ratio of the upright side of the figure to the transverse.

Let there be an hyperbola or ellipse or circumference of a circle whose diameter is the straight line *AB,* and let the center of it be the point *F,* and let the straight line *CD* be drawn tangent to the section, and the straight line *CE* be dropped ordinatewise.

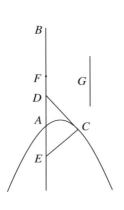 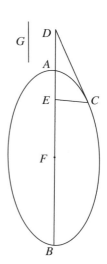

I say that

        CE : FE :: the upright : the transverse comp. *ED : EC,*

and

        CE : ED :: the upright : the transverse comp. *FE : EC.*

For let

        rect. *FE, ED* = rect. *EC, G.*

And since

        rect. *FE, ED* : sq. *CE ::* the transverse : the upright (I. 37),

and

        rect. *FE, ED* = rect. *CE, G,*

therefore

        rect. *CE, G* : sq. *CE :: G : CE ::* the transverse : the upright.

And since

        rect. *FE, ED* = rect. *CE, G,*

hence

        FE : EC :: G : ED.

And since

$$CE : ED :: CE : G \text{ comp. } G : ED,$$

but

$$CE : G :: \text{ the upright : the transverse,}$$

therefore

$$CE : ED :: \text{ the upright : the transverse comp. } FE : EC.$$

PROPOSITION 40

If a straight line touching an hyperbola or ellipse or circumference of a circle meets the second diameter, and if from the point of contact a straight line is dropped to the same diameter parallel to the other diameter, then whichever of the two straight lines is taken [along the second diameter]—of which one is the straight line between the dropped straight line and the center of the section, and the other is between the dropped straight line and the tangent—the dropped straight line will have to it the ratio compounded of the ratio of the transverse side to the upright and of the ratio of the other of the two straight lines to the dropped straight line.

Let there be an hyperbola or ellipse or circumference of a circle AB, and its diameter the straight line BFC, and its second diameter the straight line DFE, and let the straight line HLA be drawn tangent, and the straight line AG parallel to the straight line BC.

I say that
$$AG : HG :: \text{ the transverse : the upright comp. } FG : GA,$$
and
$$AG : FG :: \text{ the transverse : the upright comp. } HG : GA.$$

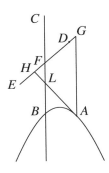

Let
$$\text{rect. } GA, K = \text{rect. } HG, GF.$$
And since
the upright : the transverse :: rect. HG, GF : sq. GA (I. 38),
and
$$\text{rect. } GA, K = \text{rect. } HG, GF,$$
therefore also
$$\text{rect. } GA, K : \text{sq. } GA :: K : AG :: \text{ the upright : the transverse.}$$
And since
$$AG : GF :: AG : K \text{ comp. } K : GF,$$

but
$$AG : K :: \text{the transverse : the upright,}$$
and
$$K : GF :: HG : GA$$
because
$$\text{rect. } HG, GF = \text{rect. } AG, K,$$
therefore
$$AG : GF :: \text{the transverse : the upright comp. } GH : GA.$$

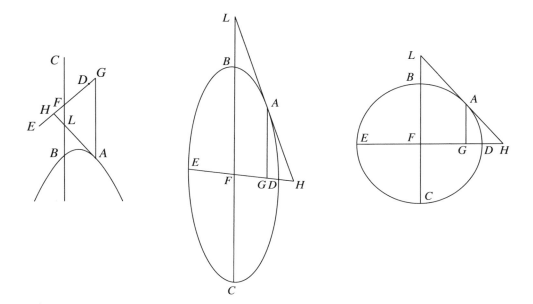

PROPOSITION 41

If in an hyperbola or ellipse or circumference of a circle a straight line is dropped ordinatewise to the diameter, and if equiangular parallelogrammic figures are described both on the ordinate and on the radius, and if the ordinate side has to the remaining side of the figure the ratio compounded of the ratio of the radius to the remaining side of its figure, and of the ratio of the upright side of the section's figure to the transverse, then the figure on the straight line between the center and the ordinate, similar to the figure on the radius, is in the case of the hyperbola greater than the figure on the ordinate by the figure on the radius, and, in the case of the ellipse and circumference of a circle, together with the figure on the ordinate is equal to the figure on the radius.

Let there be an hyperbola or ellipse or circumference of a circle whose diameter is the straight line *AB,* and center the point *E,* and let the straight line *CD* be dropped ordinatewise, and on the straight lines *EA* and *CD* let the equiangular figures *AF* and *DG* be described, and let

$$CD : CG :: AE : EF \text{ comp. the upright : the transverse.}$$

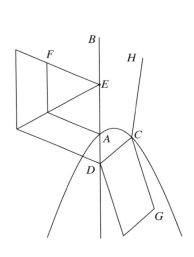
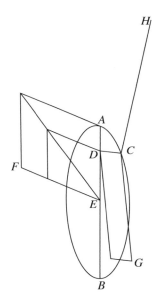

I say that, with the figure on *ED* similar to *AF,* in the case of the hyperbola,

figure on *ED* = *AF* + *GD,*

and in the case of the ellipse and circle,

figure on *ED* + *GD* = *AF.*

For let it be contrived that

the upright : the transverse :: *DC* : *CH.*

And since

DC : *CH* :: the upright : the transverse,

but

DC : *CH* :: sq. *DC* : rect. *DC, CH*

and

the upright : the transverse :: sq. *DC* : rect. *BD, DA* (I. 21),

therefore

rect. *BD, DA* = rect. *DC, CH.*

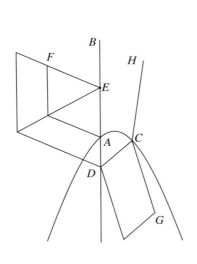
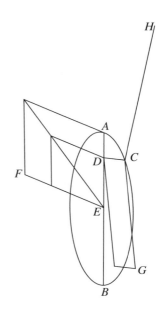

And since

DC : CG :: AE : EF comp. the upright : the transverse

or

DC : CG :: AE : EF comp. DC : CH,

and further

DC : CG :: DC : CH comp. CH : CG,

therefore

ratio *AE* : *EF* comp. *DC* : *CH* :: ratio *DC* : *CH* comp. *CH* : *CG.*

Let the common ratio, *DC* : *CH*, be taken away; therefore

AE : EF :: CH : CG.

But

HC : CG :: rect. HC, CD : rect. GC, CD,

and

AE : EF :: sq. AE : rect. AE, EF;

therefore

rect. *HC, CD* : rect. *GC, CD* :: sq. *AE* : rect. *AE, EF.*

And it has been shown that

rect. *HC, CD* = rect. *BD, DA*;

therefore
$$\text{rect. } BD, DA : \text{rect. } GC, CD :: \text{sq. } AE : \text{rect. } AE, EF.$$
Alternately
$$\text{rect. } BD, DA : \text{sq. } AE :: \text{rect. } GC, CD : \text{rect. } AE, EF.$$
And
$$\text{rect. } GC, CD : \text{rect. } AE, EF :: \text{pllg. } DG : \text{pllg. } FA;$$
for they are equiangular and have to one another the ratio compounded of their sides, $GC : AE$ and $CD : EF$ (Eucl. VI. 23); and therefore
$$\text{rect. } BD, DA : \text{sq. } EA :: \text{pllg. } DG : \text{pllg. } FA.$$

Moreover in the case of the hyperbola we are to say : *componendo*
$$\text{rect. } BD, DA + \text{sq. } AE : \text{sq. } AE :: \text{pllg. } GD + \text{pllg. } AF : \text{pllg. } AF,$$
or
$$\text{sq. } DE : \text{sq. } EA :: \text{pllg. } GD + \text{pllg. } AF : \text{pllg. } AF \text{ (Eucl. II. 6)}.$$
And as the square on DE is to the square on EA, so is the figure described on ED, similar and similarly situated to the parallelogram AF, to the parallelogram AF (Eucl. VI. 20 porism); therefore, with the figure on ED similar to the parallelogram AF,
$$\text{pllg. } GD + \text{pllg. } AF : \text{pllg. } AF :: \text{figure on } ED : \text{pllg. } AF.$$
Therefore
$$\text{figure on } ED = \text{pllg. } GD + \text{pllg. } AF,$$
the figure on ED being similar to the parallelogram AF.

And in the case of the ellipse and of the circumference of a circle we shall say : since then
$$\text{whole sq. } AE : \text{whole pllg. } AF ::$$
$$\text{rect. } AD, DB \text{ subtracted} : \text{pllg. } DG \text{ subtracted},$$
also remainder is to remainder as whole to whole (Eucl. V. 19).
And
$$\text{sq. } AE - \text{rect. } BD, DA = \text{sq. } DE \text{ (Eucl. II. 5)};$$
therefore
$$\text{sq. } DE : \text{pllg. } AF - \text{pllg. } DG :: \text{sq. } AE : \text{pllg. } AF.$$
But
$$\text{sq. } AE : \text{pllg. } AF :: \text{sq. } DE : \text{figure on } DE \text{ (Eucl. VI. 20 porism)}$$
the figure on DE being similar to the parallelogram AF. Therefore, the figure on DE being similar to the parallelogram AF,
$$\text{sq. } DE : \text{pllg. } AF - DG :: \text{sq. } DE : \text{figure on } DE.$$
Therefore, the figure on DE being similar to the parallelogram AF,
$$\text{figure on } DE = \text{pllg. } AF - \text{pllg. } DG.$$
Therefore
$$\text{figure on } DE + \text{pllg. } DG = \text{pllg. } AF.$$

PROPOSITION 42

If a straight line touching a parabola meets the diameter, and if from the point of contact a straight line is dropped ordinatewise to the diameter, and if some point being taken on the section, two straight lines are dropped to the diameter, one of them parallel to the tangent, and the other parallel to the straight line dropped from the point of contact, then the triangle resulting from them [i.e., from the diameter and the two straight lines dropped from the random point] is equal to the parallelogram contained by the straight line dropped from the point of contact and by the straight line cut off by the parallel from the vertex of the section.

Let there be a parabola, whose diameter is the straight line *AB,* and let the straight line *AC* be drawn tangent to the section, and let the straight line *CH* be dropped ordinatewise, and from some point at random let the straight line *DF* be dropped ordinatewise, and through the point *D* let the straight line *DE* be drawn parallel to the straight line *AC,* and through the point *C* the straight line *CG* parallel to the straight line *BF,* and through the point *B* the straight line *BG* parallel to the straight line *HC.*

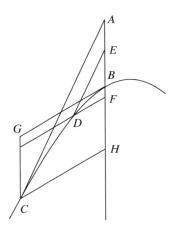

I say that

$$\text{trgl. } DEF = \text{pllg. } GF.$$

For since the straight line *AC* touches the section, and the straight line *CH* has been dropped ordinatewise,

$$AB = BH \text{ (I. 35)};$$

therefore
$$AH = 2BH.$$

Therefore
$$\text{trgl. } AHC = \text{pllg. } BC \text{ (Eucl. I. 41).}$$

And since
$$\text{sq. } CH : \text{sq. } DF :: HB : BF$$

because of the section (I. 20), but
$$\text{sq. } CH : \text{sq. } DF :: \text{trgl. } ACH : \text{trgl. } EDF \text{ (Eucl. VI. 20 porism),}$$

and
$$HB : BF :: \text{pllg. } GH : \text{pllg. } GF \text{ (Eucl. VI. 1),}$$

therefore
$$\text{trgl. } ACH : \text{trgl. } EDF :: \text{pllg. } HG : \text{pllg. } FG.$$

Therefore alternately
$$\text{trgl. } AHC : \text{pllg. } BC :: \text{trgl. } EDF : \text{pllg. } GF.$$

But
$$\text{trgl. } ACH = \text{pllg. } GH;$$

therefore
$$\text{trgl. EDF} = \text{pllg. } GF.$$

PROPOSITION 43

If a straight line touching an hyperbola or ellipse or circumference of a circle meets the diameter, and if from the point of contact a straight line is dropped ordinatewise to the diameter, and if through the vertex a parallel to [the ordinatewise line] is drawn meeting the straight line drawn through the point of contact and the center [or meeting this line extended], and if, some [random] point being taken on the section, two straight lines are drawn to the diameter, one of which is parallel to the tangent and the other parallel to the straight line dropped [ordinatewise] from the point of contact, then, in the case of the hyperbola, the triangle resulting from them [i.e., the two lines drawn through the random point to the diameter], will be less than the triangle cut off by the straight line through the center to the point of contact [and by the ordinatewise line through the random point] by the triangle on the radius* similar to the triangle cut off, and, in the case of the ellipse and

---

* The radius is defined in Definition 9, page 36. (Ed.)

the circumference of the circle, [the triangle resulting from the two lines through the random point to the diameter, taken] together with the triangle cut off [by the line] from the center [to the point of contact and by the ordinatewise line through the random point], will be equal to the triangle on the radius similar to the triangle cut off.

Let there be an hyperbola or ellipse or circumference of a circle whose diameter is the straight line *AB*, and center the point *C*, and let the straight line *DE* be drawn tangent to the section, and let the straight line *CE* be joined, and let the straight line *EF* be dropped ordinatewise, and let some point *G* be taken on the section, and let the straight line *GH* be drawn parallel to the tangent, and let the straight line *GK* be dropped ordinatewise [and extended to meet *CE* at *M*], and through *B* let the straight line *BL* be erected ordinatewise.

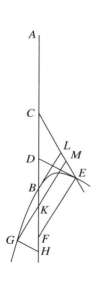
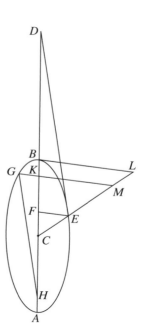

I say that triangle *KMC* differs from triangle *CLB* by triangle *GKH*.

For since the straight line *ED* touches, and the straight line *EF* has been dropped, hence

$EF : FD :: CF : FE$ comp. the upright : the transverse (I. 39).

But

$EF : FD :: GK : KH,$

and

$$CF : FE :: CB : BL \text{ (Eucl. VI. 4)};$$

therefore

$$GK : KH :: BC : BL \text{ comp. the upright : the transverse.}$$

And through those things shown in the forty-first theorem (I. 41), triangle *CKM* differs from triangle *BCL* by triangle *GHK*; for the same things have also been shown in the case of the parallelograms, their doubles.

PROPOSITION 44

If a straight line touching one of the opposite sections meets the diameter, and if from the point of contact some straight line is dropped ordinatewise to the diameter, and if a parallel to it is drawn through the vertex of the other section meeting the straight line drawn through the point of contact and the center, and, if some point is taken at random on the section and [from it] straight lines are dropped to the diameter, one of which is parallel to the tangent and the other parallel to the straight line dropped ordinatewise from the point of contact, then the triangle resulting from them will be less than the triangle the dropped straight line cuts off from the center of the section, by the triangle on the radius similar to the triangle cut off.

Let there be the opposite sections *AF* and *BE,* and let their diameter be the straight line *AB,* and center the point *C,* and from some point *F* of those on the section *FA* let the straight line *FG* be drawn tangent to the section, and the straight line *FO* ordinatewise, and let the straight line *CF* be joined and produced, as *CE* (I. 29), and through *B* let the straight line *BL* be drawn parallel to the straight line *FO,* and let some point *N* be taken on the section *BE,* and from *N* let the straight line *NH* be dropped ordinatewise, and let the straight line *NK* be drawn parallel to the straight line *FG.*

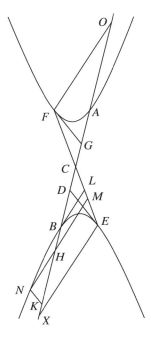

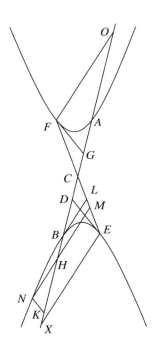

I say that

trgl. HKN + trgl. CBL = trgl. CMH.

For through E let the straight line ED be drawn tangent to the section $BE,$ and let the straight line EX be drawn ordinate-wise. Since then FA and BE are opposite sections whose diameter is $AB,$ and the straight line through whose center is $FCE,$ and FG and ED are tangents to the section, hence DE is parallel to $FG.$[*] And the straight line NK is parallel to FG; therefore NK is also parallel to $ED,$ and the straight line MH to $BL.$ Since then BE is an hyperbola, whose diameter is the straight line $AB,$ and whose center is $C,$ and the straight line DE is tangent to the section, and EX drawn ordinate-wise, and BL is parallel to $EX,$ and N has been taken on the section as the point from which NH has been dropped

---

[*] Eutocius, commenting, says : "For since AF is an hyperbola, and FG a tangent, and FO an ordinate,

$$\text{rect. } OC, CG = \text{sq. } CA \text{ (I. 37)};$$

likewise then also

$$\text{rect. } XC, CD = \text{sq. } CB.$$

But

$$\text{sq. } AC = \text{sq. } CB;$$

therefore also

$$\text{rect. } OC, CG = \text{rect. } XC, CD.$$

And

$$OC = CX \text{ (I. 14, 30)};$$

and therefore

$$GC = CD;$$

and also

$$FC = CE \text{ (I. 30)};$$

therefore

$$FC = EC, CG = CD.$$

And they contain equal angles at the point C; for they are vertical. And so also

$$FG = ED$$

and

$$\text{angle } CFG = \text{angle } CED.$$

And they are alternate; therefore the straight line FG is parallel to the straight line $ED.$"

(Tr.) (Two superfluous lines deleted by Ed.)

ordinatewise, and *KN* has been drawn parallel to *DE,* therefore

trgl. *NHK* + trgl. *BCL* = trgl. *HMC*;

for this has been shown in the forty-third theorem (I. 43).

PROPOSITION 45

If a straight line touching an hyperbola or ellipse or circumference of a circle meets the second diameter, and if from the point of contact some straight line is dropped to the same diameter parallel to the other diameter, and if through the point of contact and the center a straight line is produced, and, if some point is taken at random on the section, and [from it] two straight lines are drawn to the second diameter one of which is parallel to the tangent and the other parallel to the dropped straight line, then, in the case of the hyperbola, the triangle resulting from them is greater than the triangle that the dropped straight line cuts off from the center, by the triangle whose base is the tangent and vertex is the center of the section, and, in the case of the ellipse and circle, [the triangle resulting from the two straight lines drawn to the second diameter,] together with the triangle cut off, will be equal to the triangle whose base is the tangent and whose vertex is the center of the section.

Let there be an hyperbola or ellipse or circumference of a circle *ABC,* whose diameter is the straight line *AH,* and

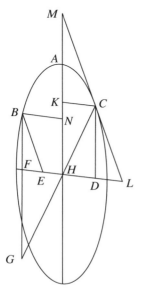

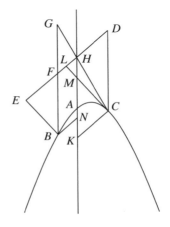

second diameter *HD,* and center *H,* and let the straight line *CML* touch it at *C,* and let the straight line *CD* be drawn parallel to *AH,* and let the straight line *HC* be joined and produced, and let some point *B* be taken at random on the section, and from *B* let the straight lines *BE* and *BF* be drawn from *B* parallel to the straight lines *LC* and *CD.*

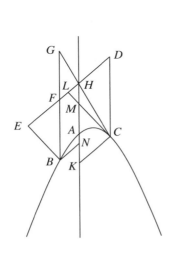

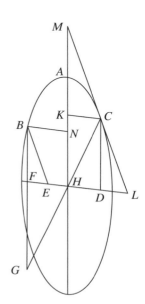

I say that, in the case of the hyperbola,

trgl. *BEF* = trgl. *GHF* + trgl. *LCH,*

and, in the case of the ellipse and circle,

trgl. *BEF* + trgl. *FGH* = trgl. *CLH.*

For let the straight lines *CK* and *BN* be drawn parallel to *DH.* Since then the straight line *CM* is tangent, and the straight line *CK* has been dropped ordinatewise, hence

CK : KH :: MK : KC comp. the upright : the transverse (I. 39),

and

MK : KC :: CD : DL (Eucl. VI. 4);

therefore

CK : KH :: CD : DL comp. the upright : the transverse.

And triangle *CDL* is the figure on *KH*; and triangle *CKH,* that is triangle *CDH,* is the figure on *CK,* that is on *DH*; therefore, in the case of the hyperbola,

trgl. *CDL* = trgl. *CKH* + trgl. on *AH* similar to trgl. *CDL,*

and, in the case of the ellipse and the circle,

trgl. *CDH* + trgl. *CDL* = trgl. on *AH* similar to trgl. *CDL*;

for this was also shown in the case of their doubles in the forty-first theorem (I. 41).

Since then triangle *CDL* differs either from triangle *CKH* or from triangle *CDH* by the triangle on *AH* similar to triangle *CDL,* and it also differs by triangle *CHL,* therefore

trgl. *CHL* = trgl. on *AH* similar to trgl. *CDL.*

Since then triangle *BFE* is similar to triangle *CDL,* and triangle *GFH* to triangle *CDH,* therefore they have the same ratio.[*] And triangle *BFE* is described on *NH* between the ordinate and the center, and triangle *GFH* on the ordinate *BN,* that is on *FH*; and by things already shown (I. 41) triangle *BFE* differs from triangle *GHF* by the triangle on *AH* similar to *CDL,* and so also by triangle *CHL.*

PROPOSITION 46

If a straight line touching a parabola meets the diameter, then the straight line drawn through the point of contact parallel to the diameter in the direction of the section bisects the straight lines drawn in the section parallel to the tangent.

---

[*] That is (Eucl. VI. 4),

BF : *FE* :: *CD* : *DL*

and

GF : *FH* :: *CD* : *DH* :: *CK* : *KH.*

Therefore these first ratios can be substituted in the central proportion of the theorem, namely,

CK : *KH* :: *CD* : *DL* comp. the upright : the transverse,

and so satisfy I. 41. (Tr.)

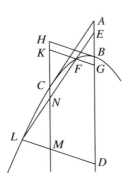

Let there be a parabola whose diameter is the straight line *ABD*, and let the straight line *AC* touch the section (I. 24), and through *C* let the straight line *HCM* be drawn parallel to the straight line *AD* (I. 26), and let some point *L* be taken at random on the section, and let the straight line *LNFE* (I. 18, 22) be drawn parallel to *AC*.

I say that
$$LN = NF.$$

Let the straight lines *BH, KFG,* and *LMD* be drawn ordinatewise. Since then by the things already shown in the forty-second theorem (I. 42)
$$\text{trgl. } ELD = \text{pllg. } BM,$$
and
$$\text{trgl. } EFG = \text{pllg. } BK,$$
therefore the remainders
$$\text{pllg. } GM = \text{quadr. } LFGD.$$
Let the common pentagon *MDGFN* be subtracted; therefore the remainders
$$\text{trgl. } KFN = \text{trgl. } LMN.$$
And KF is parallel to *LM*; therefore
$$FN = LN \text{ (Heath's Lemma I to Eucl. VI. 22).}$$

PROPOSITION 47

If a straight line touching an hyperbola or ellipse or circumference of a circle meets the diameter, and if through the point of contact and the center a straight line is drawn in the direction of the section, then it bisects the straight lines drawn in the section parallel to the tangent.

Let there be an hyperbola or ellipse or circumference of a circle whose diameter is the straight line *AB* and center *C,* and let the straight line *DE* be drawn tangent to the section, and let the straight line *CE* be joined and produced,

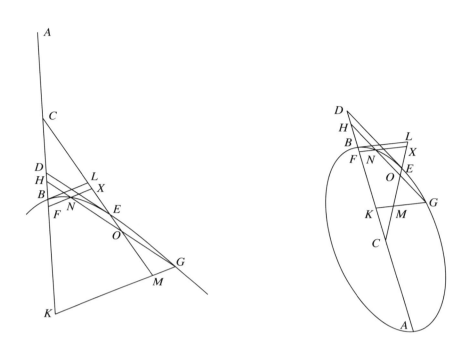

and let a point N be taken at random on the section, and through N let the straight line $HNOG$ be drawn parallel.

I say that

$$NO = OG.$$

For let the straight lines XNF, BL, and GMK be dropped ordinatewise. Therefore by things already shown in the forty-third theorem (I. 43)

$$\text{trgl. } HNF = \text{quadr. } LBFX,$$

and

$$\text{trgl.} GHK = \text{quadr. } LBKM.$$

Therefore the remainders

$$\text{quadr. } NGKF = \text{quadr. } MKFX.$$

Let the common pentagon $ONFKM$ be subtracted; therefore the remainders

$$\text{trgl. } OMG = \text{trgl. } NXO.$$

And the straight line MG is parallel to the straight line NX; therefore

$$NO = OG \text{ (Heath's Lemma I to Eucl. VI. 22)}.$$

PROPOSITION 48

If a straight line touching one of the opposite sections meets the diameter, and if through the point of contact and the center a straight line produced cuts the other section, then whatever line is drawn in the other section parallel to the tangent, will be bisected by the straight line produced.

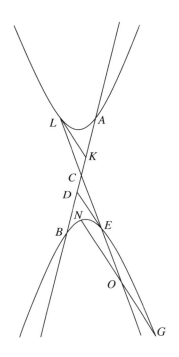

Let there be opposite sections whose diameter is the straight line *AB* and center *C*, and let the straight line *KL* touch the section *A*, and let the straight line *LC* be joined and produced (I. 29), and let some point *N* be taken on the section *B*, and through *N* let the straight line *NG* be drawn parallel to the straight line *LK*.

I say that
$$NO = OG.$$

For let the straight line *ED* be drawn through *E* tangent to the section; therefore *ED* is parallel to *LK* (I. 44, note). And so also to *NG*. Since then *BNG* is an hyperbola whose center is *C* and tangent *DE*, and since *CE* has been joined and a point *N* has been taken on the section and through it *NG* has been drawn parallel to *DE*, by a theorem already shown (I. 47) for the hyperbola
$$NO = OG.$$

PROPOSITION 49

If a straight line touching a parabola meets the diameter, and if through the point of contact a parallel to the diameter is drawn, and if from the vertex a straight line is drawn parallel to an ordinate, and if it is contrived that as the segment of the tangent between the [ordinatewise] erected straight line and the point of contact is to the segment of the parallel between the point of contact and the [ordinatewise] erected straight line, so is some [found] straight line to the double of the tangent, then whatever straight line is drawn [parallel to the tangent] from the section to the straight line drawn through the point of contact parallel to the diameter, will equal in square the rectangle contained by the straight line found [i.e., the upright side] and by the straight line cut off by it [i.e., the line parallel to the tangent] from the point of contact.

Let there be a parabola whose diameter is the straight line *MBC*, and *CD* its tangent, and through *D* let the straight line *FDN* be drawn parallel to the straight line *BC*, and let the straight line *FB* be erected ordinatewise (I. 17), and let it be contrived that

$$ED : DF :: \text{some straight line } G : 2CD,$$

and let some point *K* be taken on the section, and let the straight line *KLP* be drawn through *K* parallel to *CD*.

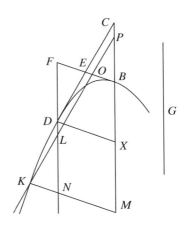

I say that

$$\text{sq. } KL = \text{rect. } G, DL;$$

that is that, with the straight line *DL* as diameter, the straight line *G* is the upright side.

For let the straight lines *DX* and *KNM* be dropped ordinatewise. And since the straight line *CD* touches the section, and the straight line *DX* has been dropped ordinatewise, then

$$CB = BX \text{ (I. 35)}.$$

But

$$BX = FD$$

And therefore

$$CB = FD.$$

And so also

$$\text{trgl. } ECB = \text{trgl. } EFD.$$

Let the common figure *DEBMN* be added; therefore

$$\text{quadr. } DCMN = \text{pllg. } FM$$
$$= \text{trgl. } KPM \text{ (I. 42)}.$$

Let the common quadrilateral *LPMN* be subtracted; therefore the remainders

$$\text{trgl. } KLN = \text{pllg. } LC.$$

And

$$\text{angle } DLP = \text{angle } KLN;$$

therefore

$$\text{rect. } KL, LN = 2 \text{ rect. } LD, DC.^{*}$$

And since

$$ED : DF :: G : 2CD,$$

and

$$ED : DF :: KL : LN,$$

therefore also

$$G : 2CD :: KL : LN.$$

But

$$KL : LN :: \text{sq. } KL : \text{rect. } KL, LN,$$

and

$$G : 2CD :: \text{rect. } G, DL : 2 \text{ rect. } LD, DC;$$

---

* Eutocius, commenting, says: "For let the triangle KLN and the parallelogram $DLPC$ be set out. And since

$$\text{trgl. } KLN = \text{pllg. } DP,$$

let the straight line NR be drawn through N parallel to LK, and through K, KR parallel to LN; therefore LR is a parallelogram and

$$\text{pllg. } LR = 2 \text{ trgl. } KLN;$$

and so also

$$\text{pllg. } LR = 2 \text{ pllg. } DP.$$

Then let the straight lines DC and LP be produced to S and T, and let CS be made equal to DC, and PT to LP, and let ST be joined; therefore

$$\text{pllg. } DT = 2 \text{ pllg. } DP;$$

and so

$$\text{pllg. } LR = \text{pllg. } LS.$$

But it is also equiangular with it because of the angles at L being vertical; but in equal and equiangular parallelograms the sides about the equal angles are reciprocally proportional; therefore

$$KL : LT \text{ or } DS :: DL : LN,$$

and

$$\text{rect. } KL, LN = \text{rect. } LD, DS.$$

And since

$$DS = 2DC,$$

hence

$$\text{rect. } KL, LN = 2 \text{ rect. } LD, DC.$$

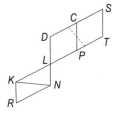

"And if DC is parallel to LP, and CP is not parallel to LD, it is clear $DCPL$ is a trapezoid, and so I say that

$$\text{rect. } KL, LN = \text{rect. } DL, (CD + LP).$$

For if LP is filled out, as we have said before, and the straight lines DC and LP are produced, and CS is made equal to LP, and PT to DC, and the straight line ST is joined, then

$$\text{pllg. } DT = 2DP,$$

and the same demonstration will fit. And this will be useful in what follows (I. 50)." (Tr.)

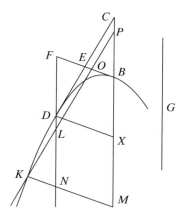

therefore

       sq. *KL* : rect. *KL, LN* :: rect. *G, DL* : 2 rect. *CD, DL*.

And alternately; but

        rect. *KL, LN* = 2 rect. *CD, DL*;

therefore also

         sq. *KL* = rect. *G, DL*.

PROPOSITION 50

If a straight line touching an hyperbola or ellipse or circumference of a circle meets the diameter, and if a straight line is produced through the point of contact and the center, and if from the vertex a straight line erected parallel to an ordinate meets the straight line drawn through the point of contact and the center, and if it is contrived that as the segment of the tangent between the point of contact and the straight line erected [ordinatewise from the vertex] is to the segment of the straight line drawn through the point of contact and the center—[the segment] between the point of contact and the straight line erected [ordinatewise from the vertex]—so some [found] straight line is to the double of the tangent, then any straight line parallel to the tangent and drawn from the section to the straight line drawn through the point of contact and the center will equal in square a rectangular area applied to the straight line found, having as breadth the straight line cut off [of the diameter] by it [i.e., the ordinatewise line] from the point of contact, and projecting beyond, in the case of the hyperbola, by a figure similar to the rectangle contained by the double of the straight line between the center and the point of contact and by the straight line found; but, in the case of the ellipse and circle, defective by this same figure.

Let there be an hyperbola or ellipse or circumference of a circle whose diameter is the straight line *AB*, and center *C*, and let the straight line *DE* be a tangent, and let the straight line *CE* be joined and produced both ways, and let the straight line *CK* be made equal to the straight line *EC*, and through *B* let the straight line *BFG* be erected ordinatewise, and through *E* let the straight line *EH* be drawn perpendicular to *EC*, and let it be that

$$FE : EG :: EH : 2ED,$$

and let the straight line *HK* be joined and produced, and let some point *L* be taken on the section, and through it let the straight line *LMX* be drawn parallel to *ED*, and the straight line *LRN* parallel to *BG*, and the straight line *MP* parallel to *EH*.

I say that

$$\text{sq. } LM = \text{rect. } EM,\ MP$$

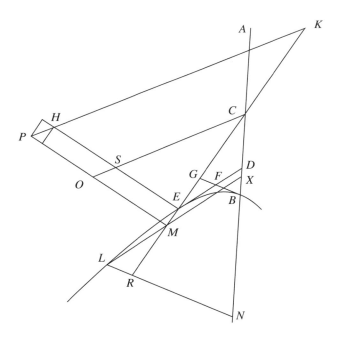

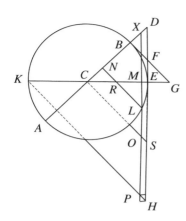 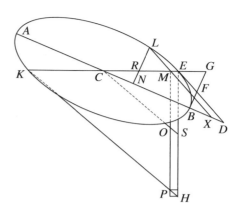

For let the straight line *CSO* be drawn through *C* parallel to *KP*. And since

$$EC = CK,$$

and

$$EC : KC :: ES : SH,$$

therefore also

$$ES = SH.$$

And since

$$FE : EG :: HE : 2ED,$$

and

$$2ES = EH,$$

therefore also

$$FE : EG :: SE : ED.$$

And

$$FE : EG :: LM : MR \text{ (Eucl. VI. 4)};$$

therefore

$$LM : MR :: SE : ED.$$

And since it was shown (I. 43) that, in the case of the hyperbola,

$$\text{trgl. } RNC = \text{trgl. } LNX + \text{trgl. } GBC$$
$$= \text{trgl. } LNX + \text{trgl. } CDE,^*$$

---

* That

$$\text{trgl. } GBC = \text{trgl. } CDE$$

is proved by Apollonius in the course of another proof of I. 43, reported by Eutocius. It is also proved in III. 1, without the help of intervening propositions. (Tr.)

and, in the case of the ellipse and circle,
$$\text{trgl. } RNC + \text{trgl. } LNX = \text{trgl. } GBC$$
$$= \text{trgl. } CDE;$$
therefore, in the case of the hyperbola with the common triangle *ECD* and the common quadrilateral *NRMX* subtracted, and in the case of the ellipse and circle with the common triangle *MXC* subtracted,[*]
$$\text{trgl. } LMR = \text{quadr. } MEDX.$$

And *MX* is parallel to *DE*, and
$$\text{angle } LMR = \text{angle } EMX;$$
therefore
$$\text{rect. } LM, MR = \text{rect. } EM, (ED + MX) \quad \text{(I. 49, note, para. 2).}$$
And since
$$MC : CE :: MX : ED$$
and
$$MC : CE :: MO : ES,$$

---

[*] The position of point *L* furnishes different cases which at times, as in the present theorem, require a change in the course of the proof.

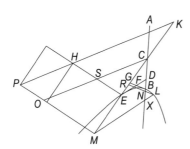

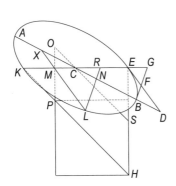

For the hyperbola illustrated here, instead of the subtraction as in the theorem, we have
$$\text{trgl. } RNC = \text{trgl. } LNX + \text{trgl. } CDE$$
$$\text{quadr. } MCNL = \text{quadr. } MCNL.$$
Subtracting the first equals from the second identity, we have
$$\text{trgl. } LMR = \text{quadr. } MEDX.$$
The rest of the proof is the same.

For the ellipse and circle in this configuration, we have as in the theorem above
$$\text{trgl. } RNC + \text{trgl. } LNX = \text{trgl. } CDE,$$
and subtracting the common triangle *CMX*,
$$\text{trgl. } LMR = \text{trgl. } CDE - \text{trgl. } CMX;$$
therefore
$$\text{rect. } LM, MR = \text{rect. } EC, ED - \text{rect, } MC, MX.$$
$$= \text{rect. } ME, (ED - MX).$$

The proof is similar to that in the note to Prop. 49.

These cases will come up again in Book III, and in general it is convenient to think of quadrilateral *MEDX* as standing for the difference of the two triangles when one pair of its sides cross each other. (Tr. with changes by Ed.)

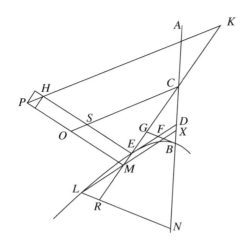
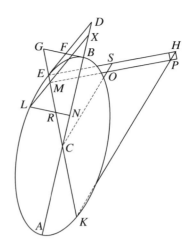

therefore

$$MO : ES :: MX : ED.$$

And *componendo*

$$MO + ES : ES :: MX + ED : ED;$$

alternately

$$MO + ES : MX + ED :: ES : ED.$$

But

$$MO + ES : MX + ED :: \text{rect. } MO + ES, EM : \text{rect. } MX + ED, EM,$$

and

$$ES : ED :: LM : MR :: FE : EG$$
(by prev. para.)

or

$$ES : ED :: \text{sq. } LM : \text{rect. } LM, MR;$$

therefore

rect. $MO + ES, ME$: rect. $MX + ED, EM$ :: sq. LM : rect. $LM, MR.$

And alternately

rect. $MO + ES, ME$: sq. LM :: rect. $MX + ED, EM$: rect. $LM, MR.$

But

$$\text{rect. } LM, MR = \text{rect. } ME, MX + ED \text{ (above)};$$

therefore

$$\text{sq. } LM = \text{rect. } EM, MO + ES.$$

And

$$SE = SH,$$

and

$$SH = OP;$$

therefore

$$\text{sq. } LM = \text{rect. } EM, MP.$$

PROPOSITION 51

If a straight line touching either of the opposite sections meets the diameter, and if through the point of contact and the center some straight line is produced to the other section, and if from the vertex a straight line is erected parallel to an ordinate and meets the straight line drawn through the point of contact and the center, and if it is contrived that, as the segment of the tangent between the straight line erected and the point of contact is to the segment of the straight line—drawn through the point of contact and the center—between the point of contact and the straight line erected, so is some [found] straight line to the double of the tangent, then whatever straight line in the other of the sections is drawn to the straight line through the point of contact and the center, parallel to the tangent, will equal in square the rectangle applied to the straight line found, having as breadth the straight line cut off by it from the point of contact and projecting beyond by a figure similar to the rectangle contained by the straight line between the opposite sections and the straight line found.

Let there be opposite sections whose diameter is the straight line *AB* and center *E,* and let the straight line *CD* be drawn tangent to the section *B* and the straight line *CE* joined and produced (I. 29), and let the straight line *BLG* be drawn ordinatewise (I. 17), and let it be contrived that

$$LC : CG :: \text{some straight line } K : 2CD.$$

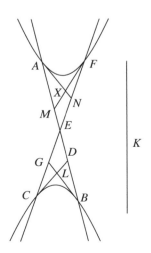

Now it is evident that the straight lines in the section *BC,* parallel to *CD* and drawn to *EC* produced are equal in square to the areas applied to *K,* having as breadths the straight line cut off by them from the point of contact, and projecting beyond by a figure similar to the rectangle *CF, K;* for

$$FC = 2CE.$$

I say then that in section *FA* the same thing will come about.

For let the straight line *MF* be drawn through *F* tangent to the section *AF,* and let the straight line *AXN* be erected ordinatewise. And since *BC* and *AF* are opposite sections, and *CD* and *MF* are tangents to them, therefore *CD* is equal and parallel to *MF* (I. 44, note). But also
$$CE = EF;$$
therefore also
$$ED = EM.$$
And since
$$LC : CG :: K : 2CD \text{ or } 2MF,$$
therefore also
$$XF : FN :: K : 2MF.$$
Since then *AF* is an hyperbola whose diameter is *AB* and tangent *MF,* and *AN* has been drawn ordinatewise, and
$$XF : FN :: K : 2FM,$$
hence any lines drawn from the section to *EF* produced, parallel to *FM,* will equal in square the rectangle contained by the straight line *K* and the line cut off by them from *F,* projecting beyond by a figure similar to the rectangle *CF, K* (I. 50).

[Porism]

And with these things shown, it is at once evident that in the parabola each of the straight lines drawn off parallel to the original diameter is a diameter (I. 46), but in the hyperbola and ellipse and opposite sections each of the straight lines drawn through the center is a diameter (I. 47–48); and that in the parabola the straight lines dropped to each of the diameters parallel to the tangents will equal in square the rectangles applied to it (I. 49), but in the hyperbola and opposite sections they will equal in square the areas applied to the diameter and projecting beyond by the same figure (I. 50–51), but in the ellipse the areas applied to the diameter and defective by the same figure (I. 50); and that all the things which have been already proved about the sections as following when the principal diameters are used,* will also, those very same things, follow when the other diameters are taken.

---

* The principal diameter ($\delta\iota\acute{\alpha}\mu\epsilon\tau\rho\sigma\varsigma$ $\dot{\alpha}\rho\chi\iota\kappa\acute{\eta}$) is that whose being is established in I.7 porism. (Tr.)

PROPOSITION 52 (PROBLEM)

Given a straight line in a plane bounded at one point, to find in the plane the section of a cone called parabola, whose diameter is the given straight line, and whose vertex is the end of the straight line, and where whatever straight line is dropped from the section to the diameter at a given angle, will equal in square the rectangle contained by the straight line cut off by it from the vertex of the section and by some other given straight line.

Let there be the straight line *AB* given in position and bounded at the point *A*, and another straight line *CD* given in magnitude, and first let the given angle be a right angle; it is required then to find a parabola in the plane of reference whose diameter is the straight line *AB* and whose vertex is the point *A*, and whose upright side is the straight line *CD*, and where the straight lines dropped ordinatewise will be dropped at a right angle, that is so that *AB* is the axis (Def. 7).

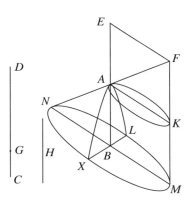

Let *AB* be produced to *E*, and let *CG* be taken as the fourth part of *CD*, and let

$$EA > CG,$$

and let

$$CD : H :: H : EA.$$

Therefore

$$CD : EA :: \text{sq. } H : \text{sq. } EA,$$

and

$$CD < 4EA;$$

therefore also

$$\text{sq. } H < 4 \text{ sq. } EA.$$

Therefore

$$H < 2EA;$$

and so the two straight lines *EA* are greater than *H*. It is therefore possible for a triangle to be constructed from *H* and two straight lines *EA*. Then let the triangle *EAF* be constructed on *EA* at right angles to the plane of reference so that

$$EA = AF,$$

and

$$H = FE,$$

and let the straight line AK be drawn parallel to FE, and FK to EA, and let a cone be conceived whose vertex is the point F and whose base is the circle about diameter KA, at right angles to the plane through AFK. Then the cone will be a right cone (Def. 3); for

$$AF = FK.$$

And let the cone be cut by a plane parallel to the circle KA, and let it make as a section the circle MNX (I. 4), at right angles clearly to the plane through MFN, and let the straight line MN be the common section of the circle MNX and of the triangle MFN; therefore it is the diameter of the circle. And let the straight line XL be the common section of the plane of reference and of the circle. Since then circle MNX is at right angles to triangle MFN, and the plane of reference also at right angles to triangle MFN, therefore the straight line LX, their common section, is at right angles to triangle MFN, that is to triangle KFA (Eucl. XI. 19); and therefore it is perpendicular to all the straight lines touching it and in the triangle; and so it is perpendicular to both MN and AB.

Again since a cone, whose base is the circle MNX and whose vertex is the point F, has been cut by a plane at right angles to the triangle MFN and makes as a section circle MNX, and since it has also been cut by another plane, the plane of reference, cutting the base of the cone in a straight line XL at right angles to MN which is the common section of the circle MNX and the triangle MFN, and the common section of the plane of reference and of the triangle MFN, the straight line AB, is parallel to the side of the cone FKM, therefore the resulting section of the cone in the plane of reference is a parabola, and its diameter AB (I. 11), and the straight lines dropped ordinatewise from the section to AB will be dropped at right angles; for they are parallel to XL which is perpendicular to AB. And since

$$CD : H :: H : EA,$$

and

$$EA = AF = FK$$

and

$$H = EF = AK,$$

therefore

$$CD : AK :: AK : AF.$$

And therefore

$$CD : AF :: \text{sq. } AK : \text{sq. } AF \text{ or rect. } AF, FK.$$

Therefore CD is the upright side of the section; for this has been shown in the eleventh theorem (I. 11).

PROPOSITION 53 (PROBLEM)

With the same things supposed, let the given angle not be right, and let the angle *HAE* be made equal to it and let

$$AH = \text{half } CD,$$

and from *H* let the straight line *HE* be drawn perpendicular to *AE*, and through *E* let the straight line *EL* be drawn parallel to *BH*, and from *A* let the straight line *AL* be drawn perpendicular to *EL*, and let *EL* be bisected at *K*, and from *K* let the straight line *KM* be drawn perpendicular to *EL* and produced to *F* and *G*, and let rect. *LK*, *KM* = sq. *AL*. And given the two straight lines *LK* and *KM*, *KL* in position and bounded at *K*, and *KM* in magnitude, and let a parabola be described with a right angle whose diameter is the straight line *KL*, and whose vertex is the point *K*, and whose upright side is the straight line *KM*, as has been shown before (I. 52); and it will pass through the point *A* because

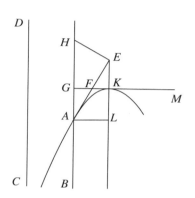

$$\text{sq. } AL = \text{rect. } LK, KM \text{ (I. 11)},$$

and the straight line *EA* will touch the section because

$$EK = KL \text{ (I. 33)}.$$

And *HA* is parallel to *EKL*; therefore *HAB* is the diameter of the section, and the straight lines dropped to it parallel to *AE* will be bisected by *AB* (I. 46). And they will be dropped at angle *HAE*. And since

$$\text{angle } AEH = \text{angle } AGF,$$

and angle at *A* is common, therefore triangle *AHE* is similar to triangle *AGF*.

Therefore

$$HA : EA :: FA : AG;$$

therefore

$$2AH : 2AE :: FA : AG.$$

But

$$CD = 2AH;$$

therefore

$$FA : AG :: CD : 2AE.$$

Then by things already shown in the forty-ninth theorem (I. 49) the straight line *CD* is the upright side.

PROPOSITION 54 (PROBLEM)

Given two bounded straight lines perpendicular to each other, one of them being produced on the side of the right angle, to find on the straight line produced the section of a cone called hyperbola in the same plane with the straight lines, so that the straight line produced is a diameter of the section and the point at the angle is the vertex, and where whatever straight line is dropped from the section to the diameter, making an angle equal to a given angle, will equal in square the rectangle applied to the other straight line having as breadth the straight line cut off by the dropped straight line begin-ning with the vertex and projecting beyond by a figure similar and similarly situated to that contained by the original straight lines.

Let there be the two bounded straight lines *AB* and *BC* perpendicular to each other, and let *AB* be produced to *D*; it is required then to find in the plane through the lines *AB, BC* an hyperbola whose diameter will be the straight line *ABD* and vertex *B*, and upright side the straight line *BC*, and where the

straight lines dropped from the section to *BD* at the given angle will equal in square the rect-angles applied to *BC* having as breadths the straight lines cut off by them from *B* and projecting beyond by a figure similar and similarly situated to the rectangle *AB, BC*.

First let the given angle be a right angle, and on *AB* let a plane be erect-ed at right angles to the plane of reference, and let the circle *AEBF* be described in it about *AB*, so that the segment of the circle's diameter within the sector *AEB*

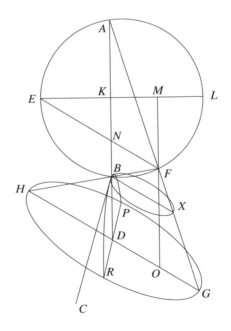

has to the segment of the diameter within the sector *AFB* a ratio not greater than that of *AB* to *BC*,[*] and let *AEB* be bisected at *E*, and let the straight line *EK* be drawn perpendicular from *E* to the straight line *AB* and let it

_____

[*] Eutocius, commenting, adds: "Let there be two straight lines *AB* and *BC*, and let it be required to describe a circle on *AB* so that its diameter is cut by *AB* in such a way that the part of it on the side of *C* has to the remainder a ratio not greater than that of *AB* to *BC*.

"Now let it be supposed that they have the same ratio, and let *AB* be bisected at *D*, and through it let the straight line *EDF* be drawn perpendicular to *AB*, and let it be contrived that *AB* : *BC* :: *ED* : *DF*,
and let *EF* be bisected; then it is clear that if

$$AB = BC$$

and

$$ED = DF,$$

the point *D* will be the midpoint of *EF*, and if

$$AB > BC$$

and

$$ED > DF,$$

the midpoint will be below *D*, and if

$$AB < BC,$$

it will be above *D*.

"And now [assuming *AB* > *BC*] let it be below as *G*, and with center *G* and radius *GF* let a circle be described; then it will have to pass either within or without the points *A* and *B*. And if it should pass through the points *A* and *B*, what was enjoined would be done; but let it fall beyond the points *A* and *B*, and let the straight line *AB*, produced both ways, meet the circumference at *H* and *K*, and let *FH*, *HE*, *EK* and *KF* be joined, and let *MB* be drawn through *B* parallel to *FK*, and *BL* parallel to *KE*, and let *MA* and *AL* be joined; then these will also be parallel to *FH* and *HE* because

$$AD = DB$$

and

$$DH = DK$$

and *FDE* is perpendicular to *HK*. And since the angle at *K* is a right angle, and *MB* and *BL* are parallel to *FK* and *KE*, therefore the angle at *B* is a right angle; then for the same reasons also the angle at *A*. And so the circle described on *ML* will pass through the points *A* and *B* (Eucl. III. 31). Let the circle *MALB* be described. And since *MB* is parallel to *FK*,

$$FD : DM :: KD : DB.$$

Then likewise also

$$KD : DB :: ED : DL.$$

And therefore

$$FD : DM :: ED : DL.$$

And alternately,

$$ED : DF :: AB : BC :: LD : DM.$$

"And likewise if the circle described on *FE* cuts *AB*, the same thing could be shown." (Tr.)

be produced to *L*; therefore the straight line *EL* is a diameter (Eucl. III. 1). If then

$$AB : BC :: EK : KL$$

we use point *L*, but if not, let it be contrived (Eucl. VI. 12) that

$$AB : BC :: EK : KM$$

with

$$KM < KL,$$

and through *M* let *MF* be drawn parallel to *AB*, and let *AF*, *EF*, and *FB* be joined, and through *B* let *BX* be drawn parallel to *FE*. Since then

  angle AFE = angle *EFB*,

but

  angle AFE = angle *AXB*,

and

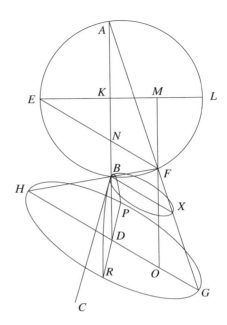

$$\text{angle } EFB = \text{angle } XBF,$$

therefore also

$$\text{angle XBF} = \text{angle } FXB;$$

therefore also

$$FB = FX.$$

Let a cone be conceived whose vertex is the point *F* and whose base is the circle about diameter *BX* at right angles to triangle *BFX*. Then the cone will be a right cone; for

$$FB = FX.$$

Then let the straight lines *BF*, *FX*, and *MF* be produced, and let the cone be cut by a plane parallel to the circle *BX*; then the section will be a circle (I. 4). Let it be the circle *GPR*; and so *GH* will be the diameter of the circle (I. 4, end). And let the straight line *PDR* be the common section of circle *GH* and of the plane of reference; then *PDR* will be perpendicular to both of the straight lines *GH* and *DB*; for both of the circles *XB* and *HG* are perpendicular to triangle *FGH*, and the plane of reference is perpendicular to triangle *FGH*; and therefore their common section, the straight line *PDR*, is perpendicular to triangle *FGH*; therefore it makes right angles also with all the straight lines touching it and in the same plane.

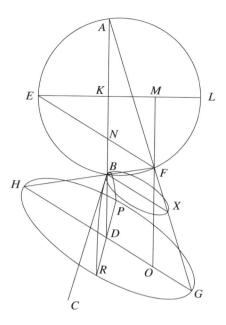

And since a cone whose base is circle *GH* and vertex *F*, has been cut by a plane perpendicular to triangle *FGH,* and has also been cut by another plane, the plane of reference, in the straight line *PDR* perpendicular to the straight line *GDH,* and the common section of the plane of reference and of triangle *GFH,* that is the straight line *DB,* produced in the direction of *B,* meets the straight line *GF* at *A,* therefore by things already shown before (I. 12) the section *PBR* will be an hyperbola whose vertex is the point *B,* and where the straight lines dropped ordinatewise to *BD* will be dropped at a right angle; for they are parallel to straight line *PDR.* And since

$$AB : BC :: EK : KM,$$

and

$$EK : KM :: EN : NF :: \text{rect. } EN, NF : \text{sq. } NF,$$

therefore

$$AB : BC :: \text{rect. } EN, NF : \text{sq. } NF.$$

And

$$\text{rect. } EN, NF = \text{rect. } AN, NB \text{ (Eucl. III. 35)};$$

therefore

$$AB : CB :: \text{rect. } AN, NB : \text{sq. } NF.$$

But

$$\text{rect. } AN, NB : \text{sq. } NF :: AN : NF \text{ comp. } BN : NF;$$

but
$$AN : NF :: AD : DG :: FO : OG,$$
and
$$BN : NF :: FO : OH;$$
therefore
$$AB : BC :: FO : OG \text{ comp. } FO : OH,$$
that is
$$\text{sq. } FO : \text{rect. } OG, OH.$$
Therefore
$$AB : BC :: \text{sq. } FO : \text{rect. } OG, OH.$$
And the straight line FO is parallel to the straight line AD; therefore the straight line AB is the transverse side, and BC the upright side; for these things have been shown in the twelfth theorem (I. 12).

PROPOSITION 55 (PROBLEM)

Then let the given angle not be a right angle, and let there be the two given straight lines AB and AC, and let the given angle be equal to angle BAH; then it is required to describe an hyperbola whose diameter will be the straight line AB, and upright side AC, and where the ordinates will be dropped at angle HAB.

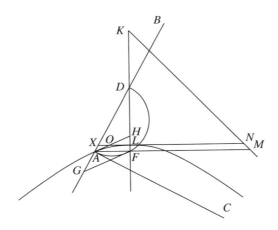

Let the straight line AB be bisected at D, and let the semicircle AFD be described on AD, and let some straight line FG, parallel to AH, be drawn to the semicircle making

$$\text{sq. } FG : \text{rect. } DG, GA :: AC : AB,^*$$

---

* Eutocius, commenting, gives this construction : "Let there be the semicircle *ABC* on the diameter *AC*, and the given ratio *EF* to *FG*, and let it be required to do what is proposed.

"Let *FH* be made equal to *EF*, and let *HG* be bisected at *K*, and let the straight line *CB* be drawn in the semicircle at angle *ACB* (the required angle), and from the center *L* let the straight line *LS* be drawn perpendicular to it, and produced let it meet the circumference at *N*, and through *N* let *NM* be drawn parallel to *CB*; therefore it will touch the circle. And let it be contrived that

$$FH : HK :: MX : XN,$$

and let *NO* be made equal to *XN*, and let the straight lines *LX* and *LO* cutting the semicircle at *P* and *R* be joined, and let the straight line *PRD* be joined.

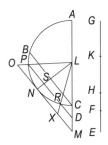

Since then

$$XN = NO,$$

and *NL* is common and perpendicular, therefore

$$LO = LX.$$

And also

$$LP = LR;$$

and therefore the remainders

$$PO = RX.$$

Therefore *PRD* is parallel to *MO*. And

$$FH : HK :: MX : NX;$$

and

$$HK : HG :: NX : XO;$$

therefore *ex aequali*

$$FH : HG :: MX : XO;$$

inversely

$$HG : FH :: XO : MX;$$

componendo

$$GF : FH :: OM : MX$$

or

$$GF : FE :: PD : DR.$$

And

$$PD : DR :: \text{rect. } PD, DR : \text{sq. } DR,$$

but

$$\text{rect. } PD, DR = \text{rect. } AD, DC \text{ (Eucl. III. 36)};$$

therefore

$$GF : FE :: \text{rect. } AD, DC : \text{sq. } DR.$$

Therefore inversely

$$FE : GF :: \text{sq. } DR : \text{rect. } AD, DC."$$

(Tr.)

and let the straight line *FHD* be joined and produced to *D*,[*] and let

FD : DL :: DL : DH,

and let *DK* be made equal to *DL*, and let

rect. *LF, FM* = sq. *AF*,

and let *KM* be joined, and through *L* let *LN* be drawn perpendicular to *KF* and let it be produced towards *X*. And with two given bounded straight lines *KL* and *LN* perpendicular to each other, let an hyperbola be described whose transverse side is *KL*, and upright side *LN*,

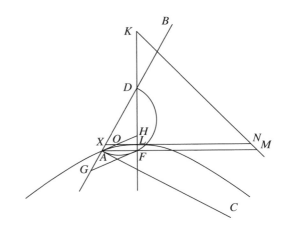

and where the straight lines dropped from the section to the diameter will be dropped at a right angle and will equal in square the rectangle applied to *LN* having as breadths the straight lines cut off by them from *L* and projecting beyond by a figure similar to rectangle *KL, LN* (I. 54); and the section will pass through *A*; for

sq. *AF* = rect. *LF, FM* (I. 12).

And *AH* will touch it; for

rect. *FD, DH* = sq. *DL* (I. 37 converse).

And so *AB* is a diameter of the section (I. 47 and Def. 4). And since

CA : 2AD or AB :: sq. FG : rect. DG, GA,

but

CA : 2AD :: CA : 2AH comp. 2AH : 2AD

or

CA : 2AD :: CA : 2AH comp. AH : AD

and

AH : AD :: FG : GD,

therefore

CA : AB :: CA : 2AH comp. FG : GD.

But also

sq. FG : rect. DG, GA :: FG : GD comp. FG : GA;

therefore

---

[*]Although this last phrase "and produced to *D*" may seem unnecessary or even incorrect, it does appear in the Greek text. (Ed.)

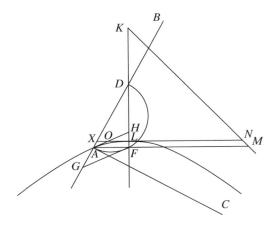

ratio $CA : 2AH$ comp. $FG : GD$ = ratio $FG : GA$ comp. $FG : GD$.
Let the common ratio $FG : GD$ be taken away; therefore
$$CA : 2AH :: FG : GA.$$
But
$$FG : GA :: OA : AX,$$
therefore
$$CA : 2AH :: OA : AX.$$

But whenever this is so, the straight line AC is a parameter; for this has been shown in the fiftieth theorem (I. 50).

PROPOSITION 56 (PROBLEM)

Given two bounded straight lines perpendicular to each other, to find about one of them as diameter and in the same plane with the two straight lines the section of a cone called ellipse, whose vertex will be the point at the right angle, and where the straight lines dropped ordinatewise from the section to the diameter at a given angle will equal in square the rectangles applied to the other straight line, having as breadth the straight line cut off by them from the vertex of the section and defective by a figure similar and similarly situated to the rectangle contained by the given straight lines.

Let there be two given straight lines AB and AC perpendicular to each other, of which the greater is the straight line AB; then it is required to describe in the plane of reference an ellipse whose diameter will be the straight line *AB* and vertex *A*, and upright side *AC*, and where the ordinates will be dropped from the section to the diameter at a given angle and will equal in square the rectangles applied to *AC* having as breadths the straight lines cut off by them from *A* and defective by a figure similar and similarly situated to rectangle *BA, AC*.

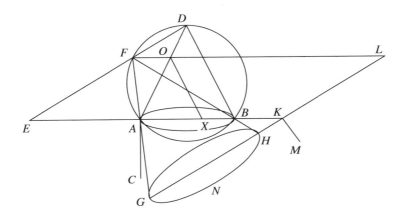

And first let the given angle be a right angle, and let a plane be erected from *AB* at right angles to the plane of reference, and in it, on *AB*, let the sector of a circle *ADB* be described, and its midpoint be *D*, and let the straight lines *DA* and *DB* be joined, and let the straight line *AX* be made equal to *AC*, and through *X* let the straight line *XO* be drawn parallel to *DB*, and through *O* let *OF* be drawn parallel to *AB*, and let *DF* be joined and let it meet *AB* produced at *E*; then we will have

$$AB : AC :: AB : AX :: DA : AO :: DE : EF.$$

And let the straight lines *AF* and *FB* be joined and produced, and let some point *G* be taken at random on *FA*, and through it let the straight line *GL* be drawn parallel to *DE* and let it meet *AB* produced at *K*; then let *FO* be produced and let it meet *GK* at *L*. Since then

$$\text{arc } AD = \text{arc } DB,$$
$$\text{angle } ABD = \text{angle } DFB \text{ (Eucl. III. 27).}$$

And since

$$\text{angle } EFA = \text{angle } FDA + \text{angle } FAD,$$

but

$$\text{angle } FAD = \text{angle } FBD,$$

and

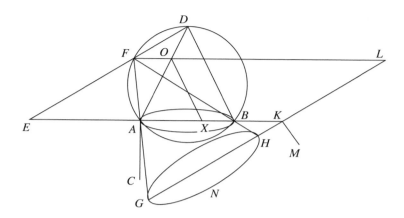

$$\text{angle } FDA = \text{angle } FBA,$$
therefore also
$$\text{angle } EFA = \text{angle } DBA = \text{angle } DFB.$$
And also DE is parallel to LG; therefore
$$\text{angle } EFA = \text{angle } FGH,$$
and
$$\text{angle } DFB = \text{angle } FHG.$$
And so also
$$\text{angle FGH} = \text{angle } FHG,$$
and
$$FG = FH.$$

Then let circle GHN be described about HG at right angles to triangle HGF, let a cone be conceived whose base is the circle GHN, and whose vertex is the point F; then the cone will be a right cone because
$$FG = FH.$$
And since the circle GHN is at right angles to plane HGF, and the plane of reference is also at right angles to the plane through GH and HF, therefore their common section will be at right angles to the plane through GH and HF. Then let their common section be the straight line KM; therefore the straight line KM is perpendicular to both of the straight lines AK and KG.

And since a cone whose base is the circle GHN, and whose vertex is the point F, has been cut by a plane through the axis and makes as a section triangle *GHF*, and has been cut also by another plane through *AK* and *KM*, which is the plane of reference, in the straight line *KM* which is perpendicular to *GK*, and the plane meets the sides of the cone *FG* and *FH*, therefore the resulting section is an ellipse whose diameter is *AB* and where the ordinates will be dropped at a right angle (I. 13); for they are parallel to *KM*. And since

$$DE : EF :: \text{rect. } DE, EF \text{ or rect. } BE, EA : \text{sq. } EF,$$

and

$$\text{rect. } BE, EA : \text{sq. } EF :: BE : EF \text{ comp. } AE : EF,$$

but

$$BE : EF :: BK : KH \ [:: FL : LH\ ],$$

and

$$AE : EF :: AK : KG :: FL : LG,$$

therefore

$$BA : AC :: FL : LG \text{ comp. } FL : LH \text{ (see above)},$$

which is the same as

$$\text{sq. } FL : \text{rect. } GL, LH;$$

therefore

$$BA : AC :: \text{sq. } FL : \text{rect. } GL, LH.$$

And whenever this is so, the straight line *AC* is the upright side of the figure, as has been shown in the thirteenth theorem (I. 13).

PROPOSITION 57 (PROBLEM)

With the same things supposed let the straight line *AB* be less than *AC*, and let it be required to describe an ellipse about diameter *AB* so that *AC* is the upright.

Let *AB* be bisected at *D*, and from *D* let the straight line *EDF* be drawn perpendicular to *AB*, and let

$$\text{sq. } FE = \text{rect. } BA, AC$$

so that

$$FD = DE,$$

and let *FG* be drawn parallel to *AB*, and let it be contrived that

$$AC : AB :: EF : FG;$$

therefore also

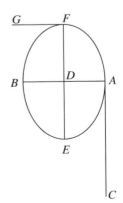

$$EF > FG.$$

And since

$$\text{rect. } CA, AB = \text{sq. } EF,$$

hence

$CA : AB :: \text{sq. } FE : \text{sq. } AB :: \text{sq. } DF : \text{sq. } DA.$

But

$$CA : AB :: EF : FG,$$

therefore

$$EF : FG :: \text{sq. } FD : \text{sq. } DA.$$

But

$$\text{sq. } FD = \text{rect. } FD, DE;$$

therefore

$EF : FG :: \text{rect. } ED, DF : \text{sq. } AD.$

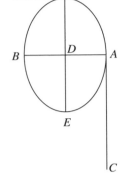

Then with two bounded straight lines situated at right angles to each other and with EF greater, let an ellipse be described whose diameter is EF and upright side FG (I. 56); then the section will pass through A because

$$\text{rect. } FD, DE : \text{sq. } DA :: EF : FG \text{ (I. 21)}.$$

And

$$AD = DB;$$

then it will also pass through B. Then an ellipse has been described about AB.

And since

$$CA : AB :: \text{sq. } FD : \text{sq. } DA,$$

and

$$\text{sq. } DA = \text{rect. } AD, DB,$$

therefore

$$CA : AB :: \text{sq. } DF : \text{rect. } AD, DB.$$

And so the straight line AC is an upright side (I. 21).

PROPOSITION 58 (PROBLEM)

But then let the given angle not be a right angle, and let the angle BAD be equal to it, and let the straight line AB be bisected at E, and let the semicircle AFE be described on AE, and in it let the straight line FG be drawn parallel to AD making

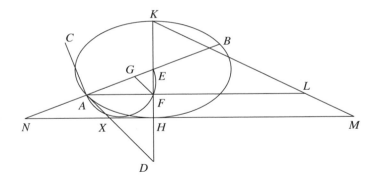

sq. *FG* : rect. *AG, GE* :: *CA* : *AB,*[*]

and let the straight lines *AF* and *EF* be joined and produced, and let

$$DE : EH :: EH : EF,$$

and let

$$EK = EH,$$

and let it be contrived that

$$\text{rect. } HF, FL = \text{sq. } AF,$$

---

[*] Eutocius, commenting, gives this construction : "Let there be the semicircle *ABC* and within it some straight line *AB* (at the required angle to *AC*), and let two un-equal straight lines *DE* and *EF* be laid down, and let *EF* be produced to *G*, and let *FG* be made equal to *DE*, and let the whole line *EG* be bisected at *H*, and let the center of the circle, *K*, be taken, and from it let a perpendicular be drawn to *AB* and let it meet the circumference at *L*, and through *L* let *LM* be drawn parallel to *AB*, and let *KA* produced meet *LM* at *M*, and let it be contrived that

$$HF : FG :: LM : MN,$$

and let

$$LX = LN,$$

and let the straight lines *NK* and *KX* be joined and produced, and let the circle, finished out, cut them at *P* and *O*, and let the straight line *ORP* be joined.

"Since then

$$FH : FG :: LM : MN,$$

componendo

$$HG : GF :: LN : NM;$$

inversely

$$FG : GH :: NM : NL,$$

and

$$FG : GE :: MN : NX;$$

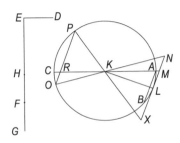

[Footnote continued on next page.]

and let the straight line *KL* be joined, and from *H* let the straight line *HMX* be drawn perpendicular to *HF* and so parallel to the straight line *AFL*; for the angle at *F* is right. And with the two given bounded straight lines *KH*, and *HM* perpendicular to each other, let an ellipse be described whose transverse diameter is *KH*, and the upright side of whose figure is *HM*, and where the ordinates to *HK* will be dropped at right angles (I. 56–57); then the section will pass through *A* because sq. *FA* = rect. *HF*, *FL* (I. 13). And since

$$HE = EK,$$

and

$$AE = EB,$$

the section will also pass through *B*, and *E* will be the center, and the straight line *AEB* the diameter. And the straight line *DA* will touch the section because

---

[Footnote continued from previous page:]

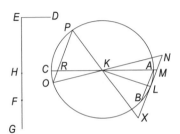

separando
$$FG : FE :: NM : MX.$$
And since
$$NL = LX,$$
and the straight line *LK* is common and at right angles, therefore also
$$KN = KX.$$
And also
$$KO = KP;$$
therefore *NX* is parallel to *OP*. Therefore triangle *KMN* is similar to triangle *OKR* and triangle *KMX* to triangle *PRK*.
Therefore
$$KM : KR :: MN : RO.$$
But also
$$KM : KR :: MX : PR;$$
and therefore
$$NM : RO :: MX : PR;$$
and alternately
$$NM : MX :: RO : RP.$$
But
$$NM : MX :: GF : FE :: DE : EF,$$
and
$$OR : RP :: sq.\ OR : rect.\ OR,\ RP;$$
and therefore
$$DE : EF :: sq.\ OR : rect.\ OR,\ RP.$$
And
$$rect.\ OR,\ RP = rect.\ AR,\ RC \ (Eucl.\ III.\ 35).$$
Therefore
$$DE : EF :: sq.\ OR : rect.\ AR,\ RC.\,"$$

(Tr.)

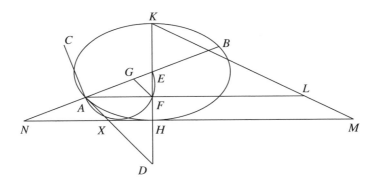

$$\text{rect. } DE, EF = \text{sq. } EH.$$

And since

$$CA : AB :: \text{sq. } FG : \text{rect. } AG, GE,$$

but

$$CA : AB :: CA : 2AD \text{ comp. } 2AD : AB \text{ or } DA : AE,$$

and

$$\text{sq. } FG : \text{rect. } AG, GE :: FG : GE \text{ comp. } FG : GA,$$

therefore

$$\text{ratio } CA : 2AD \text{ comp. } DA : AE :: \text{ratio } FG : GE \text{ comp. } FG : GA.$$

But

$$DA : AE :: FG : GE;$$

and the common ratio being taken away, we will have

$$CA : 2AD :: FG : GA$$

or

$$CA : 2AD :: XA : AN.$$

And whenever this is so, the straight line AC is the upright side of the figure (I. 50).

PROPOSITION 59 (PROBLEM)

Given two bounded straight lines perpendicular to each other, to find opposite sections whose diameter is one of the given straight lines, and whose vertices are the ends of the straight line, and where the straight lines dropped in each of the sections at a given angle will equal in square the rectangles applied to the other of the straight lines and projecting beyond by a figure similar to the rectangle contained by the given straight lines.

Let there be the two given bounded straight lines *BE* and *BH*, perpendicular to each other, and let the given angle be *G*; then it is required to describe opposite sections about one of the straight lines *BE*, *BH*, so that the ordinates are dropped at an angle *G*.

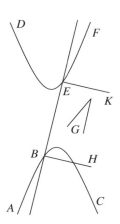

And given the two straight lines *BE* and *BH*, let an hyperbola be described whose transverse diameter will be the straight line *BE*, and the upright side of whose figure will be *HB*, and where the ordinates to *BE* produced will be at an angle *G*, and let it be the line *ABC*; for we have already described how this must be done (I. 55). Then let the straight line *EK* be drawn through *E* perpendicular to *BE* and equal to *BH*, and let another hyperbola *DEF* be likewise described whose diameter is *BE* and the upright side of whose figure is *EK*, and where the ordinates from the section will be dropped at a same angle *G*. Then it is evident that *B* and *E* are opposite sections, and there is one diameter for them, and their uprights are equal.

PROPOSITION 60 (PROBLEM)

Given two straight lines bisecting each other, to describe about each of them opposite sections, so that the straight lines are their conjugate diameters and the diameter of one pair of opposite sections is equal in square to the figure of the other pair, and likewise the diameter of the second pair of opposite sections is equal in square to the figure of the first pair.

Let there be the two given straight lines *AC* and *DE* bisecting each other; then it is required to describe opposite sections about each of them as a diameter so that the straight lines *AC* and *DE* are conjugates in them, and *DE* is equal in square to the figure about *AC*, and *AC* is equal in square to the figure about *DE*.

Let

$$\text{rect. } AC, CL = \text{sq. } DE,$$

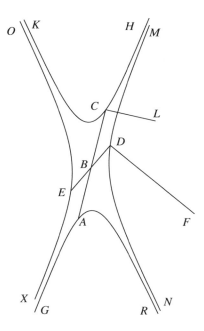

and let *LC* be perpendicular to *CA*. And given two straight lines *AC* and *CL* perpendicular to each other, let the opposite sections *RAG* and *HCK* be described whose transverse diameter will be *CA* and whose upright side will be *CL*, and where the ordinates from the sections to *CA* will be dropped at the given angle (I. 59). Then the straight line *DE* will be a second diameter of the opposite sections (Def. 11); for it is the mean proportion between the sides of the figure, and, parallel to an ordinate, it has been bisected at *B*.

Then again let

rect. *DE, DF* = sq. *AC,*

and let *DF* be perpendicular to *DE*. And given two straight lines *ED* and *DF* lying perpendicular to each other, let the opposite sections *MDN* and *OEX* be described, whose transverse diameter will be *DE* and the upright side of whose figure will be *DF*, and where the ordinates from the sections will be dropped to *DE* at the given angle (I. 59); then the straight line *AC* will also be a second diameter of the sections *MDN* and *XEO*. And so *AC* bisects the parallels to *DE* between the sections *RAG* and *HCK*, and *DE* the parallels to *AC*; and this it was required to do.

And let such sections be called conjugate.

CONICS
BOOK TWO

APOLLONIUS to EUDEMUS,

If you are well, well and good, and I, too, fare pretty well.

I have sent you my son Apollonius bringing you the second book of the conics as arranged by us. Go through it then carefully and acquaint those with it worthy of sharing in such things. And Philonides, the geometer, I introduced to you in Ephesus, if ever he happen about Pergamum, acquaint him with it too. And take care of yourself, to be well. Good-bye.

PROPOSITION 1

If a straight line touch an hyperbola at its vertex, and from it on both sides of the diameter a straight line is cut off equal in square to the fourth of the figure, then the straight lines drawn from the center of the section to the ends thus taken on the tangent will not meet the section.

Let there be an hyperbola whose diameter is the straight line *AB* and center *C*, and upright the straight line *BF*; and let the straight line *DE* touch the section at *B*, and let the squares on *BD* and *BE* each be equal to the fourth of the figure *AB*, *BF*, and the straight lines *CD* and *CE* be joined and produced.

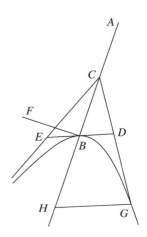

I say that they will not meet the section.

For if possible, let *CD* meet the section at *G,* and from *G* let the straight line *GH* be dropped ordinatewise; therefore it is parallel to *DB* (I. 17). Since then

$$AB : BF :: sq. \ AB : rect. \ AB, BF,$$

but

$$sq. \ CB = fourth \ sq. \ AB,$$

and

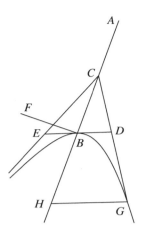

sq. BD = fourth rect. AB, BF,
therefore

$AB : BF ::$ sq. $CB :$ sq. $DB ::$ sq. $CH :$ sq. HG.[*]
And also

$AB : BF ::$ rect. AH, $HB :$ sq. HG (I. 21);
therefore

sq. $CH :$ sq. $HG ::$ rect. AH, $HB :$ sq. HG.
Therefore

rect. AH, HB = sq. CH;

and this is absurd (Eucl. II. 6). Therefore the straight line CD will not meet the section. Then likewise we could show that neither does CE; therefore the straight lines CD and CE are asymptotes ($\dot{\alpha}\sigma\acute{\upsilon}\mu\pi\tau\omega\tau\circ\varsigma$)[**] to the section.

PROPOSITION 2

With the same things it is to be shown that a straight line cutting the angle contained by the straight lines DC and CE is not another asymptote.

For if possible, let CH be it, and let the straight line BH be drawn through B parallel to CD and let it meet CH at H, and let DG be made equal to BH and let GH be joined and produced to the points K, L, and M. Since then BH and DG are equal and parallel, DB and HG are also equal and parallel. And since AB is bisected at C and a straight line BL is added to it,

rect. AL, LB + sq. CB = sq. CL (Eucl. II. 6).
Likewise then, since GM is parallel to DE, and

$DB = BE$,

---

[*] This result is used later, for example in II.10. The proportion is true, because of similar triangles, whether G falls on the section or not. (Ed.)

[**] The word $\dot{\alpha}\sigma\acute{\upsilon}\mu\pi\tau\omega\tau\circ\varsigma$ means literally "not capable of meeting" and is used in a general way in Euclid to refer to any non-secant lines or planes. In Apollonius it is also used in this way, as for instance in II. 14, porism, where it refers to any straight lines not meeting the hyperbola. The special case where in English the lines are spoken of as asymptotes is the one defined here. Book II, proposition 14, porism further declares their peculiar property and significance. (Tr.)

therefore also
$$GL = LM.$$
And since
$$GH = DB,$$
therefore
$$GK > DB.$$
And also
$$KM > BE,$$
since also
$$LM > BE;$$
therefore
$$\text{rect. } MK, KG > \text{rect. } DB, BE,$$
that is,
$$> \text{sq. } DB.$$
Since then
$$AB : BF :: \text{sq. } CB : \text{sq. } BD \text{ (II. 1)},$$
but
$$AB : BF :: \text{rect. } AL, LB : \text{sq. } LK \text{ (I. 21)},$$
and
$$\text{sq. } CB : \text{sq. } BD :: \text{sq. } CL : \text{sq. } LG,$$
therefore also
$$\text{sq. } CL : \text{sq. } LG :: \text{rect. } AL, LB : \text{sq. } LK.$$
Since then

whole sq. LC : whole sq. LG ::
part subtr. rect. AL, LB : part subtr. sq. LK,

therefore also
sq. LC : sq. LG :: remainder sq. CB : remainder rect. MK, KG (Eucl. II. 5),
that is
$$\text{sq. } CB : \text{rect. } MK, KG :: \text{sq. } CB : \text{sq. } DB.$$
Therefore
$$\text{sq. } DB = \text{rect. } MK, KG;$$
and this is absurd; for it has been shown to be greater than it. Therefore the straight line CH is not an asymptote to the section.

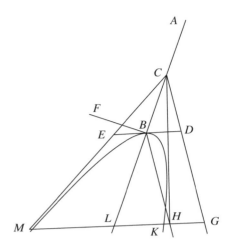

PROPOSITION 3

If a straight line touches an hyperbola, it will meet both of the asymptotes and it will be bisected at the point of contact, and the square on each of

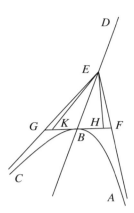

its segments will be equal to the fourth of the figure resulting on the diameter drawn through the point of contact.

Let there be the hyperbola *ABC*, and its center *E*, and asymptotes *FE* and *EG*, and let some straight line *HK* touch it at *B*.

I say that the straight line *HK* produced will meet the straight lines *FE* and *EG*.

For if possible, let it not meet them, and let *EB* be joined and produced, and let *ED* be made equal to *EB*; therefore the straight line *BD* is a diameter. Then let the squares on *HB* and *BK* each be made equal to the fourth of the figure on *BD* [and the parameter], and let *EH* and *EK* be joined. Therefore they are asymptotes (II. 1); and this is absurd (II. 2); for *FE* and *EG* are supposed asymptotes. Therefore *KH* produced will meet the asymptotes *EF* and *EG* at *F* and *G*.

I say then also that the squares on *BF* and *BG* will each be equal to the fourth of the figure on *BD*.

For let it not be, but if possible, let the squares on *BH* and *BK* each be equal to the fourth of the figure. Therefore *HE* and *EK* are asymptotes (II. 1); and this is absurd (II. 2). Therefore the squares on *FB* and *BG* will each be equal to the fourth of the figure on *BD*.

PROPOSITION 4 (PROBLEM)

Given two straight lines containing an angle and a point within the angle, to describe through the point the section of a cone called hyperbola so that the given straight lines are its asymptotes.

Let there be the two straight lines *AC* and *AB* containing a chance angle at *A*, and let some point *D* be given, and let it be required to describe through *D* an hyperbola to the asymptotes *CA* and *AB*.

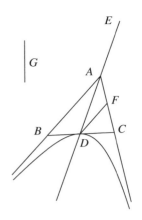

Let the straight line *AD* be joined and produced to *E*, and let *AE* be made equal to *DA*, and let the straight line *DF* be drawn through *D* parallel to *AB*, and let *FC* be made equal to *AF*, and let *CD* be joined and produced to *B*, and let it be contrived that

$$\text{rect. } DE, G = \text{sq. } CB,$$

and with *AD* extended let an hyperbola be described about it through *D* so that the ordinates equal in square the areas applied to *G* and exceeding by a figure similar to rectangle *DE*, *G*. Since then *DF* is parallel to *BA*, and

$$CF = FA,$$

therefore

$$CD = DB;$$

and so

$$\text{sq. } CB = 4 \text{ sq. } CD.$$

And

$$\text{sq. } CB = \text{rect. } DE, G;$$

therefore the squares on *CD* and *DB* are each equal to the fourth part of the figure *DE*, *G*. Therefore the straight lines *AB* and *AC* are asymptotes to the hyperbola described.

PROPOSITION 5

If the diameter of a parabola or hyperbola bisects some straight line [within the section], the tangent to the section at the end of the diameter will be parallel to the bisected straight line.

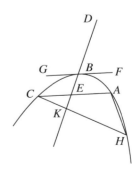

Let there be the parabola or hyperbola *ABC* whose diameter is the straight line *DBE*, and let the straight line *FBG* touch the section, and let some straight line *AEC* be drawn in the section making *AE* equal to *EC*.

I say that *AC* is parallel to *FG*.

For if not, let the straight line *CH* be drawn through *C* parallel to *FG* and let *HA* be joined. Since then *ABC* is

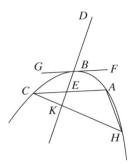

a parabola or hyperbola whose diameter is *DE*, and tangent *FG*, and *CH* is parallel to it, therefore

$$CK = KH \text{ (I. 46, 47)}.$$

But also

$$CE = EA.$$

Therefore *AH* is parallel to *KE*; and this is absurd; for produced it meets *BD* (I. 22).

PROPOSITION 6

If the diameter of an ellipse or circumference of a circle bisects some straight line not through the center, the tangent to the section at the end of the diameter will be parallel to the bisected straight line.

Let there be an ellipse or circumference of a circle whose diameter is the straight line *AB*, and let *AB* bisect *CD*, a straight line not through the center, at the point *E*.

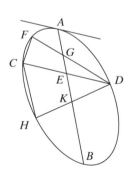

I say that the tangent to the section at *A* is parallel to *CD*.

For let it not be, but if possible, let *DF* be parallel to the tangent at *A*; therefore

$$DG = FG \text{ (I.47)}.$$

But also

$$DE = EC;$$

therefore *CF* is parallel to *GE*; and this is absurd. For if *G* is the center of the section *AB*, the straight line *CF* will meet the straight line *AB* (I. 23); and if it is not, suppose it to be *K*, and let *DK* be joined and produced to *H*, and let *CH* be joined. Since then

$$DK = KH,$$

and also

$$DE = EC,$$

therefore *CH* is parallel to *AB*. But also *CF*; and this is absurd. Therefore the tangent at *A* is parallel to *CD*.

PROPOSITION 7

If a straight line touches a section of a cone or circumference of a circle, and a parallel to it is drawn in the section and bisected, the straight line joined from the point of contact to the midpoint will be a diameter of the section.

Let there be a section of a cone or circumference of a circle *ABC*, and *FG* tangent to it, and *AC* parallel to *FG* and bisected at *E*, and let *BE* be joined.

I say that *BE* is a diameter of the section.

For let it not be, but, if possible, let *BH* be a diameter of the section. Therefore
$$AH = HC \text{ (Def. 4)};$$
and this is absurd; for
$$AE = EC.$$
Therefore *BH* will not be a diameter of the section. Then likewise we could show that there is no other than *BE*.

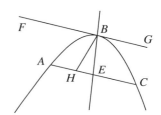

PROPOSITION 8

If a straight line meets an hyperbola in two points, produced both ways it will meet the asymptotes, and the straight lines cut off on it by the section from the asymptotes will be equal.

Let there be the hyperbola *ABC*, and the asymptotes *ED* and *DF*, and let some straight line *AC* meet *ABC*.

I say that produced both ways it will meet the asymptotes.

Let *AC* be bisected at *G* and let *DG* be joined. Therefore it is a diameter of the section (I. 47); therefore the tangent at *B* is parallel to *AC* (II. 5). Then let *HBK* be the tangent (I. 32); then it will meet *ED* and *DF* (II. 3). Since then *AC* is parallel

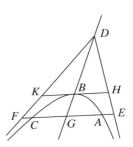

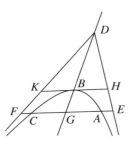

to *KH*, and *KH* meets *DK* and *DH*, therefore also *AC* will meet *DE* and *DF*.

Let it meet them at *E* and *F*; and
$$HB = BK \text{ (II. 3)};$$
therefore also
$$FG = GE.$$
And so also
$$CF = AE.$$

PROPOSITION 9

If a straight line meeting the asymptotes is bisected by the hyperbola, it will touch the section in one point only.

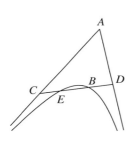

For let the straight line *CD* meeting the asymptotes *CA*, *AD* be bisected by the hyperbola at the point *E*.

I say that it touches the hyperbola at no other point.

For if possible, let it touch it at *B*. Therefore
$$CE = BD \text{ (II. 8)};$$
and this is absurd; for *CE* is supposed equal to *ED*. Therefore it will not touch the section at another point.

PROPOSITION 10

If some straight line cutting the section meet both of the asymptotes, the rect-angle contained by the straight lines cut off between the asymptotes and the section is equal to the fourth of the figure resulting on the diameter bisecting the straight lines drawn parallel to the drawn straight line.

Let there be the hyperbola *ABC,* and let *DE, EF* be its asymptotes, and let some straight line *DF* be drawn cutting the section and the asymptotes, and let *AC* be bisected at *G*, and let *GE* be joined, and let *EH* be made equal to

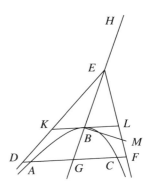

BE, and let *BM* be drawn from *B* perpendicular to
HEB; therefore *BH* is a diameter (I. 51 porism),
and *BM* is the upright side.

I say that
rect. *DA, AF =* fourth rect. *HB, BM,*
then likewise also
rect. *DC, CF =* fourth rect. *HB, BM.*

For let *KL* be drawn through *B* tangent to the
section; therefore it is parallel to *DF* (II. 5). And
since it has been shown
$$HB : BM :: \text{sq. } EB : \text{sq. } BK :: \text{sq. } EG : \text{sq. } GD \text{ (II. 1)},$$
and
$$HB : BM :: \text{rect. } HG, GB : \text{sq. } GA \text{ (I. 21)},$$
therefore
$$\text{sq. } EG : \text{sq. } GD :: \text{rect. } HG, GB : \text{sq. } GA.$$
Since then
whole sq. *EG* : whole sq. *GD* ::
part subtr. rect. *HG, GB* : part subtr. sq. *AG,*
therefore also (Eucl. II. 6, II. 5, V. 19)
remainder sq. *EB* : remainder rect. *DA, AF* :: sq. *EG* : sq. *GD,*
or
remainder sq. *EB* : remainder rect. *DA, AF* :: sq. *EB* : sq. *BK.*
Therefore
rect. *FA, AD =* sq. *BK.*

Then likewise it could be shown also that
rect. *DC, CF =* sq. *BL;*
therefore also
rect. *FA, AD =* rect. *DC, CF.*

PROPOSITION 11

*If some straight line cut each of the straight lines containing the angle that
is adjacent to the angle which contains the hyperbola, then this straight line
will meet the section in one point only, and the rectangle contained by the
straight lines cut off [on this straight line] between the containing straight*

lines and the section will be equal to the fourth part of the square on the diameter drawn parallel to the cutting straight line.

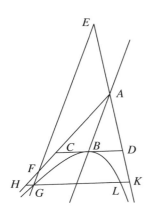

Let there be an hyperbola whose asymptotes are *CA, AD,* and let *DA* be produced to *E,* and through some point *E* let *EF* be drawn cutting *EA* and *AC* [extended as necessary].

Now it is evident that it meets the section in one point only; for the straight line drawn through *A* parallel to *EF* as *AB* will cut angle *CAD* and will meet the section (II. 2) and be its diameter (I. 51 porism); therefore *EF* will meet the section in one point only (I. 26).

Let it meet it at *G.*

I say then also that
$$\text{rect. } EG, GF = \text{sq. } AB.$$

For let the straight line *HGLK* be drawn ordinatewise through *G*; therefore the tangent through *B* is parallel to *GH* (II. 5). Let it be *CD.* Since then
$$CB = BD \text{ (II. 3),}$$
therefore
$$\text{sq. } CB \text{ or rect. } CB, BD : \text{sq. } BA :: CB : BA \text{ comp. } DB : BA.$$
But
$$CB : BA :: HG : GF,$$
and
$$DB : BA :: GK : GE;$$
therefore
$$\text{sq. } CB : \text{sq. } BA :: HG : GF \text{ comp. } KG : GE.$$
But also
$$\text{rect. } KG, GH : \text{rect. } EG, GF :: HG : GF \text{ comp. } KG : GE;$$
therefore
$$\text{rect. } KG, GH : \text{rect. } EG, GF :: \text{sq. } CB : \text{sq. } BA.$$
Alternately
$$\text{rect. } KG, GH : \text{sq. } CB :: \text{rect. } EG, GF : \text{sq. } BA.$$
But it was shown
$$\text{rect. } KG, GH = \text{sq. } CB \text{ (II. 10);}$$
therefore also
$$\text{rect. } EG, GF = \text{sq. } AB.$$

PROPOSITION 12

If two straight lines at chance angles are drawn to the asymptotes from some point of those on the section, and parallels are drawn to the two straight lines from some point of those on the section, then the rectangle contained by the parallels will be equal to that contained by those straight lines to which they were drawn parallel.

Let there be an hyperbola whose asymptotes are *AB* and *BC*, and let some point *D* be taken on the section, and from it let *DE* and *DF* be dropped [at chance angles] to *AB* and *BC*, and let some other point on the section *G* be taken, and through *G* let *GH* and *GK* be drawn parallel to *ED* and *DF*.

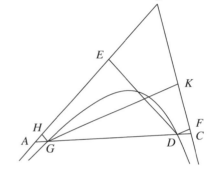

I say that

rect. *ED, DF* = rect. *HG, GK*.

For let *DG* be joined and produced to *A* and *C*. Since then

rect. *AD, DC* = rect. *AG, GC* (II. 8),

therefore

AG : AD :: DC : CG.

But

AG : AD :: GH : ED,

and

DC : CG :: DF : GK;

therefore

GH : DE :: DF : GK.

Therefore

rect. *ED, DF* = rect. *HG, GK*.

PROPOSITION 13

If in the place bounded by the asymptotes and the section some straight line is drawn parallel to one of the asymptotes, it will meet the section in one point only.

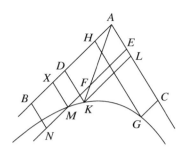

Let there be an hyperbola whose asymptotes are *CA* and *AB*, and let some point *E* be taken, and through it let *EF* be drawn parallel to *AB*.

I say that it will meet the section.

For if possible, let it not meet it, and let some point *G* on the section be taken, and through *G* let *GC* and *GH* be drawn parallel to *CA* and *AB*, and let

$$\text{rect. } CG, GH = \text{rect. } AE, EF,$$

and let *AF* be joined and produced; then it will meet the section (II. 2). Let it meet it at *K*, and through *K* parallel to *CA* and *AB* let *KL* and *KD* be drawn; therefore

$$\text{rect. } CG, GH = \text{rect. } LK, KD \text{ (II. 12)}.$$

And it is supposed that also

$$\text{rect. } CG, GH = \text{rect. } AE, EF;$$

therefore

$$\text{rect. } LK, KD \text{ or rect. } KL, LA = \text{rect. } AE, EF;$$

and this is impossible; for both

$$KL > EF$$

and

$$LA > AE.$$

Therefore *EF* will meet the section.

Let it meet it at *M*.

I say then that it will not meet it at any other point.

For if possible, let it also meet it at *N*, and through *M* and *N* let *MX* and *NB* be drawn parallel to *CA*. Therefore

$$\text{rect. } EM, MX = \text{rect. } EN, NB \text{ (II. 12)};$$

and this is impossible. Therefore it will not meet the section in another point.

PROPOSITION 14

The asymptotes and the section, if produced indefinitely, draw nearer to each other; and they reach a distance less than any given distance.

Let there be an hyperbola whose asymptotes are AB and AC, and a given distance K.

I say that AB and AC and the section, if produced, draw nearer to each other and will reach a distance less than K.

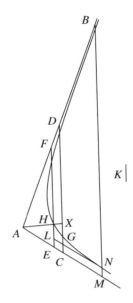

For let EHF and CGD be drawn parallel to the tangent, and let AH be joined and produced to X. Since then

   rect. CG, GD = rect. FH, HE (II. 10),

therefore

$$DG : FH :: HE : CG.$$

But

$$DG > FH \text{ (Eucl. VI.4, CN 5)};$$

therefore also

$$HE > CG.$$

Then likewise we could show that the succeeding straight lines are less.

Then let the distance EL be taken less than K, and through L let LN be drawn parallel to AC; therefore it will meet the section (II. 13). Let it meet it at N, and through N let MNB be drawn parallel to EF. Therefore

$$MN = EL$$

and so

$$MN < K.$$

PORISM

Then from this it is evident that the straight lines AB and AC are nearer than all the asymptotes to the section, and the angle contained by BA, AC is clearly less than that contained by other asymptotes to the section.

PROPOSITION 15

The asymptotes of opposite sections are common.

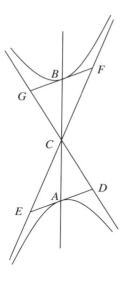

Let there be opposite sections whose diameter is *AB* and center *C*.

I say that the asymptotes of the sections *A* and *B* are common.

Let the straight lines *DAE* and *FBG* be drawn tangent to the sections through the points *A* and *B*; they are therefore parallel (I. 44, note). Then let each of the straight lines *DA*, *AE*, *FB*, and *BG* be cut off equal in square to the fourth of the figure applied to *AB*; therefore

$$DA = AE = FB = BG.$$

Then let *CD*, *CE*, *CF*, and *CG* be joined. Then it is evident that *DC* is in a straight line with *CG*, and *CE* with *CF*, because of the parallels. Since then it is an hyperbola whose diameter is *AB* and tangent *DE*, and *DA* and *AE* are each equal in square to the fourth of the figure applied to *AB*, therefore *DC* and *CE* are asymptotes (II. 1). For the same reasons then *FC* and *CG* are also asymptotes to section *B*. Therefore the asymptotes of opposite sections are common.

PROPOSITION 16

If in opposite sections some straight line is drawn cutting each of the straight lines containing the angle adjacent to the angles containing the sections, it will meet each of the opposite sections in one point only, and the straight lines cut off on it by the sections from the asymptotes will be equal.

For let there be the opposite sections *A* and *B* whose center is *C* and asymptotes *DCG* and *ECF*, and let some straight line *HK* be drawn through, cutting each of the straight lines *DC* and *CF*.

I say that produced it will meet each of the sections in one point only.

For since *DC* and *CE* are asymptotes of section *A*, and some straight line *HK* has been drawn across cutting both of the straight lines containing the adjacent angle *DCF*, therefore *HK* produced will meet the section (II. 11).

Then likewise also *B*.

Let it meet them at *L* and *M*.

Let the straight line *ACB* be drawn through *C*
parallel to *LM*; therefore

rect. *KL, LH* = sq. *AC* (II. 11)

and

rect. *HM, MK* = sq. *CB* (II. 11).

And so also

rect. *KL, LH* = rect. *HM, MK*,

and

LH = *KM*.

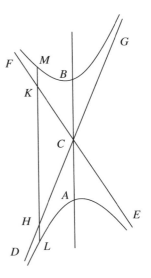

PROPOSITION 17

The asymptotes of conjugate opposite sections are common.

Let there be conjugate opposite sections whose conjugate diameters are *AB*
and *CD*, and whose center is *E*.

I say that their asymptotes are common.

For let the straight lines *FAG, GDH, HBK*, and *KCF*
be drawn through the points *A, B, C*, and *D* touching the sections; therefore *FGHK* is a parallelogram
(I. 44, note). Then let *FEH* and *KEG* be joined;
therefore they are straight lines (II. 15) and diagonals
of the parallelogram, and they are all bisected at the
point *E*. And since the figure on *AB* is equal to the
square on *CD* (I. 60), and

CE = *ED*,

therefore each of the squares on *FA, AG, KB*, and *BH*
is equal to a fourth of the figure on *AB*. Therefore
the straight lines *FEH* and *KEG* are asymptotes of
the sections *A* and *B* (II. 1). Then likewise we could

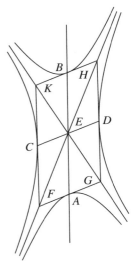

show that the same straight lines are also asymptotes of the sections *C* and
D. Therefore the asymptotes of conjugate opposite sections are common.

PROPOSITION 18

If a straight line meeting one of the conjugate opposite sections, when produced both ways, falls outside the section, it will meet both of the adjacent sections in one point only.

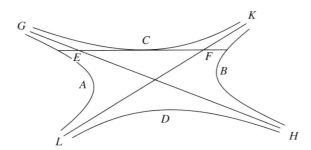

Let there be the conjugate opposite sections *A, B, C,* and *D,* and let some straight line *EF* meet the section *C* and produced both ways fall outside the section.

I say that it will meet both of the sections *A* and *B* in one point only.

For let *GH* and *KL* be asymptotes of the sections. Therefore *EF* meets both *GH* and *KL* (II. 3). Then it is evident that it will also meet the sections *A* and *B* in one point only (II. 16).

PROPOSITION 19

If some straight line is drawn touching some one of the conjugate opposite sections at random, it will meet the adjacent sections and will be bisected at the point of contact.

Let there be the conjugate opposite sections *A, B, C,* and *D,* and let some straight line *ECF* touch it at *C.*

I say that produced it will meet sections *A* and *B* and will be bisected at *C*.

It is evident now that it will meet sections *A* and *B* (II. 18); let it meet them at *G* and *H*.

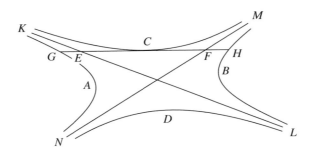

I say that

$$CG = CH.$$

For let the asymptotes of the sections *KL* and *MN* be drawn. Therefore
$$EG = FH \text{ (II. 16)},$$

and

$$CE = CF \text{ (II. 3)},$$

and

$$CG = CH.$$

PROPOSITION 20

If a straight line touches one of the conjugate opposite sections, and two straight lines are drawn through their center, one through the point of contact, and one parallel to the tangent until it meet one of the adjacent sections, then the straight line touching the section at the point of meeting will be parallel to the straight line drawn through the point of contact and the center, and those through the points of contact and the center will be conjugate diameters of the opposite sections.

Let there be conjugate opposite sections whose conjugate diameters are the straight lines *AB* and *CD*, and center *Y*, and let *EF* be drawn touching the section *A*, and produced let it meet *CY* at *T*, and let *EY* be joined and produced to *X*, and through *Y* let *YG* be drawn parallel to *EF*, and through *G* let *HG* be drawn touching the section.

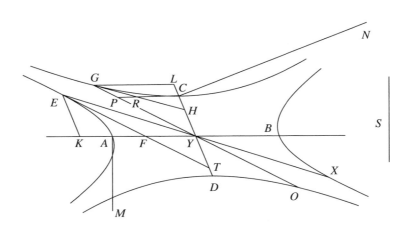

I say that *HG* is parallel to *YE*, and *GO* and *EX* are conjugate diameters.

For let the straight lines *KE, GL,* and *CRP* be drawn ordinatewise, and let *AM* and *CN* be the parameters. Since then
$$BA : AM :: NC : CD \text{ (I. 60)},$$
but
$$BA : AM :: \text{rect. } YK, KF : \text{sq. } KE \text{ (I. 37)},$$
and
$$NC : CD :: \text{sq. } GL : \text{rect. } YL, LH \text{ (I. 37)},$$
therefore also
$$\text{rect. } YK, KF : \text{sq. } EK :: \text{sq. } GL : \text{rect. } YL, LH.$$
But
$$\text{rect. } YK, KF : \text{sq. } EK :: YK : KE \text{ comp. } FK : KE,$$
and
$$\text{sq. } GL : \text{rect. } YL, LH :: GL : LY \text{ comp. } GL : LH;$$
therefore
$$\text{ratio } YK : KE \text{ comp. } FK : KE = \text{ratio } GL : LY \text{ comp. } GL : LH;$$

and of these
$$FK : KE :: GL : LY;$$
for each of the straight lines *EK, KF,* and *FE* is parallel to each of the straight lines *YL, LG,* and *GY* respectively. Therefore as remainder
$$YK : KE :: GL : LH.$$
Also the sides about the equal angles at *K* and *L* are proportional; therefore triangle *EKY* is similar to triangle *GHL* and will have equal the angles the corresponding sides subtend. Therefore
$$\text{angle } EYK = \text{angle } LGH.$$
But also
$$\text{angle } KYG = \text{angle } LGY;$$
and therefore
$$\text{angle } EYG = \text{angle } HGY.$$
Therefore *EY* is parallel to *GH*.

Then let it be contrived that
$$PG : GR :: HG : S;$$
therefore *S* is a half of the parameter of the ordinates to the diameter *GO* in sections *C* and *D* (I. 51). And since *CD* is the second diameter of the sections *A* and *B*, and *ET* meets it, therefore
$$\text{rect. } TY, EK = \text{sq. } CY;$$
for if we draw from *E* a parallel to *KY*, the rectangle contained by *TY* and the straight line cut off by the parallel will be equal to the square on *CY* (I. 38). And therefore
$$TY : EK :: \text{sq. } TY : \text{sq. } YC \text{ (Eucl. VI. 20).}$$
But
$$TY : EK :: TF : FE$$
or
$$TY : EK :: \text{trgl. } TYF : \text{trgl. } EFY \text{ (Eucl. VI. 1),}$$
and
$$\text{sq. } TY : \text{sq. } CY :: \text{trgl. } YTF : \text{trgl. } YCP \text{ (Eucl. VI. 19)}$$
or
$$\text{sq. } TY : \text{sq. } CY :: \text{trgl. } YTF : \text{trgl. } GHY \text{ (III. 1).}$$
Therefore
$$\text{trgl. } TYF : \text{trgl. } EFY :: \text{trgl. } TFY : \text{trgl. } YGH.$$
Therefore
$$\text{trgl. } GHY = \text{trgl. } YEF.$$
But they also have
$$\text{angle } HGY = \text{angle } YEF;$$
for *EY* is parallel to *GH*, and *EF* to *GY*. Therefore the sides about the equal angles are reciprocally proportional (Eucl. VI. 15). Therefore

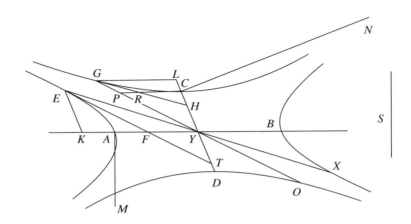

$$GH : EY :: EF : GY;$$

therefore

$$\text{rect. } HG, GY = \text{rect. } YE, EF.$$

And since

$$S : HG :: RG : GP,$$

and

$$RG : GP :: YE : EF;$$

for they are parallel; therefore also

$$S : HG :: YE : EF.$$

But, with YG taken as common height,

$$S : HG :: \text{rect. } S, YG : \text{rect. } HG, GY,$$

and

$$YE : EF :: \text{sq. } YE : \text{rect. } YE, EF.$$

And therefore

$$\text{rect. } S, YG : \text{rect. } HG, GY :: \text{sq. } YE : \text{rect. } YE, EF.$$

Alternately

$$\text{rect. } S, GY : \text{sq. } EY :: \text{rect. } HG, GY : \text{rect. } FE, EY.$$

But

$$\text{rect. } HG, GY = \text{rect. } YE, EF \text{ (above)},$$

therefore also

$$\text{rect. } S, GY = \text{sq. } EY.$$

And rectangle S, GY is a fourth of the figure on GO; for

$$GY = \text{half } GO,$$

and S is the parameter; and
$$\text{sq. } EY = \text{fourth sq. } EX;$$
for
$$EY = YX.$$
Therefore the square on EX is equal to the figure on GO. Then likewise we could show also that GO is equal in square to the figure on EX. Therefore EX and GO are conjugate diameters of the opposite sections A, B, C, and D.

PROPOSITION 21

The same things being supposed it is to be shown that the point of meeting of the tangents is on one of the asymptotes.

Let there be conjugate opposite sections, whose diameters are the straight lines AB and CD, and let the straight lines AE and EC be drawn tangent.

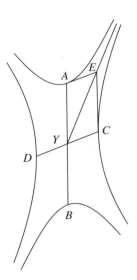

I say that the point E is on the asymptote.

For since the square on CY is equal to the fourth of the figure on AB (I. 60), and
$$\text{sq. } AE = \text{sq. } CY \text{ (II. 17),}$$
therefore also the square on AE is equal to the fourth part of the figure on AB. Let EY be joined; therefore EY is an asymptote (II. 1); therefore the point E is on the asymptote.

PROPOSITION 22

If in conjugate opposite sections a radius is drawn to any one of the sections, and a parallel is drawn to it meeting one of the adjacent sections and meeting the asymptotes, then the rectangle contained by the segments produced between the section and the asymptotes on the straight line drawn is equal to the square on the radius.

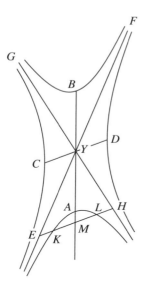

Let there be the conjugate opposite sections A, B, C and D, and let there be the asymptotes of the sections YEF and YGH, and from the center Y let some straight line YCD be drawn across, and let HE be drawn parallel to it cutting both the adjacent section and the asymptotes.

I say that

$$\text{rect. } EK, KH = \text{sq. } CY.$$

Let KL be bisected at M, and let MY be joined and produced; therefore AB is the diameter of the sections A and B (I. 51 porism). And since the tangent at A is parallel to EH (II. 5), therefore EH has been dropped ordinatewise to AB (I. 17). And the center is Y; therefore AB and CD are conjugate diameters (Def. 6). Therefore the square on CY is equal to the fourth of the figure on AB (I. 60). And the rectangle HK, KE is equal to the fourth part of the figure on AB (II. 10); therefore also

$$\text{rect. } HK, KE = \text{sq. } CY.$$

PROPOSITION 23

If in conjugate opposite sections some radius is drawn to any one of the sections, and a parallel is drawn to it meeting the three adjacent sections, then the rectangle contained by the segments produced between the three sections on the straight line drawn is twice the square on the radius.

Let there be the conjugate opposite sections A, B, C, and D, and let the center of the section be Y, and from the point Y let some straight line CY be drawn to meet any one of the sections, and let KL be drawn parallel to CY cutting the three adjacent sections.

I say that

$$\text{rect. } KM, ML = 2 \text{ sq. } CY.$$

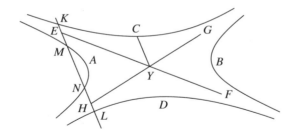

Let the asymptotes to the sections, *EF* and *GH*, be drawn; therefore

$$\text{sq. } CY = \text{rect. } HM, ME \text{ (II. 22)} = \text{rect. } HK, KE \text{ (II. 11)}.$$

And

$$\text{rect. } HM, ME + \text{rect. } HK, KE = \text{rect. } LM, MK$$

because of the straight lines on the ends being equal (II. 8, 16). Therefore also

$$\text{rect. } LM, MK = 2 \text{ sq. } CY.$$

PROPOSITION 24

If two straight lines meet a parabola each at two points, and if a point of meeting of neither one of them is contained by the points of meeting of the other, then the straight lines will meet each other outside the section.

Let there be the parabola *ABCD*, and let the two straight lines *AB* and *CD* meet *ABCD*, and let a point of meeting of neither of them be contained by the points of meeting of the other.

I say that the straight lines produced will meet each other.

Let the diameters of the section, *EBF* and *GCH*, be drawn through the points *B* and *C*; therefore they are parallel (I. 51 porism) and

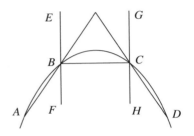

each one cuts the section in one point only (I. 26). Then let *BC* be joined; therefore

angle *EBC* + angle *BCG* = 2 rt. angles,

and *DC* and *BA* produced make angles less than two right angles. Therefore they will meet each other outside the section (I. 10; Eucl. Post. 5).

PROPOSITION 25

If two straight lines meet an hyperbola each at two points, and if a point of meeting of neither of them is contained by the points of meeting of the other, then the straight lines will meet each other outside the section, but within the angle containing the section.

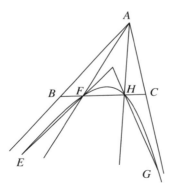

Let there be an hyperbola, whose asymptotes are *AB* and *AC,* and let the two straight lines *EF* and *GH* cut the section, and let a point of meeting of neither of them be contained by the points of meeting of the other.

I say that the straight lines *EF* and *GH* produced will meet outside the section, but within the angle *CAB*.

For let the straight lines *AF* and *AH* be joined and produced, and let *FH* be joined. And since the straight lines *EF* and *GH* produced cut the angles *AFH* and *AHF,* and the said angles are less than two right angles (Eucl. I. 17), the straight lines *EF* and *GH* produced will meet each other outside the section, but within the angle *BAC*.

[Porism]

Then we could likewise show it, even if the straight lines *EF* and *GH* are tangents to the sections.

PROPOSITION 26

If in an ellipse or circumference of a circle two straight lines not through the center cut each other, then they do not bisect each other.

For if possible, in the ellipse or circumference of a circle let the two straight lines *CD* and *EF* not through the center bisect each other at *G,* and let the point *H* be the center of the section, and let *GH* be joined and produced to *A* and *B*.

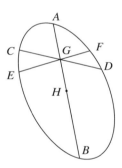

Since then the straight line *AB* is a diameter bisecting *EF,* therefore the tangent at *A* is parallel to *EF* (II. 6). We could then likewise show that it is also parallel to *CD*. And so also *EF* is parallel to *CD*. And this is impossible. Therefore *CD* and *EF* do not bisect each other.

PROPOSITION 27

If two straight lines touch an ellipse or circumference of a circle, and if the straight line joining the points of contact is through the center of the section, the tangents will be parallel; but if not, they will meet on the same side of the center.

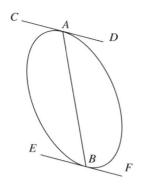

Let there be the ellipse or circumference of a circle *AB*, and let the straight lines *CAD* and *EBF* touch it, and *AB* be joined, and first let it be through the center.

I say that *CD* is parallel to *EF*.

For since *AB* is a diameter of the section, and *CD* touches it at *A*, therefore *CD* is parallel to the ordinates to *AB* (I. 17). Then for the same reasons *BF* is also parallel to the same ordinates. Therefore *CD* is also parallel to *EF*.

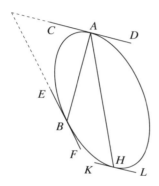

Then let *AB* not be through the center, as in the second drawing, and let the diameter *AH* be drawn, and let *KHL* be drawn tangent through *H*; therefore *KL* is parallel to *CD*. Therefore *EF* produced will meet *CD* on the same side of the center as *AB*.

PROPOSITION 28

If in a section of a cone or circumference of a circle some straight line bisects two parallel straight lines, then it will be a diameter of the section.

For let *AB* and *CD*, two parallel straight lines in a conic section, be bisected at *E* and *F*, and let *EF* be joined and produced.

I say that it is a diameter of the section.

For if not, let the straight line *GFH* be so if possible. Therefore the tangent at G is parallel to AB (II. 5, 6). And so the same straight line is parallel to CD. And GH is a diameter; therefore

$$CH = HD \text{ (Def. 4)}$$

and this is impossible; for it is supposed

$$CE = ED.$$

Therefore GH is not a diameter. Then likewise we could show that there is no other except EF. Therefore EF will be a diameter of the section.

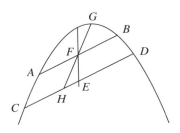

PROPOSITION 29

If in a section of a cone or circumference of a circle two tangents meet, the straight line drawn from their point of meeting to the midpoint of the straight line joining the points of contact is a diameter of the section.

Let there be a section of a cone or circumference of a circle to which let the straight lines AB and AC, meeting at A, be drawn tangent, and let BC be joined and bisected at D, and let AD be joined.

I say that it is a diameter of the section.

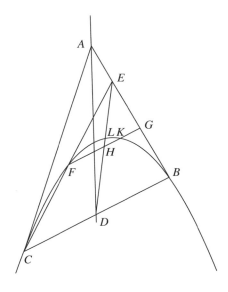

For if possible, let DE be a diameter, and let EC be joined; then it will cut the section (I. 35, 36). Let it cut it at F, and through F let FKG be drawn parallel to CDB. Since then

$$CD = DB$$

also

$$FH = HG.$$

And since the tangent at L is parallel to BC (II. 5, 6), and FG is also parallel to BC, therefore also FG is parallel to the tangent at L. Therefore

$$FH = HK \text{ (I. 46, 47);}$$

and this is impossible. Therefore DE is not a diameter. Then likewise we could show that there is no other except AD.

PROPOSITION 30

If two straight lines tangent to a section of a cone or to a circumference of a circle meet, the diameter drawn from the point of meeting will bisect the straight line joining the points of contact.

Let there be the section of a cone or circumference of a circle *BC*, and let two tangents *BA* and *AC* be drawn to it meeting at *A*, and let *BC* be joined and let *AD* be drawn through *A* as a diameter of the section.

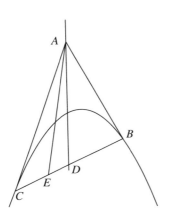

I say that
$$DB = DC.$$
For let it not be, but if possible, let
$$BE = EC,$$
and let *AE* be joined; therefore *AE* is a diameter of the section (II. 29). But *AD* is also a diameter; and this is absurd. For if the section is an ellipse, the point *A* at which the diameters meet each other, will be a center outside the section; and this is impossible; and if the section is a parabola, the diameters meet each other (I. 51 porism); and if it is an hyperbola, and the straight lines *BA* and *AC* meet the section without containing one another's points of meeting, then the center is within the angle containing the hyperbola (II. 25); but it is also on it, for it has been supposed a center since *DA* and *AE* are diameters (I. 51 porism); and this is absurd. Therefore *BE* is not equal to *EC*.

PROPOSITION 31

If two straight lines touch each of the opposite sections, then if the straight line joining the points of contact falls through the center, the tangents will be parallel, but if not, they will meet on the same side as the center.

Let there be the opposite sections *A* and *B*, and let the straight lines *CAD* and

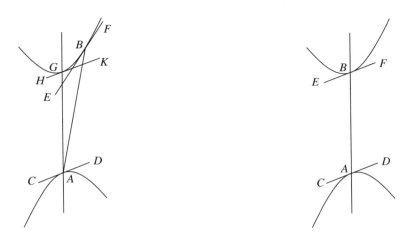

EBF be tangent to them at *A* and *B*, and let the straight line joined from *A* to *B* fall first through the center of the sections.

I say that *CD* is parallel to *EF*.

For since they are opposite sections of which *AB* is a diameter, and *CD* touches one of them at *A*, therefore the straight line drawn through *B* parallel to *CD* touches the section (I. . 44, note). But *EF* also touches it; therefore *CD* is parallel to *EF*.

Then let the straight line from *A* to *B* not be through the center of the sections, and let *AG* be drawn as a diameter of the sections, and let *HK* be drawn tangent to the section; therefore *HK* is parallel to *CD*, and since the straight lines *EF* and *HK* touch an hyperbola, therefore they will meet (II. 25 porism). And *HK* is parallel to *CD*; therefore also the straight lines *CD* and *EF* produced will meet. And it is evident they are on the same side as the center.

PROPOSITION 32

If straight lines meet each of the opposite sections, in one point when touching or in two points when cutting, and, when produced, the straight lines meet, then their point of meeting will be in the angle adjacent to the angle containing the section.

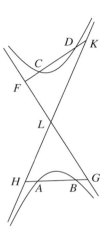

Let there be opposite sections and the straight lines *AB* and *CD* either touching the opposite sections in one point or cutting them in two points, and let them meet when produced.

I say that their point of meeting will be in the angle adjacent to the angle containing the section.

Let *FG* and *HK* be asymptotes to the sections; therefore *AB* produced will meet the asymptotes (II. 8). Let it meet them at *H* and *G*. And since *FK* and *HG* are supposed as meeting, it is evident that either they will meet in the place under angle *HLF* or in that under angle *KLG*. And likewise also, if they touch (II. 3).

PROPOSITION 33

If a straight line meeting one of the opposite sections, when produced both ways, falls outside the section, it will not meet the other section, but will fall through the three places of which one is that contained by the angle containing the section, and two are those contained by the angle adjacent to the angle containing the section.

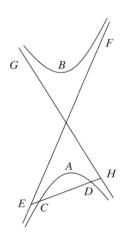

Let there be the opposite sections *A* and *B*, and let some straight line *CD* cut *A*, and, when produced both ways, let it fall outside the section.

I say that the straight line *CD* does not meet the section *B*.

For let *EF* and *GH* be drawn as asymptotes to the sections; therefore *CD* produced will meet the asymptotes (II. 8). And it only meets them in the points *E* and *H*. And so it will not meet the section *B*.

And it is evident that it will fall through the three places. For if some straight line meets both of the opposite sections, it will meet neither of the opposite sections in two points. For if it meets it in two points, by what has just been proved it will not meet the other section.

PROPOSITION 34

If some straight line touch one of the opposite sections and a parallel to it be drawn in the other section, then the straight line drawn from the point of contact to the midpoint of the parallel will be a diameter of the opposite sections.

Let there be the opposite sections *A* and *B*, and let some straight line *CD* touch one of them *A*, at *A*, and let *EF* be drawn parallel to *CD* in the other section, and let it be bisected at *G*, and let *AG* be joined.

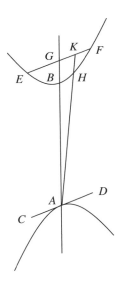

I say that *AG* is a diameter of the opposite sections.

For if possible, let *AHK* be. Therefore the tangent at *H* is parallel to *CD* (II. 31). But *CD* is also parallel to *EF*; and therefore the tangent at *H* is parallel to *EF*. Therefore

$$EK = KF \text{ (I. 47);}$$

and this is impossible; for

$$EG = GF.$$

Therefore *AH* is not a diameter of the opposite sections.

Therefore *AB* is.

PROPOSITION 35

If a diameter in one of the opposite sections bisects some straight line, the straight line touching the other section at the end of the diameter will be parallel to the bisected straight line.

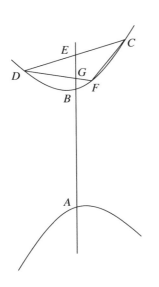

Let there be the opposite sections A and B, and let their diameter AB bisect the straight line CD in section B at E.

I say that the tangent to the section at A is parallel to CD.

For if possible, let DF be parallel to the tangent to the section at A; therefore
$$DG = GF \text{ (I. 48).}$$
But also
$$DE = EC.$$
Therefore CF is parallel to EG; and this is impossible; for produced it meets it (I. 22). Therefore DF is not parallel to the tangent to the section at A nor is any other straight line except CD.

PROPOSITION 36

If parallel straight lines are drawn, one in each of the opposite sections, then the straight line joining their midpoints will be a diameter of the opposite sections.

Let there be the opposite sections A and B, and let the straight lines CD and EF be drawn, one in each of them, and let them be parallel, and let them both be bisected at points G and H, and let GH be joined.

I say that GH is a diameter of the opposite sections.

For if not, let GK be one. Therefore the tangent to A is parallel to CD (II. 5);

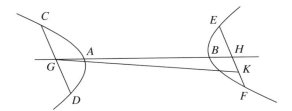

and so also to *EF*. Therefore
$$EK = KF \text{ (I. 48)};$$
and this is impossible, since also
$$EH = HF.$$
Therefore *GK* is not a diameter of the opposite sections. Therefore *GH* is.

PROPOSITION 37

If a straight line not through the center cuts the opposite sections, then the straight line joined from its midpoint to the center is a so-called upright diameter of the opposite sections, and the straight line drawn from the center parallel to the bisected straight line is a transverse diameter conjugate to it.

Let there be the opposite sections *A* and *B*, and let some straight line *CD* not through the center cut the sections *A* and *B* and let it be bisected at *E*, and let *Y* be the center of the sections, and let *YE* be joined, and through *Y* let *AB* be drawn parallel to *CD*.

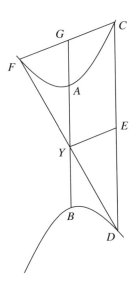

I say that the straight lines *AB* and *EY* are conjugate diameters of the sections.

For let *DY* be joined and produced to *F*, and let *CF* be joined. Therefore
$$DY = YF \text{ (I. 30).}$$
But also
$$DE = EC;$$

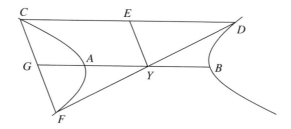

therefore EY is parallel to FC. Let BA be produced to G. And since

$$DY = YF,$$

therefore also

$$EY = FG;$$

and so also

$$CG = FG.$$

Therefore the tangent at A is parallel to CF (II. 5); and so also to EY. Therefore EY and AB are conjugate diameters (I. 16).

PROPOSITION 38

If two straight lines meeting touch opposite sections, the straight line joined from the point of meeting to the midpoint of the straight line joining the points of contact will be a so-called upright diameter of the opposite sections, and the straight line drawn through the center parallel to the straight line joining the points of contact is a transverse diameter conjugate to it.

Let there be the opposite sections A and B, and CY and YD touching the sections, and let CD be joined and bisected at E, and let EY be joined.

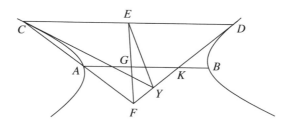

I say that the diameter *EY* is a so-called upright, and the straight line drawn through the center parallel to *CD* is a transverse diameter conjugate to it.

For if possible, let *EF* be a diameter, and let *F* be a point taken at random; therefore *DY* will meet *EF*. Let it meet it at *F,* and let *CF* be joined; therefore *CF* will hit the section (I. 32). Let it hit it at *A,* and through *A* let *AB* be drawn parallel to *CD*. Since then *EF* is a diameter, and bisects *CD*, it also bisects the parallels to it (Def. 4). Therefore

$$AG = GB.$$

And since

$$CE = ED,$$

and is on triangle *CFD,* therefore also

$$AG = GK.$$

And so also

$$GK = GB;$$

and this is impossible. Therefore *EF* will not be a diameter.

PROPOSITION 39

If two straight lines meeting touch opposite sections, the straight line drawn through the center and the point of meeting of the tangents bisects the straight line joining the points of contact.

Let there be the opposite sections *A* and *B*, and let two straight lines *CE* and *ED* be drawn touching *A* and *B,* and let *CD* be joined, and let *EF* be drawn as a diameter.

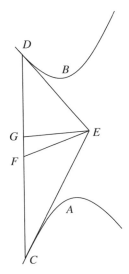

I say that

$$CF = FD.$$

For if not, let *CD* be bisected at *G*, and let *GE* be joined; therefore *GE* is a diameter (II. 38). But *EF* is also; therefore *E* is the center (I. 31 porism). Therefore the point of meeting of the tangents is at the center of the sections; and this is absurd (II. 32). Therefore *CF* is not unequal to *FD*. Therefore equal.

PROPOSITION 40

If two straight lines touching opposite sections meet, and through the point of meeting a straight line is drawn, parallel to the straight line joining the points of contact, and meeting the sections, then the straight lines drawn from the points of meeting to the midpoint of the straight line joining the points of contact touch the sections.

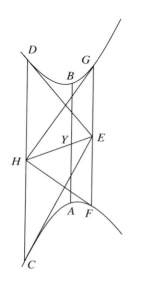

Let there be the opposite sections *A* and *B,* and let two straight lines *CE* and *ED* be drawn touching *A* and *B,* and let *CD* be joined, and through *E* let *FEG* be drawn parallel to *CD,* and let *CD* be bisected at *H,* and let *FH* and *HG* be joined.

I say that *FH* and *HG* touch the sections.

Let *EH* be joined; therefore *EH* is an upright diameter, and the straight line drawn through the center parallel to *CD* a transverse diameter conjugate to it (II. 38). And let the center *Y* be taken, and let *AYB* be drawn parallel to *CD;* therefore *HE* and *AB* are conjugate diameters. And *CH* has been drawn ordinatewise to the second diameter, and *CE* has been drawn touching the section and meeting the second diameter. Therefore the rectangle *EY, YH* is equal to the square on the half of the second diameter (I. 38), that is to the fourth part of the figure on *AB* (Def. 10). And since *FE* has been drawn ordinatewise and *FH* joined, therefore *FH* touches the section *A* (I. 38). Likewise then also *GH* touches section *B.* Therefore *FH* and *HG* touch sections *A* and *B.*

PROPOSITION 41

If in opposite sections two straight lines not through the center cut each other, then they do not bisect each other.

Let there be the opposite sections *A* and *B,* and in *A* and *B* let the two straight lines *CB* and *AD* not through the center cut each other at *E.*

I say that they do not bisect each other.

For if possible, let them bisect each other, and let Y be the center of the sections, and let EY be joined; therefore EY is a diameter (II. 37). Let YF be drawn through Y parallel to BC; therefore YF is a diameter and conjugate to EY (II. 37). Therefore the tangent at F is parallel to EY (Def. 6). Then for the same reasons, with HK drawn parallel to AD, the tangent at H is parallel to EY; and so also the tangent at F is parallel to the tangent at H; and this is absurd; for it has been shown it also meets it (II. 31). Therefore the straight lines CB and AD not being through the center do not bisect each other.

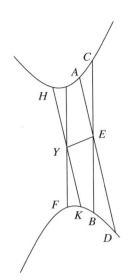

PROPOSITION 42

If in conjugate opposite sections two straight lines not through the center cut each other, they do not bisect each other.

Let there be the conjugate opposite sections A, B, C and D, and in A, B, C and D let the two straight lines not through the center, EF and GH, cut each other at K.

I say that they do not bisect each other.

For if possible, let them bisect each other, and let the center of the sections be Y, and let AB be drawn parallel to EF and CD to HG, and let KY be joined; therefore KY and AB are conjugate diameters (II. 37). Likewise YK and CD are also conjugate diameters. And so also the tangent at A is parallel to the tangent at C; and this is impossible; for it meets it, since the tangent at C cuts the sections A and B (II. 19), and the tangent at A sections C and D; it is evident also that their point of meeting is in the place under angle AYC (II. 21). Therefore the straight lines EF and GH not being through the center do not bisect each other.

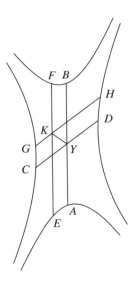

PROPOSITION 43

If a straight line cuts one of the conjugate opposite sections in two points, and through the center one straight line is drawn to the midpoint of the cutting straight line and another straight line is drawn parallel to the cutting straight line, they will be conjugate diameters of the opposite sections.

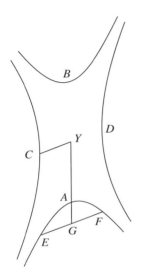

Let there be the conjugate opposite sections *A, B, C* and *D*, and let some straight line cut section *A* at the two points *E* and *F*, and let *FE* be bisected at *G*, and let *Y* be center, and let *YG* be joined, and let *CY* be drawn parallel to *EF*.

I say that *AY* and *YC* are conjugate diameters.

For since *AY* is a diameter, and bisects *EF*, the tangent at *A* is parallel to *EF* (II. 5); and so also to *CY*. Since then they are opposite sections, and a tangent has been drawn to one of them, *A*, at *A*, and from the center *Y* one straight line *YA* is joined to the point of contact, and another *CY* has been drawn parallel to the tangent, therefore *YA* and *CY* are conjugate diameters; for this has been shown before (II. 20).

PROPOSITION 44 (PROBLEM)

Given a section of a cone, to find a diameter.

Let there be the given conic section on which are the points *A, B, C, D* and *E*. Then it is required to find a diameter.

Let it have been done, and let it be *CH*. Then with *DF* and *EH* drawn ordinatewise and produced

$$DF = FB,$$

and

$$EH = HA \text{ (Def. 4).}$$

If then we fix the straight lines *BD* and *EA* in position to be parallel, the points *H* and *F* will be given. And so *HFC* will be given in position.

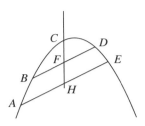

Then it will be constructed (συντεθήσεται) thus: let there be the given conic section on which are the points *A, B, C, D* and *E*, and let the straight lines *BD* and *AE* be drawn parallel and be bisected at *F* and *H*. And the straight line *FH* joined will be a diameter of the section (II.28). And in the same way we could also find an indefinite number of diameters.

PROPOSITION 45 (PROBLEM)

Given an ellipse or hyperbola, to find the center.

And this is evident; for if two diameters of the section, *AB* and *CD*, are drawn through (II. 44), the point at which they cut each other will be the center of the section, as indicated.

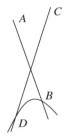

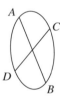

PROPOSITION 46 (PROBLEM)

Given a section of a cone, to find the axis.

Let the given section of a cone first be a parabola, on which are the points *F, C* and *E*. Then it is required to find its axis.

For let *AB* be drawn as a diameter of it (II. 44). If then *AB* is an axis, what was enjoined would have been done; but if not, let it have been done, and let *CD* be the axis; therefore the axis *CD* is parallel to

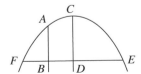

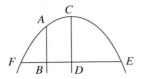

AB (I. 51 porism) and bisects the straight lines drawn perpendicular to it (Def. 7). And the perpendiculars to *CD* are also perpendiculars to *AB*; and so *CD* bisects the perpendiculars to *AB*. If then I fix *EF*, a perpendicular to *AB*, it will be given in position, and therefore

$$ED = DF;$$

therefore the point *D* is given. Therefore through the given point *D*, *CD* has been drawn parallel to *AB* which is given in position; therefore *CD* is given in position.

Then it will be constructed thus: let there be the given parabola on which are the points *F, E* and *A,* and let *AB*, a diameter of it, be drawn (II. 44), and let *BE* be drawn perpendicular to it and let it be produced to *F*. If then

$$EB = BF,$$

it is evident that *AB* is the axis (Def. 7); but if not, let *EF* be bisected by *D*, and let *CD* be drawn parallel to *AB*. Then it is evident that *CD* is the axis of the section; for being parallel to a diameter, that is being a diameter (I. 51 porism), it bisects *EF* at right angles. Therefore *CD* has been found as the axis of the given parabola (Def. 7).

And it is evident that the parabola has only one axis. For if there is another, as *AB,* it will be parallel to *CD* (I. 51 porism). And it cuts *EF*, and so it also bisects it (Def. 4).

Therefore

$$BE = BF;$$

and this is absurd.

PROPOSITION 47 (PROBLEM)

Given an hyperbola or ellipse, to find the axis.

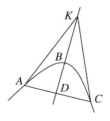

Let there be the hyperbola or ellipse *ABC*; then it is required to find its axis.

Let it have been found and let it be *KD,* and *K* the center of the section; therefore *KD* bisects the ordinates to itself and at right angles (Def. 7).

Let the perpendicular *CDA* be drawn, and
let *KA* and *KC* be joined. Since then
$$CD = DA,$$
therefore
$$CK = KA.$$

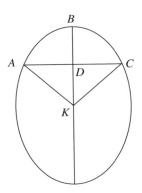

If then we fix the given point *C*, *CK* will
be given. And so the circle described with
center *K* and radius *KC* will also pass
through *A* and will be given in position.
And the section *ABC* is also given in position; therefore the point *A* is given. But the
point *C* is also given; therefore *CA* is given
in position. Also
$$CD = DA,$$
therefore the point *D* is given. But also *K*
is given; therefore *DK* is given in position.

Then it will be constructed thus: let there be the given hyperbola or ellipse
ABC, and let *K* be taken as its center;

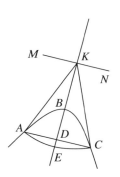

and let a point *C* be taken at random
on the section, and let the circle *CEA*,
with center *K* and radius *KC*, be
described, and let *CA* be joined and
bisected at *D*, and let *KC*, *KD*, and
KA be joined, and let *KD* be drawn
through to *B*.

Since then
$$AD = DC$$
and *DK* is common, therefore the
two straight lines *CD* and *DK* are
equal to the two straight lines *AD*
and *DK*, and
$$\text{base } KA = \text{base } KC.$$
Therefore *KBD* bisects *ADC* at right
angles. Therefore *KD* is an axis
(Def. 7).

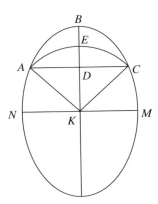

Let *MKN* be drawn through *K* parallel to *CA*; therefore *MN* is the axis of
the section conjugate to *BK* (Def. 8).

PROPOSITION 48 (PROBLEM)

Then with these things shown, let it be next in order to show that there are no other axes of the same sections.

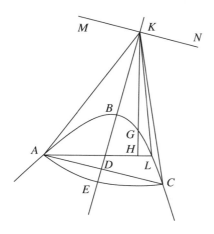 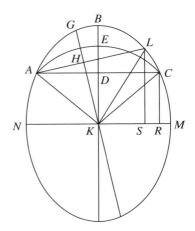

For if possible, let there also be another axis *KG*. Then in the same way as before, with *AH* drawn perpendicular,

$$AH = HL \text{ (Def. 4)};$$

and so also

$$AK = KL.$$

But also

$$AK = KC;$$

therefore

$$KL = KC;$$

and this is absurd.

Now that the circle *AEC* does not hit the section also in another point between the points *A, B* and *C,* is evident in the case of the hyperbola; and in the case of the ellipse let the perpendiculars *CR* and *LS* be drawn. Since then

$$KC = KL;$$

for they are radii; also

$$\text{sq. } KC = \text{sq. } KL.$$

But

$$\text{sq. } CR + \text{sq. } RK = \text{sq. } CK,$$

and

$$\text{sq. } KS + \text{sq. } SL = \text{sq. } LK;$$

therefore

$$\text{sq. } CR + \text{sq. } RK = \text{sq. } KS + \text{sq. } SL.$$

Therefore

difference between sq. CR and sq. $SL =$
difference between sq. KS and sq. RK.

Again since

$$\text{rect. } MR, RN + \text{sq. } RK = \text{sq. } KM,$$

and also

$$\text{rect. } MS, SN + \text{sq. } SK = \text{sq. } KM \text{ (Eucl. II. 5)},$$

therefore

$$\text{rect. } MR, RN + \text{sq. } RK = \text{rect. } MS, SN + \text{sq. } SK.$$

Therefore

difference between sq. SK and sq. $KR =$
difference between rect. MR, RN and rect. MS, SN.

And it was shown that

difference between sq. SK and sq. $KR =$
difference between sq. CR and sq. SL;

therefore

difference between sq. CR and sq. $SL =$
difference between rect. MR, RN and rect. MS, SN.

And since CR and LS are ordinates

$$\text{sq. } CR : \text{rect. } MR, RN :: \text{sq. } SL : \text{rect. } MS, SN \text{ (I. 21)}.$$

But the same difference was also shown for both; therefore

$$\text{sq. } CR = \text{rect. } MR, RN,$$

and

$$\text{sq. } SL = \text{rect. } MS, SN \text{ (Eucl. V. 16, 17, 9)}.$$

Therefore the line LCM is a circle; and this is absurd; for it is supposed an ellipse.

PROPOSITION 49 (PROBLEM)

Given a section of a cone and a point not within the section, to draw from the point a straight line touching the section in one point.

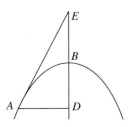

Let the given section of a cone first be a parabola whose axis is *BD*. Then it is required to draw a straight line as prescribed from the given point which is not within the section.

Then the given point is either on the line or on the axis or somewhere else outside.

Now let it be on the line, and let it be *A,* and let it have been done, and let it be *AE,* and let *AD* be drawn perpendicular; then it will be given in position. And

$$BE = BD \text{ (I. 35)};$$

and *BD* is given; therefore *BE* is also given. And the point *B* is given; therefore *E* is also given. But *A* also; therefore *AE* is given in position.

Then it will be constructed thus: let *AD* be drawn perpendicular from *A,* and let *BE* be made equal to *BD,* and let *AE* be joined. Then it is evident that it touches the section (I. 33).

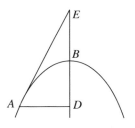

Again let the given point *E* be on the axis, and let it have been done, and let *AE* be drawn tangent, and let *AD* be drawn perpendicular; therefore

$$BE = BD \text{ (I. 35)}.$$

And *BE* is given; therefore also *BD* is given. And the point *B* is given; therefore *D* is also given. And *DA* is perpendicular; therefore *DA* is given in position. Therefore the point *A* is given. But also *E*; therefore *AE* is given in position.

Then it will be constructed thus: let *BD* be made equal to *BE,* and from *D* let *DA* be drawn perpendicular to *ED,* and let *AE* be joined. Then it is evident that *AE* touches (I. 33).

And it is evident also that, even if the given point is the same as *B,* the straight line drawn from *B* perpendicular touches the section (I. 17).

Then let *C* be the given point, and let it have been done, and let *CA* be it, and through *C* let *CF* be drawn parallel to the axis, that is to *BD*; therefore *CF* is given in position. And from *A* let *AF* be drawn ordinatewise to *CF*; then

$$CG = FG \text{ (I. 35).}$$

And the point *G* is given; therefore *F* is also given. And *FA* has been erected ordinatewise, that is, parallel to the tangent at *G* (I. 32); therefore *FA* is given in position. Therefore *A* is also given; but also *C*. Therefore *CA* is given in position.

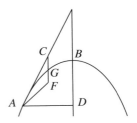

It will be constructed thus: let *CF* be drawn through *C* parallel to *BD*, and let *FG* be made equal to *CG*, and let *FA* be drawn parallel to the tangent at *G* (above), and let *AC* be joined. It is evident then that this will do the problem (I. 33).

Again let it be an hyperbola whose axis is *DBC* and center *H*, and asymptotes *HE* and *HF*. Then the given point will be given either on the section or on the axis or within angle *EHF* or in the adjacent place or on one of the asymptotes containing the section or in the place between the straight lines containing the angle vertical to angle *EHF*.

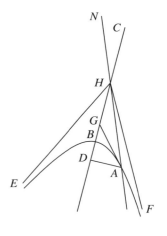

Let the point *A* first be on the section, and let it have been done, and let *AG* be tangent, and let *AD* be drawn perpendicular, and let *BC* be the transverse side of the figure; then

$$CD : DB :: CG : GB \text{ (I. 36).}$$

And the ratio of *CD* to *DB* is given; for both the straight lines are given; therefore also the ratio of *CG* to *GB* is given. And *BC* is given; therefore point *G* is given. But also *A*; therefore *AG* is given in position.

It will be constructed thus: let *AD* be drawn perpendicular from *A*, and let

$$CG : GB :: CD : DB;$$

and let *AG* be joined. Then it is evident that *AG* touches the section (I. 34).

Then again let the given point *G* be on the axis, and let it have been done, and let *AG* be drawn tangent, and let *AD* be drawn perpendicular. Then for the same reasons

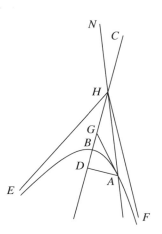

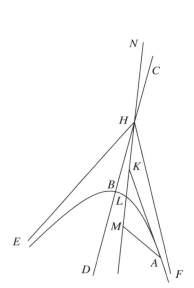

$CG : GB :: CD : DB$ (I. 36).
And BC is given; therefore the point D is given. And DA is perpendicular; therefore DA is given in position. And also the section is given in position; therefore the point A is given. But also G; therefore AG is given in position.

Then it will be constructed thus: let the other things be supposed the same, and let it be contrived that

$$CG : GB :: CD : DB,$$

and let DA be drawn perpendicular, and let AG be joined. Then it is evident that AG does the problem (I. 34), and that from G another tangent to the section could be drawn on the other side.

With the same things supposed let the given point K be in the place inside angle EHF, and let it be required to draw a tangent to the section from K. Let it have been done, and it be KA, and let KH be joined and produced, and let HN be made equal to LH, therefore they are all given. Then also LN will be given. Then let AM be drawn ordinatewise to MN;
then also

$$NK : KL :: MN : ML.$$

And the ratio of NK to KL is given; therefore also the ratio of NM to ML is given. And the point L is given, therefore also M is given. And MA has been erected parallel to the tangent at L; therefore MA is given in position. And also the section ALB is given in position; therefore the point A is given. But K is also given; therefore AK is given.

Then it will be constructed thus: let the other things be supposed the same, and the given point K, and KH be joined and produced, and let HN be made equal to HL, and let it be contrived that

$$NK : KL :: NM : ML$$

and let MA be drawn parallel to the tangent at L (above), and let KA be joined; therefore KA touches the section (I. 34).

And it is evident that a tangent to the section could also be drawn to the other side.

With the same things supposed let the given point F be on one of the asymptotes containing the section, and let it be required to draw from F a tangent to the section. And let it have been done, and let it be FAE, and through A let AD be drawn parallel to EH; then
$$DH = DF,$$
since also
$$FA = AE \text{ (II. 3).}$$
And FH is given; therefore also point D is given. And through the given point D, DA has been drawn parallel in position to EH; therefore DA is given in position. And the section is also given in position; therefore the point A is given. But F is also given; therefore the straight line FAE is given in position.

Then it will be constructed thus: let there be the section AB, and asymptotes EH and HF, and the given point F on one of the asymptotes containing the section, and let FH be bisected at D, and through D let DA be drawn parallel to HE, and let FA be joined. And since
$$FD = DH,$$
therefore also
$$FA = AE.$$
And so by things shown before, the straight line FAE touches the section (II. 9).

With the same things supposed, let the given point be in the place under the angle adjacent to the straight lines containing the section, and let it be K; it is required then to draw a tangent to the section from K. And let it have been done, and let it be KA, and let KH be joined and produced; then it will be given in position. If then a given point C is taken on the section, and through C, CD is drawn parallel to KH, it will be given in position. And if CD is bisected at E, and HE is joined and produced, it will be, in position, a diameter conjugate to KH (Def. 6). Then let HG be made equal to BH, and through A let AL be drawn parallel to BH; then because KL and BG are conjugate diameters, and AK a tangent, and AL a straight line drawn parallel to

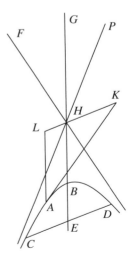

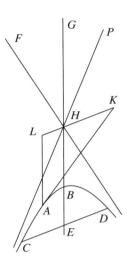

BG, therefore rectangle *KH, HL* is equal to the fourth part of the figure on *BG* (I. 38). Therefore rectangle *KH, HL* is given. And *KH* is given; therefore *HL* is also given. But it is also given in position; and the point *H* is given; therefore *L* is also given. And through *L*, *LA* has been drawn parallel in position to *BG*; therefore *LA* is given in position. And the section is also given in position; therefore the point *A* is given. But also *K*; therefore *AK* is given in position.

Then it will be constructed thus: let the other things be supposed the same, and let the given point *K* be in the aforesaid place, and let *KH* be joined and produced, and let some point *C* be taken, and let *CD* be drawn parallel to *KH*, and let *CD* be bisected by *E* and let *EH* be joined and produced, and let *HG* be made equal to *BH*; therefore *GB* is a transverse diameter conjugate to *KHL* (Def. 6). Then let rectangle *KH, HL* be made equal to the fourth of the figure on *BG*, and through *L* let *LA* be drawn parallel to *BG*, and let *KA* be joined; then it is clear that *KA* touches the section by the converse of the theorem (I. 38).

And if it is given in the place between the straight lines *FH* and *HP*, the problem is impossible. For the tangent will cut *GH*. And so it will meet both *FH* and *HP*; and this is impossible by the things shown in the thirty-first theorem of the first book (I. 31) and in the third of this book (II. 3).

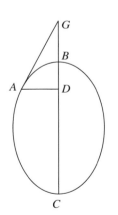

With the same things supposed let the section be an ellipse, and the given point *A* on the section, and let it be required to draw from *A* a tangent to the section. Let it have been done, and let it be *AG*, and let *AD* be drawn from *A* ordinatewise to the axis *BC*; then the point *D* will be given, and
$$CD : DB :: CG : GB \text{ (I. 36)}.$$
And the ratio of *CD* to *DB* is given; therefore the ratio of *CG* to *GB* is also given. Therefore the point *G* is given. But also *A* ; therefore *AG* is given in position.

Then it will be constructed thus: let *AD* be drawn perpendicular, and let
$$CG : GB :: CD : DB,$$

and let *AG* be joined. Then it is evident that *AG* touches, as also in the case of the hyperbola (I. 34).

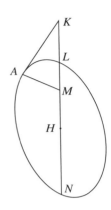

Then again let the given point be *K,* and let it be required to draw a tangent. Let it have been done, and let it be *KA,* and let the straight line *KLH* be joined to the center *H* and produced to *N*; then it will be given in position. And if *AM* is drawn ordinatewise, then

NK : KL :: NM : ML (I. 36).

And the ratio of *NK : KL* is given; therefore the ratio of *MN* to *LM* is also given. Therefore the point *M* is given. And *MA* has been erected ordinatewise; for it is parallel to the tangent at *L*; therefore *MA* is given in position. Therefore the point *A* is given. But also *K*; therefore *KA* is given in position.

And the construction ($\sigma\acute{\upsilon}\nu\theta\epsilon\sigma\iota\varsigma$) is the same as for the preceding.

PROPOSITION 50 (PROBLEM)

Given the section of a cone, to draw a tangent which will make with the axis, on the same side as the section, an angle equal to a given acute angle.

Let the section of a cone first be a parabola whose axis is *AB*; then it is required to draw a tangent to the section which will make with the axis *AB*, on the same side as the section, an angle equal to the given acute angle.

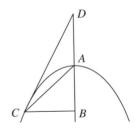

Let it have been done, and let it be *CD*; therefore angle *BDC* is given. Let *BC* be drawn perpendicular; then the angle at *B* is also given. Therefore the ratio of *DB* to *BC* is given. But the ratio of *BD* to *BA* is given; therefore also the ratio of *AB* to *BC* is given. And the angle at *B* is given; therefore angle *BAC* is also given. And it is [given] with respect to *BA* which is given in position, and with respect to the given point *A*; therefore *CA* is given in position. And the section is also given in position; therefore the point *C* is given. And *CD* touches; therefore *CD* is given in position.

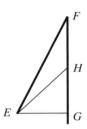

Then the problem will be constructed thus: let the given section of a cone first be a parabola whose axis is *AB,* and the given acute angle, angle *EFG,* and let some point *E* be taken on *EF,* and let *EG* be drawn perpendicular, and let *FG* be bisected by *H,* and let *HE* be joined, and let angle *BAC* be constructed equal to angle *GHE,* and let *BC* be drawn perpendicular, and let *AD* be made equal to *BA,* and let *CD* be joined. Therefore *CD* is tangent to the section (I. 33).

I say then that

$$\text{angle } CDB = \text{angle } EFG.$$

For since

$$FG : GH :: DB : BA$$

and

$$HG : GE :: AB : BC,$$

therefore *ex aequali*

$$FG : GE :: DB : BC.$$

And the angles at *G* and *B* are right angles, therefore

$$\text{angle at } F = \text{angle at } D.$$

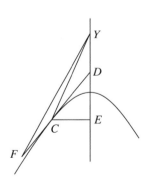

Let the section be an hyperbola, and let it have been done, and let *CD* be tangent, and let the center of the section *Y* be taken, and let *CY* be joined, and let *CE* be perpendicular; therefore the ratio of rectangle *YE, ED* to the square on *CE* is given; for it is the same as the transverse to the upright (I. 37). And the ratio of the square on *CE* to the square on *ED* is given; for each of the angles *CDE* and *DEC* is given. Therefore the ratio of rectangle *YE, ED* to the square on *ED* is given; and so also the ratio of *YE* to *ED* is given. And the angle at *E* is given; therefore the angle at *Y* is also given. Then some straight line *CY* has been drawn across in position with respect to the straight line *YE* and to the given point *Y* at a given angle; therefore *CY* is given in position. And the section is also given in position; therefore the point *C* is given. And *CD* has been drawn across as tangent; therefore *CD* is given in position.

Let the asymptote to the section *YF* be drawn; therefore *CD* produced will
meet the asymptote (II. 3). Let it meet it at *F*. Therefore
$$\text{angle } FDE > \text{angle } FYD.$$
Therefore, for the construction, the given acute angle will have to be greater
than half the angle contained by the asymptotes.

Then the problem will be constructed
thus: let there be the given hyperbola
whose axis is *AB*, and asymptote *YF*, and
the given acute angle *KHG* greater than
angle *AYF*, and let
$$\text{angle } KHL = \text{angle } AYF,$$
and let *AF* be drawn from *A* perpendicular
to *AB*, and let some point *G* be taken on
GH, and let *GK* be drawn from it perpendicular to *HK*. Since then
$$\text{angle } FYA = \text{angle } LHK,$$
and also the angles at *A* and *K* are right,
therefore

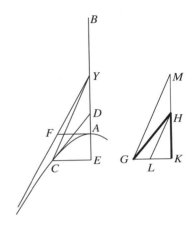

$$YA : AF :: HK : KL$$
$$HK : KL > HK : KG;$$
therefore also
$$YA : AF > HK : KG.$$

And so also
$$\text{sq. } YA : \text{sq. } AF > \text{sq. } HK : \text{sq. } KG.$$

But
$$\text{sq. } YA : \text{sq. } AF :: \text{transverse} : \text{upright (II. 1)};$$
therefore also
$$\text{transverse} : \text{upright} > \text{sq. } HK : \text{sq. } KG.$$
If then we shall contrive that
$$\text{sq. } YA : \text{sq. } AF :: \text{some other} : \text{sq. } KG,$$
it will be greater than the square on *HK*. Let it be the rectangle *MK*, *KH*; and
let *GM* be joined. Since then
$$\text{sq. } MK > \text{rect. } MK, KH,$$
therefore
$$\text{sq. } MK : \text{sq. } KG > \text{rect. } MK, KH : \text{sq. } KG$$
$$> \text{sq. } YA : \text{sq. } AF.$$

And if we shall contrive that
$$\text{sq. } MK : \text{sq. } KG :: \text{sq. } YA : \text{some other},$$
it will be to a magnitude less than the square on *AF*; and the straight line

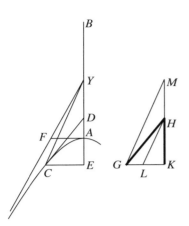

joined from Y to the point taken will make similar triangles, and therefore

angle $FYA >$ angle GMK.

Let angle AYC be made equal to angle GMK; therefore YC will cut the section (II. 2). Let it cut it at C, and from C let CD be drawn tangent to the section (II. 49), and CE drawn perpendicular; therefore triangle CYE is similar to triangle GMK. Therefore

sq. YE : sq. EC :: sq. MK : sq. KG.

But also

transverse : upright :: rect. YE, ED : sq. EC (I. 37),

and

transverse : upright :: rect. MK, KH : sq. KG.

And inversely

sq. CE : rect. YE, ED :: sq. GK : rect. MK, KH;

therefore *ex aequali*

sq. YE : rect. YE, ED :: sq. MK : rect. MK, KH.

And therefore

$YE : ED :: MK : KH$.

But also we had

$CE : EY :: GK : KM$;

therefore *ex aequali*

$CE : ED :: GK : KH$.

And the angles at E and K are right angles; therefore

angle at $D =$ angle GHK.

Let the section be an ellipse whose axis is AB. Then it is required to draw a tangent to the section which with the axis will contain, on the same side as the section, an angle equal to the given acute angle.

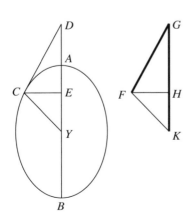

Let it have been done, and let it be CD. Therefore angle CDA is given. Let CE be drawn perpendicular; therefore the ratio of the square on DE to the square on EC is given. Let Y be the center of the section, and let CY be joined. Then the ratio of the square on CE to the rectangle DE, EY is given; for it is the same as the ratio of the upright to the transverse (I. 37), and therefore the ratio of the square on *DE* to rectangle *DE, EY* is given; and therefore the ratio of *DE* to *EY* is given. And of *DE* to *EC*; therefore also the ratio of *CE* to *EY* is given. And the angle at *E* is right; therefore the angle at *Y* is given. And it is given with respect to a straight line given in position and to a given point; therefore the point *C* is given. And from the given point *C* let *CD* be drawn tangent; therefore *CD* is given in position.

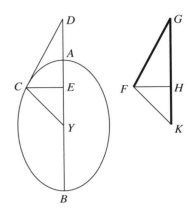

Then the problem will be constructed thus: let there be the given acute angle *FGH*, and let some point *F* be taken on *FG*, and let *FH* be drawn perpendicular, and let it be contrived that

upright : transverse :: sq. *FH* : rect. *GH, HK*,

and let *KF* be joined, and let *Y* be the center of the section, and let angle *AYC* be constructed equal to angle *GKF*, and let *CD* be drawn tangent to the section (II. 49).

I say that *CD* does the problem, that is,

angle *CDE* = angle *FGH*.

For since

YE : EC :: KH : FH,

therefore also

sq. *YE* : sq. *EC* :: sq. *KH* : sq. *FH*.

But also

sq. *EC* : rect. *DE, EY* :: sq. *FH* : rect. *KH, HG*;

for each is the same ratio as that of the upright to the transverse (I. 37, and above). And *ex aequali*; therefore

sq. *YE* : rect. *DE, EY* :: sq. *KH* : rect. *KH, HG*.

And therefore

YE : ED :: KH : HG.

But also

YE : EC :: KH : FH;

ex aequali, therefore

DE : EC :: HG : FH.

And the sides about the right angles are proportional; therefore

angle *CDE* = angle *FGH*.

Therefore *CD* does the problem.

PROPOSITION 51 (PROBLEM)

Given a section of a cone, to draw a tangent which with the diameter drawn through the point of contact will contain an angle equal to a given acute angle.

Let the given section of a cone first be a parabola whose axis is *AB,* and the given angle *H;* then it is required to draw a tangent to the parabola which with the diameter from the point of contact will contain an angle equal to the angle at *H.*

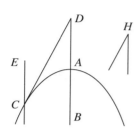

Let it have been done, and let *CD* be drawn a tangent making with the diameter *EC* drawn through the point of contact angle *ECD* equal to angle *H,* and let *CD* meet the axis at *D* (I. 24). Since then *AD* is parallel to *EC* (I. 51 porism),
angle *ADC* = angle *ECD.*
But angle *ECD* is given; for it is equal to angle *H;* therefore angle *ADC* is also given.

Then it will be constructed thus: let there be a parabola whose axis is *AB,* and the given angle *H.* Let *CD* be drawn a tangent to the section making with the axis the angle *ADC* equal to angle *H* (II. 50), and through *C* let *EC* be drawn parallel to *AB.*
Since then

$$\text{angle } H = \text{angle } ADC,$$

and

$$\text{angle } ADC = \text{angle } ECD,$$

therefore also

$$\text{angle } H = \text{angle } ECD.$$

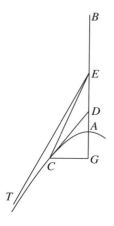

Let the section be an hyperbola whose axis is *AB,* and center *E,* and asymptote *ET,* and the given acute angle *Q,* and let *CD* be tangent, and let *CE* be joined doing the problem, and let *CG* be drawn perpendicular. Therefore the ratio of the transverse to the upright is given; and so also the ratio of rectangle *EG, GD* to the square on *CG* (I. 37). Then let some given straight line *FH* be laid out, and on it let there be described a segment of a circle admitting an angle equal to angle *Q* (Eucl. III. 33); therefore it will be greater than a semicircle (Eucl. III. 31). And from some point *K* of those on the circumference let *KL* be drawn perpendicular making

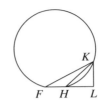

rect. *FL, LH* : sq. *LK* :: transverse : upright,

and let *FK* and *KH* be joined. Since then

angle *FKH* = angle *ECD,*

but also

rect. *EG, GD* : sq. *GC* :: transverse : upright,

and

rect. *FL, LH* : sq. *LK* :: transverse : upright,

therefore triangle *KFL* is similar to triangle *ECG,* and triangle *FHK* to triangle *ECD.*[*]

And so

angle *HFK* = angle *CED.*

---

[*] Pappus, in lemma IX to this book: "Let triangle *ABC* be similar to triangle *DEF,* and triangle *AGB* to *DEH;* the result is

rect. *BC, CG* : sq. *CA* :: rect. *EF, FH* : sq. *DF.*

"For since because of similarity

whole angle *A* = whole angle *D,*

and

angle *BAG* = angle *EDH,*

therefore

remaining angle *GAC* = remaining angle *HDF.*

But also

angle *C* = angle *F;*

therefore

GC : CA :: HF : FD.

But also

BC : CA :: EF : FD;

therefore also compounded ratio is the same with compounded. Therefore

rect. *BC, CG* : sq. *CA* :: rect. *EF, FH* : sq. *FD.* "

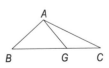

(Tr.)

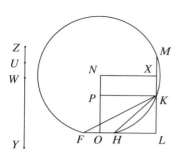

Then it will be constructed thus: let there be the given hyperbola *AC*, and axis *AB*, and center *E*, and given acute angle *Q*, and let the given ratio of the transverse to the upright be the same as *YZ* to *YW*, and let *WZ* be bisected at *U*, and let a given straight line *FH* be laid out, and on it let there be described a segment of a circle, greater than semicircle and admitting an angle equal to angle *Q* (Eucl. III. 31, 33), and let it be *FKH*, and let the center of the circle *N* be taken, and from *N* let *NO* be drawn perpendicular to *FH*, and let *NO* be cut at *P* in the ratio of *UW* to *WY*, and through *P* let *PK* be drawn parallel to *FH*, and from *K* let *KL* be drawn perpendicular to *FH* produced, and let *FK* and *KH* be joined, and let *LK* be produced to *M*, and from *N* let *NX* be drawn perpendicular to it; therefore it is parallel to *FH*. And therefore

$$NP : PO \text{ or } UW : WY :: XK : KL.$$

And doubling the antecedents

$$ZW : WY : MK : KL;$$

componendo

$$ZY : YW :: ML : LK.$$

But

$$ML : LK :: \text{rect. } ML, LK : \text{sq. } LK;$$

therefore

$$ZY : YW :: \text{rect. } ML, LK : \text{sq. } LK :: \text{rect. } FL, LH : \text{sq. } LK \text{ (Eucl. III. 36).}$$

But

$$ZY : YW :: \text{transverse : upright;}$$

therefore also

$$\text{rect. } FL, LH : \text{sq. } LK :: \text{transverse : upright.}$$

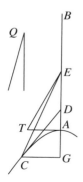

Then let *AT* be drawn from *A* perpendicular to *AB*. Since then

  sq. *EA* : sq. *AT* :: transverse : upright (II. 1),
and also

  transverse : upright :: rect. *FL, LH* : sq. *LK*,
and

  sq. *FL* : sq. *LK* > rect. *FL, LH* : sq. *LK*,
therefore also

  sq. *FL* : sq. *LK* > sq. *EA* : sq. *AT*.
And the angles at *A* and *L* are right angles;

therefore
angle F < angle E.

Then let angle AEC be constructed equal to angle LFK; therefore EC will meet the section (II. 2). Let it meet it at C. Then let CD be drawn tangent from C (II. 49), and let CG be drawn perpendicular; then

transverse : upright :: rect. EG, GD : sq. CG (I. 37).

Therefore also

rect. FL, LH : sq. LK :: rect. EG, GD : sq. CG;

therefore triangle KFL is similar to triangle ECG, and triangle KHL to triangle CGD, and triangle KFH to triangle CED. And so

angle ECD = angle FKH = angle Q.

And if the ratio of the transverse to the upright is equal to equal, KL touches the circle FKH (Eucl. III. 37), and the straight line joined from the center to K will be parallel to FH and itself will do the problem.

PROPOSITION 52

If a straight line touches an ellipse making an angle with the diameter drawn through the point of contact, it is not less than the angle adjacent to the one contained by the straight lines deflected at the middle of the section.

Let there be an ellipse whose axes are AB and CD, and center E, and let AB be the major axis, and let the straight line GFL touch the section, and let AC, CB, and FE be joined, and let BC be produced to L.

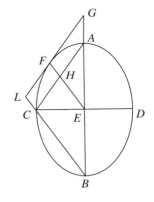

I say that angle LFE is not less than angle LCA.

For FE is either parallel to LB or not.

Let it first be parallel; and

$AE = EB$;

therefore also

$AH = HC$.

And FE is a diameter; therefore the tangent at F is parallel to AC (II. 6). But also FE is parallel to LB; therefore $FHCL$ is a parallelogram, and therefore

angle LFH = angle LCH.

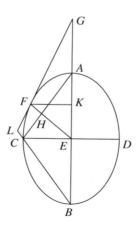

And since *AE* and *EB* are each greater than *EC*, angle *ACB* is obtuse; therefore angle *LCA* is acute. And so also angle *LFE*. And therefore angle *GFE* is obtuse.

Then let *EF* not be parallel to *LB*, and let *FK* be drawn perpendicular; therefore *LBE* is not equal to angle *FEA*. But

rt. angle at E = rt. angle at K;

therefore it is not true that

sq. BE : sq. EC :: sq. EK : sq. KF.

But

sq. BE : sq. EC :: rect. AE, EB : sq. EC :: transverse : upright (I. 21)

and

transverse : upright :: rect. GK, KE : sq. KF (I. 37).

Therefore it is not true that

rect. GK, KE : sq. KF :: sq. KE : sq. KF.

Therefore *GK* is not equal to *KE*.

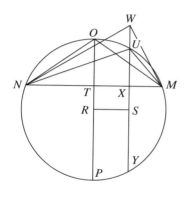

Let there be laid out a segment of a circle *MUN* admitting an angle equal to angle *ACB* (Eucl. III. 33); and angle *ACB* is obtuse; therefore *MUN* is a segment less than a semicircle (Eucl. III. 31). Then let it be contrived that

GK : KE :: NX : XM,

and from *X* let *UXY* be drawn at right angles, and let *NU* and *UM* be joined, and let *MN* be bisected at *T*, and let *OTP* be drawn at right angles; therefore it is a diameter. Let the center be *R*, and from it let *RS* be drawn perpendicular, and *ON* and *OM* be joined. Since then

$$\text{angle } MON = \text{angle } ACB,$$

and *AB* and *MN* have been bisected, the one at *E* and the other at *T*, and the angles at *E* and *T* are right angles, therefore triangles *OTN* and *BEC* are similar. Therefore

$$\text{sq. } TN : \text{sq. } TO :: \text{sq. } BE : \text{sq. } EC.$$

And since

$$TR = SX,$$

and

$$RO > SU,$$

therefore

$$RO : TR > SU : SX;$$

and *convertendo*

$$RO : OT < SU : UX.$$

And, doubling the antecedents, therefore

$$PO : TO < YU : UX.$$

And *separando*

$$PT : TO < YX : UX.$$

But

$$PT : TO :: \text{sq. } TN : \text{sq. } TO :: \text{sq. } BE : \text{sq. } EC :: \text{transverse : upright (I. 21),}$$

and

$$\text{transverse : upright :: rect. } GK, KE : \text{sq. } KF \text{ (I. 37);}$$

therefore

$$\text{rect. } GK, KE : \text{sq. } KF < YX : XU$$
$$< \text{rect. } YX, XU : \text{sq. } XU$$
$$< \text{rect. } NX, XM : \text{sq. } XU.$$

If then we contrive it that

$$\text{rect. } GK, KE : \text{sq. } KF :: \text{rect. } MX, XN : \text{some other,}$$

it will be greater than the square on *XU*. Let it be to the square on *XW*. Since then

$$GK : KE :: NX : XM,$$

and *KF* and *YW* are perpendicular to *GE* and *MN*, respectively; and

$$\text{rect. } GK, KE : \text{sq. } KF :: \text{rect. } MX, XN : \text{sq. } XW,$$

therefore

$$\text{angle } GFE = \text{angle } MWN.$$

Therefore

$$\text{angle } MUN \text{ or angle } ACB > \text{angle } GFE,$$

and the adjacent angle *LFH* is greater than angle *LCH*.

Therefore angle *LFH* is not less than angle *LCH*.

PROPOSITION 53 (PROBLEM)

Given an ellipse, to draw a tangent which will make with the diameter drawn through the point of contact an angle equal to a given acute angle; then it is required that the given acute angle be not less than the angle adjacent to the angle contained by the straight lines deflected at the middle of the section.

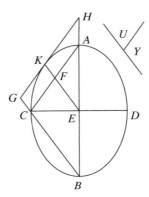

Let there be the given ellipse whose major axis is *AB* and minor axis *CD*, and center *E*, and let *AC* and *CB* be joined, and let angle *U* be the given angle not less than angle *ACG*; and so also angle *ACB* is not less than angle *Y*.

Therefore angle *U* is either greater than or equal to angle *ACG*.

Let it first be equal; and through *E* let *EK* be drawn parallel to *BC*, and through *K* let *KH* be drawn tangent to the section (II. 49). Since then
$$AE = EB,$$
and
$$AE : EB :: AF : FC,$$
therefore
$$AF = FC.$$
And *KE* is a diameter; therefore the tangent to the section at *K*, that is *HKG*, is parallel to *CA* (II. 6). And also *EK* is parallel to *GB*; therefore *KFCG* is a parallelogram; and therefore
$$\text{angle } GKF = \text{angle } GCF.$$
And angle *GCF* is equal to the given angle, that is *U*; therefore also
$$\text{angle } GKE = \text{angle } U.$$

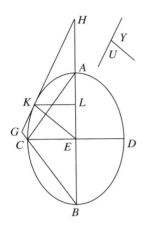
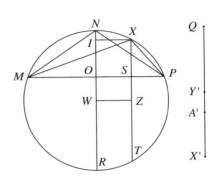

Then let

angle U > angle ACG;

then inversely

angle Y < angle ACB.

Let a circle be laid out, and let a segment be taken from it, and let it be MNP admitting an angle equal to angle Y, and let MP be bisected at O, and from O let NOR be drawn at right angles to MP, and let NM and NP be joined; therefore

angle MNP < angle ACB.

But

angle MNO = half angle MNP,

and

angle ACE = half angle ACB;

therefore

angle MNO < angle ACE.

And the angles at E and O are right angles, therefore

$AE : EC > OM : ON$.

And so also

sq. AE : sq. EC > sq. MO : sq. NO.

But

sq. AE = rect. AE, EB,

and

sq. MO = rect. MO, OP = rect. NO, OR (Eucl. III. 35);

therefore

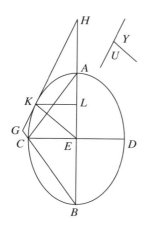
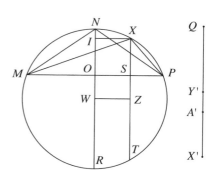

rect. *AE, EB* : sq. *EC* or transverse : upright (I. 21) > *RO* : *ON*.
Then let it be that

$$\text{transverse : upright :: } QA' : A'X',$$

and let *QX'* be bisected at *Y'*. Since then

$$\text{transverse : upright} > RO : ON,$$

also

$$QA' : A'X' > RO : ON.$$

And *componendo*

$$QX' : X'A' > RN : NO.$$

Let the center of the circle be *W*; and so also

$$Y'X' : X'A' > WN : NO.$$

And *separando*

$$A'Y' : A'X' > WO : ON.$$

Then let it be contrived that

$$A'Y' : A'X' :: WO : \text{less than } ON$$

such as *IO,* and let *IX* and *XT* and *WZ* be drawn parallel.
Therefore

$$A'Y' : A'X' :: WO : OI :: ZS : SX;$$

and *componendo*

$$Y'X' : X'A' :: ZX : XS.$$

And doubling the antecedents,

$$QX' : X'A' :: TX : XS.$$

And *separando*

$$QA' : A'X' \text{ or transverse : upright :: } TS : SX.$$

Then let *MX* and *XP* be joined, and let angle *AEK* be constructed on straight
line *AE* at point *E* equal to angle *MPX*, and through *K* let *KH* be drawn

touching the section (II. 49), and let *KL* be dropped ordinatewise. Since then

$$\text{angle } MPX = \text{angle } AEK,$$

and

$$\text{rt. angle at } S = \text{rt. angle at } L,$$

therefore triangle *XSP* is equiangular with triangle *KEL*.
And

$$\text{transverse : upright :: } TS : SX$$
$$:: \text{rect. } TS, \ SX : \text{sq. } SX$$
$$:: \text{rect. } MX, \ SP : \text{sq. } SX;$$

therefore triangle *KLE* is similar to triangle *SXP*, and triangle *MXP* to triangle *KHE*, and therefore

$$\text{angle } MXP = \text{angle } HKE.$$

But

$$\text{angle } MXP = \text{angle } MNP = \text{angle } Y;$$

therefore also

$$\text{angle } HKE = \text{angle } Y.$$

And therefore

$$\text{adjacent angle } GKE = \text{adjacent angle } U.$$

Therefore *GH* has been drawn across tangent to the section and making with the diameter *KE,* drawn through the point of contact, angle *GKE* equal to the given angle *U*; and this it was required to do.

CONICS
BOOK THREE

PROPOSITION 1

If straight lines, touching a section of a cone or circumference of a circle, meet, and diameters are drawn through the points of contact meeting the tangents, the resulting vertically related triangles will be equal.

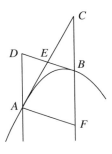 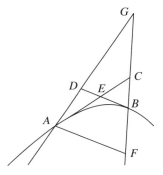 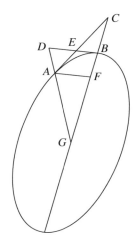

Let there be the section of a cone or circumference of a circle AB, and let AC and BD, meeting at E, touch AB, and let the diameters of the section CB and DA be drawn through A and B, meeting the tangents at C and D.

I say that

$$\text{trgl. } ADE = \text{trgl. } EBC.$$

For let AF be drawn from A parallel to BD; therefore it has been dropped ordinatewise (I. 32 converse). Then in the case of the parabola

$$\text{pllg. } ADBF = \text{trgl. } ACF \text{ (I. 42)},$$

and, with the common area $AEBF$ subtracted,

$$\text{trgl. } ADE = \text{trgl. } CBE.$$

And in the case of the others let the diameters meet at center G. Since then AF has been dropped ordinatewise, and AC touches,

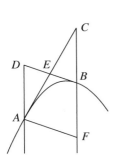
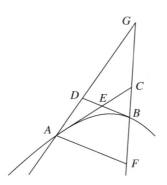
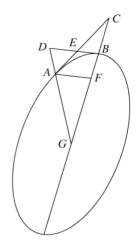

$$\text{rect. } FG, GC = \text{sq. } BG \text{ (I. 37).}$$

Therefore
$$FG : GB :: BG : GC;$$
therefore also
$$FG : GC :: \text{sq. } FG : \text{sq. } GB \text{ (Eucl. VI. 19 porism).}$$

But
$$\text{sq. } FG : \text{sq. } GB :: \text{trgl. } AGF : \text{trgl. } DGB \text{ (Eucl. VI. 19),}$$
and
$$FG : GC :: \text{trgl. } AGF : \text{trgl. } AGC;$$
therefore also
$$\text{trgl. } AGF : \text{trgl. } AGC ::$$
$$\text{trgl. } AGF : \text{trgl. } DGB.$$

Therefore
$$\text{trgl. } AGC = \text{trgl. } DGB.$$
Let the common area $AGBE$ be subtracted; therefore as remainders,
$$\text{trgl. } AED = \text{trgl. } CEB.$$

PROPOSITION 2

With the same things supposed, if some point is taken on the section or circumference of a circle, and through it parallels to the tangents are drawn as far as the diameters, then the quadrilateral produced on one of the tangents and one of the diameters will be equal to the triangle produced on the same tangent and the other diameter.

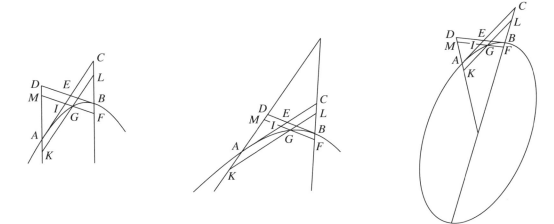

For let there be a section of a cone or circumference of a circle *AB* and let *AEC* and *BED* be tangents, and *AD* and *BC* diameters, and let some point *G* be taken on the section, and *GKL* and *GMF* be drawn parallel to the tangents.

I say that

$$\text{trgl. } AIM = \text{quadr. } CLGI.$$

For triangle *GKM* has been shown equal to quadrilateral *AL* (I. 42, 43),[*] let the common quadrilateral *IK* be added or subtracted, and

$$\text{trgl. } AIM = \text{quadr. } CG.$$

PROPOSITION 3

With the same things supposed, if two points are taken on the section or circumference of a circle, and through them parallels to the tangents are drawn as far as the diameters, the quadrilaterals produced by the straight lines drawn, and standing on the diameters as bases, are equal to each other.

---

[*] If G is not between A and B, this step is derived from a relation established in the course of the proof of I. 49 (for the parabola) or I. 50 (hyperbola and ellipse), pp. 85 and 89, respectively. (Ed.)

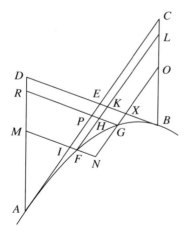

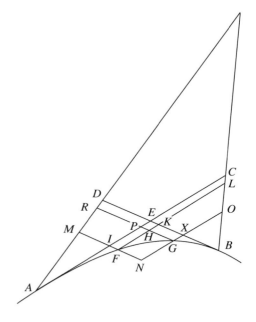

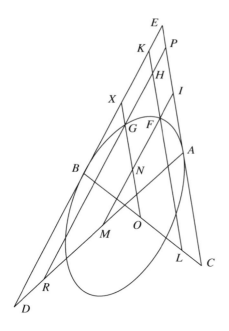

For let there be the section and tangents and diameters as said before, and let two points at random *F* and *G* be taken on the section, and through *F* let the straight lines *FHKL* and *NFIM* be drawn parallel to the tangents, and through *G* the straight lines *GXO* and *HPR*.

I say that

$$\text{quadr. } LG = \text{quadr. } MH,$$

and

$$\text{quadr. } LN = \text{quadr. } RN.$$

For since it has already been shown that

$$\text{trgl. } RPA = \text{quadr. } CG \text{ (III. 2)},$$

and

$$\text{trgl. } AMI = \text{quadr. } CF \text{ (III. 2)},$$

and

$$\text{trgl. } RPA = \text{trgl. } AMI + \text{quadr. } PM,$$

therefore also

$$\text{quadr. } CG = \text{quadr. } CF + \text{quadr. } PM;$$

and so

$$\text{quadr. } CG = \text{quadr. } CH + \text{quadr. } RF.$$

Let the common quadrilateral CH be subtracted; therefore as remainders

$$\text{quadr. } LG = \text{quadr. } HM.$$

And therefore as wholes

$$\text{quadr. } LN = \text{quadr. } RN.$$

PROPOSITION 4

If two straight lines touching opposite sections meet each other, and diameters are drawn through the points of contact meeting the tangents, then the triangles at the tangents will be equal.

Let there be the opposite sections A and B, and let the tangents to them, AC and BC, meet at C, and let D be the center of the sections and let AB and CD be joined, and CD produced to E, and let DA and BD also be joined and produced to F and G.

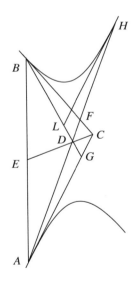

I say that

$$\text{trgl. } AGD = \text{trgl. } BDF,$$

and

$$\text{trgl. } ACF = \text{trgl. } BCG.$$

For let HL be drawn through H tangent to the section; therefore it is parallel to AG (I. 44, note). And since

$$AD = DH \text{ (I. 30),}$$

$$\text{trgl. } AGD = \text{trgl. } DHL \text{ (Eucl. VI. 19).}$$

But

$$\text{trgl. } DHL = \text{trgl. } BDF \text{ (III. 1);}$$

therefore also

$$\text{trgl. } AGD = \text{trgl. } BDF.$$

And so also

$$\text{trgl. } ACF = \text{trgl. } BCG.$$

PROPOSITION 5

If two straight lines touching opposite sections meet, and some point is taken on either of the sections, and from it two straight lines are drawn, the one parallel to the tangent, the other parallel to the line joining the points of contact, then the triangle produced by them on the diameter drawn through the point of meeting differs from the triangle cut off at the point of meeting of the tangents by the triangle cut off on the tangent and the diameter drawn through the point of contact.

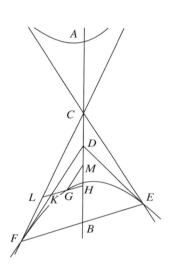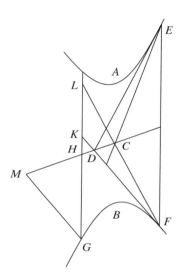

Let there be the opposite sections *A* and *B* whose center is *C*, and let tangents *ED* and *DF* meet at *D*, and let *EF* and *CD* be joined, and let *CD* be produced, and let *FC* and *EC* be joined and produced, and let some point *G* be taken on the section, and through it let *HGKL* be drawn parallel to *EF*, and *GM* parallel to *DF*.

I say that triangle *GHM* differs from triangle *KHD* by triangle *KLF*.

For since *CD* has been shown to be a diameter of the opposite sections (II. 39, 38), and *EF* to be an ordinate to it (II. 38; Def. 5), and *GH* has been drawn parallel to *EF*, and *MG* parallel to *DF*, therefore triangle *GHM* differs from triangle *CLH* by triangle *CDF* (I. 45, or I. 44, according to the case). And so triangle *GHM* differs from triangle *KHD* by triangle *KLF.*

And it is evident that

$$\text{trgl. } KLF = \text{quadr. } MGKD.$$

PROPOSITION 6

With the same things supposed, if some point is taken on one of the opposite sections, and from it parallels to the tangents are drawn meeting the tangents and the diameters, then the quadrilateral produced by them on one of the tangents and on one of the diameters will be equal to the triangle produced on the same tangent and the other diameter.

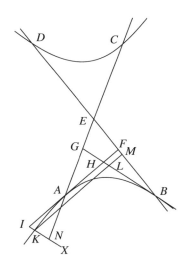

Let there be opposite sections of which *AEC* and *BED* are diameters, and let *AF* and *BG* touch the section *AB* meeting each other at *H,* and let some point *K* be taken on the section, and from it let *KML* and *KNX* be drawn parallel to the tangents.

I say that

$$\text{quadr. } KF = \text{trgl. } AIN.$$

Now since *AB* and *CD* are opposite sections, and *AF,* meeting *BD,* touches section *AB,* and *KL* has been drawn parallel to *AF,* therefore

$$\text{trgl. } AIN = \text{quadr. } KF \text{ (III. 2).}^{*}$$

PROPOSITION 7

With the same things supposed, if points are taken on each of the sections, and from them parallels to the tangents are drawn meeting the tangents and the diameters, then the quadrilaterals produced by the straight lines drawn and standing on the diameters as bases will be equal to each other.

For let the aforesaid things be supposed, and let points *K* and *L* be taken on both sections, and through them let *MKPRY* and *NSTLQ* be drawn parallel to *AF,* and *NIOKX* and *YWULZ* parallel to *BG.*

I say that what was said in the enunciation will be so.

For since

$$\text{trgl. } AOI = \text{quadr. } RO \text{ (III. 2),}$$

---

* Another and important case where the point *K* falls between *C* and *D* is given by Eutocius in his commentary to this proposition. It is as follows : ". . . and let *CPR* be drawn from *C* tangent to the section; then it is evident that it is parallel to *AF* and *ML* (I. 44, note). And since it has been shown in the second theorem (III. 2) in the figure of the hyperbola that

$$\text{trgl. } PNC = \text{quadr. } LP \text{ (III. 2, note),}$$

let the common quadrilateral *MP* be added; therefore

$$\text{trgl. } MKN = \text{quadr. } MLRC.$$

Let there be added the common triangle *CRE,* which is equal to triangle *AEF* by I. 4, note (and I. 30), therefore

$$\text{whole trgl. } MEL = \text{trgl. } MKN + \text{trgl. } AEF.$$

With common triangle *KMN* subtracted, as remainders

$$\text{trgl. } AEF = \text{quadr. } KLEN.$$

Let the common quadrilateral *FENI* be added; therefore, in whole,

$$\text{trgl. } AIN = \text{quadr. } KLFI.$$

And likewise also

$$\text{trgl. } BOL = \text{quadr. } KNGO.\text{"}$$

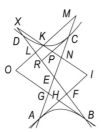

(Tr.)

let the quadrilateral *EO* be added to both;
therefore

> whole trgl. *AEF* = quadr. *KE*.

But also

> trgl. *BGE* = quadr. *LE* (III. 5, note);

and

> trgl. *AEF* = trgl. *BGE* (III. 1);

therefore

> quadr. *LE* = quadr. *IKRE*.

Let the common quadrilateral *NE* be
added; therefore as wholes

> whole quadr. *TK* = quadr. *IL,*

and also

> quadr. *KU* = quadr. *RL*.

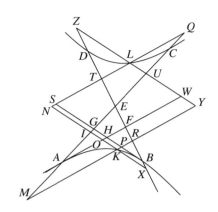

PROPOSITION 8

*With the same things supposed, instead of K and L let there be taken the
points C and D at which the diameters hit the sections, and through them let
the parallels to the tangents be drawn.*

I say that

$$\text{quadr. } DG = \text{quadr. } FC$$

and

$$\text{quadr. } XI = \text{quadr. } OT.$$

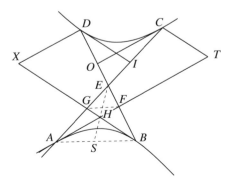

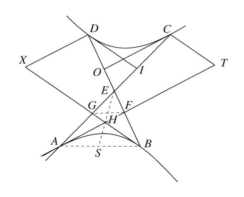

For since it was shown

trgl. AGH = trgl. HBF (III. 1),

and the straight line from A to B is parallel to the straight line from G to F,[*] therefore

$$AE : EG :: BE : EF;$$

and *convertendo*

$$EA : AG :: EB : BF.$$

And also

$$CA : AE :: DB : BE;$$

for each is double the other; therefore *ex aequali*

$$CA : AG :: DB : BF.$$

And the triangles are similar because of the parallels; therefore

trgl. CTA : trgl. AHG :: trgl. XBD : trgl. HBF (Eucl. VI. 19).

And alternately; but

trgl. AHG = trgl. HBF (III. 1);

therefore

trgl. CTA = trgl. XBD.

As parts of these it was shown

trgl. AHG = trgl. HBF;

therefore also as remainders

quadr. DH = quadr. CH.

And so also

quadr. DG = quadr. CF.

And since CO is parallel to AF,

trgl. COE = trgl. AEF.

And likewise also

trgl. DEI = trgl. BEG.

---

[*] For the point H falls within the angle AEB (II. 25), and the straight line drawn from H to the midpoint of AB, that is S, is a diameter (II. 29), and must therefore pass through E (I. 51 porism). An analogous series of propositions is found for the opposite sections: II. 32, 38, 39.
Then, since

trgl. GHA = trgl. FHB,

therefore

trgl. GFB = trgl. GFA.

Their bases are the same, therefore their heights are equal (Eucl. VI. 1). (Tr.)

But
$$\text{trgl. } BEG = \text{trgl. } AEF \text{ (III. 1)};$$
therefore also
$$\text{trgl. } COE = \text{trgl. } DEI.$$
And also
$$\text{quadr. } DG = \text{quadr. } CF \text{ (above)}.$$
Therefore, as wholes,
$$\text{quadr. } XI = \text{quadr. } OT.$$

PROPOSITION 9

With the same things supposed, if one of the points is between the diameters, as K, and the other is the same with one of the points C and D, for instance C, and the parallels are drawn, I say that
$$\text{trgl. } CEO = \text{quadr. } KE,$$
and
$$\text{quadr. } LO = \text{quadr. } LM.$$

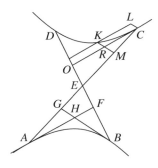

And this is evident. For since it was shown
$$\text{trgl. } CEO = \text{trgl. } AEF,$$
and
$$\text{trgl. } AEF = \text{quadr. } KE \text{ (III. 5, note)},$$
therefore also
$$\text{trgl. } CEO = \text{quadr. } KE.$$
And so also
$$\text{trgl. } CRM = \text{quadr. } KO,$$
and
$$\text{quadr. } KC = \text{quadr. } LO.$$

PROPOSITION 10

With the same things supposed, let K and L be taken not as points at which the diameters hit the sections.

Then it is to be shown that

$$quadr.\ LTRY = quadr.\ QYKI.$$

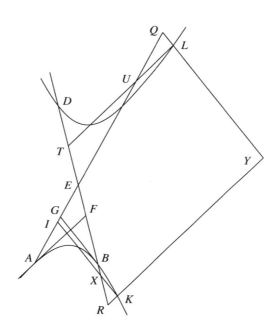

For since the straight lines *AF* and *BG* touch, and *AE* and *BE* are diameters through the points of contact, and *LT* and *KI* are parallel to the tangents,

trgl. *TUE* = trgl. *UQL* + trgl. *EFA* (I. 44).

And likewise also

trgl. *XEI* = trgl. *XRK* + trgl. *BEG*.

But

trgl. *EFA* = trgl. *BEG* (III. 1);

therefore

trgl. *TUE* − trgl. *UQL* = trgl. *XEI* − trgl. *XRK*.

Therefore
$$\text{trgl. } TUE + \text{trgl. } XRK = \text{trgl. } XEI + \text{trgl. } UQL.$$
Let the common area $KXEULY$ be added; therefore
$$\text{quadr. } LTRY = \text{quadr. } QYKI.$$

PROPOSITION 11

With the same things supposed, if some point is taken on either of the sections, and from it parallels are drawn, one parallel to the tangent and the other parallel to the straight line joining the points of contact, then the triangle produced by them on the diameter drawn through the point of meeting of the tangents differs from the triangle cut off on the tangent and the diameter drawn through the point of contact by the triangle cut off at the point of meeting of the tangents.

Let there be the opposite sections *AB* and *CD,* and let the tangents *AE* and *DE* meet at *E,* and let the center be *H,* and let *AD* and *EHG* be joined, and let some point *B* be taken at random on the section *AB,* and through it let *BFL* be drawn parallel to *AG,* and *BM* parallel to *AE.*

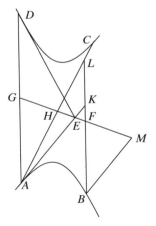

I say that triangle *BFM* differs from triangle *AKL* by triangle *KEF.*

For it is evident that *AD* is bisected by *EH* (II. 39, and II. 29), and that *EH* is a diameter conjugate to the diameter drawn through *H* parallel to *AD* (II. 38); and so *AG* is an ordinate to *EG* (Def. 6).

Since then *GE* is a diameter, and *AE* touches, and *AG* is an ordinate, and, with point *B* taken on the section, *BF* has been dropped to *EG* parallel to *AG*, and *BM* parallel to *EK,* therefore it is clear that triangle *BMF* differs from triangle *LHF* by triangle *HAE* (I. 45; I. 43).[*]

And so also triangle *BMF* differs from triangle *AKL* by triangle *KFE.*

[Porism]

And it has been proved at the same time that
$$\text{quadr. } BKEM = \text{trgl. } LKA.$$

PROPOSITION 12

With the same things being so, if on one of the sections two points are taken, and parallels are drawn from each of them, likewise the quadrilaterals produced by them will be equal.

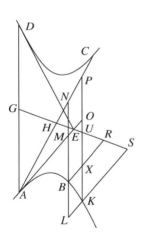

For let there be the same things as before, and let the points *B* and *K* be taken at random on section *AB,* and through them let *LBMN* and *KXOUP* be drawn parallel to *AD,* and *BXR* and *LKS* parallel to *AE.*

I say that
$$\text{quadr. } BP = \text{quadr. } KR.$$

For since it has been shown
$$\text{trgl. } AOP = \text{quadr. } KOES \text{ (III. 11 porism)}$$
and
$$\text{trgl. } AMN = \text{quadr. } BMER \text{ (III. 11 porism),}$$
therefore, as remainders, either

---

[*] That is, in the first case,
$$\text{trgl. } BMF = \text{trgl. } LHF + \text{trgl. } HAE \text{ (I. 45);}$$
in the second case, only the more general statement "differs" holds true (I. 43). It will be noticed these are different cases of I. 43 and I. 45 from those given in the text itself. (Tr.)

quadr. KR − quadr. BO = quadr. MP

or

quadr. KR + quadr. BO = quadr. MP.
And, with the common quadrilateral BO added or subtracted,
quadr. BP = quadr. XS.

PROPOSITION 13

If in conjugate opposite sections straight lines tangent to the adjacent sections meet, and diameters are drawn through the points of contact, then the triangles whose common vertex is the center of the opposite sections will be equal.

Let there be conjugate opposite sections on which there are the points A, B, C and D, and let AE and EB, meeting at E, touch the sections A and B, and let H be the center, and let AH and BH be joined and produced to D and C.

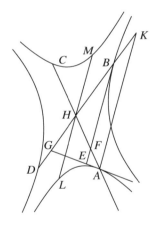

I say that
trgl. BFH = trgl. AGH.

For let AK and LHM be drawn through A and H parallel to BE. Since then BFE touches the section B, and DHB is a diameter through the point of contact, and LM is parallel to BE, LM is a diameter conjugate to diameter BD, the so-called second diameter (II. 20); and therefore AK has been drawn ordinatewise to BD. And AG touches; therefore
rect. KH, HG = sq. BH (I. 38).
Therefore
$KH : HB :: BH : GH$.

But
$KH : HB :: KA : BF :: AH : HF$;

therefore also
$AH : HF :: BH : GH$.
And the angles BHF and GHF are equal to two right angles; therefore
trgl. AGH = trgl. BHF.

PROPOSITION 14

With the same things supposed, if some point is taken on any one of the sections, and from it parallels to the tangents are drawn as far as the diameters, then the triangle produced at the center will differ from the triangle produced about the same angle by the triangle having the tangent as base, and center as vertex.

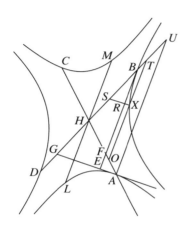

Let the other things be the same, and let some point X be taken on section B, and through it let XRS be drawn parallel to AG and XTO parallel to BE.

I say that triangle OHT differs from triangle XST by triangle HBF.

For let AU be drawn from A parallel to BF. Since then, because of the same things as before, LHM is a diameter of the section AL, and DHB is a second diameter and conjugate to it (II. 20), and AG is a tangent at A, and AU has been dropped parallel to LM, therefore (I. 40)

$AU : UG :: HU : UA$ comp. transverse side of figure on LM : upright. But

$$AU : UG :: XT : TS,$$

and

$$HU : UA :: HT : TO : HB : BF,$$

and (I. 60)

transverse side of figure on LM : upright ::
upright side of figure on BD : transverse.

Therefore

$XT : TS :: HB : BF$ comp. upright side of figure on BD : transverse

or

$XT : TS :: HT : TO$ comp. upright side of figure on BD : transverse.
And by things shown in the forty-first theorem of the first book (I. 41), triangle THO differs from triangle XTS by triangle BFH.

And so also by triangle AGH (III. 13).

PROPOSITION 15

If straight lines touching one of the conjugate opposite sections meet, and diameters are drawn through the points of contact, and some point is taken on any one of the conjugate sections, and from it parallels to the tangents are drawn as far as the diameters, then the triangle produced by them at the section is greater than the triangle produced at the center by the triangle having the tangent as base and the center of the opposite sections as vertex.[*]

Let there be conjugate opposite sections *AB*, *GS*, *T*, and *X*, whose center is *H*, and let *ADE* and *BDC* touch the section *AB*, and let the diameters *AHFW* and *BHT* be drawn through the points of contact *A* and *B*, and let some point *S* be taken on the section *GS*, and through it let *SFL* be drawn parallel to *BC* and *SU* parallel to *AE*.

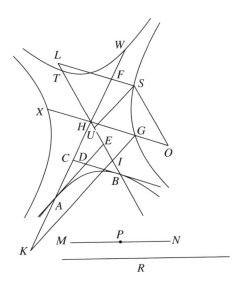

I say that
trgl. *SLU* = trgl. *HLF* + trgl. *HCB*.

For let *XHG* be drawn through *H* parallel to *BC*, and *KIG* through *G* parallel to *AE*, and *SO* parallel to *BT*; then it is evident that *XG* is a diameter conjugate to *BT* (II. 20), and that *SO* being parallel to *BT* has been dropped ordinatewise to *HGO* (Def. 6), and that *SLHO* is a parallelogram.

Since then *BC* touches, and *BH* is through the point of contact, and *AE* is another tangent, let it be contrived that
$$DB : BE :: MN : 2BC;$$

---

[*] This proposition comes as a climax to a long series, and shows that the conjugate opposite sections taken as a unit have the same property as the other conic sections. The conjugate opposite sections seem to be a sort of fifth section. (Tr.)

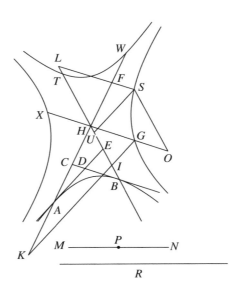

therefore *MN* is the so-called upright side of the figure on *BT* (I. 50). Let *MN* be bisected at *P*; therefore

$$DB : BE :: MP : BC.$$

Then let it be contrived that

$$XG : TB :: TB : R; \qquad\qquad (\alpha)$$

then *R* also will be the so-called upright side of the figure applied to *XG* (I. 16, 60).

Since then

$$DB : BE :: MP : BC, \qquad\qquad (\beta)$$

but

$$DB : BE :: \text{sq. } DB : \text{rect. } DB, BE,$$

and

$$MP : BC :: \text{rect. } MP, BH : \text{rect. } CB, BH,$$

therefore

$$\text{sq. } DB : \text{rect. } DB, BE :: \text{rect. } MP, BH : \text{rect. } CB, BH.$$

And

$$\text{rect. } MP, BH = \text{sq. } HG,$$

because

$$\text{sq. } XG = \text{rect. } TB, MN \text{ (I. 16)} \qquad\qquad (\gamma)$$

and

$$\text{rect. } MP, BH = \text{fourth rect. } TB, MN$$

and
$$\text{sq. } HG = \text{fourth sq. } XG;$$
therefore
$$\text{sq. } DB : \text{rect. } DB, BE :: \text{sq. } HG : \text{rect. } CB, BH.$$
Alternately
$$\text{sq. } DB : \text{sq. } HG :: \text{rect. } DB, BE : \text{rect. } CB, BH.$$
But
$$\text{sq. } DB : \text{sq. } HG :: \text{trgl. } DBE : \text{trgl. } GHI;$$
for they are similar; and
$$\text{rect. } DB, BE : \text{rect. } CB, BH :: \text{trgl. } DBE : \text{trgl. } CBH;$$
therefore
$$\text{trgl. } DBE : \text{trgl. } GHI :: \text{trgl. } DBE : \text{trgl. } CBH.$$
Therefore
$$\text{trgl. } GHI = \text{trgl. } CBH.$$

Again since
$$HB : BC :: HB : MP \text{ comp. } MP : BC,$$
but
$$HB : MP :: TB : MN : R : XG \text{ (above, } \alpha \text{ and } \gamma),$$
and
$$MP : BC :: DB : BE \text{ (above, } \beta)$$
therefore
$$HB : BC :: DB : BE \text{ comp. } R : XG.$$
And since BC is parallel to SL, and triangle HCB is similar to triangle HLF, and
$$HB : BC :: HL : LF,$$
therefore
$$HL : LF :: R : XG \text{ comp. } DB : BE$$
or
$$HL : LF :: R : XG \text{ comp. } HG : HI.$$

Since then GS is an hyperbola having XG as a diameter, and R as an upright side, and, from some point S, SO has been dropped ordinatewise, and figure HIG has been described on radius HG, and figure HLF has been described on the ordinate Sor its equal HL, and on HO the straight line between the center and the ordinate, or on SL its equal, the figure SLU has been described similar to the figure HIG described on the radius, and there are the compounded ratios as already given, therefore
$$\text{trgl. } SLU = \text{trgl. } HLF + \text{trgl. } HCB \text{ (I. 41).}$$

PROPOSITION 16

If two straight lines touching a section of a cone or circumference of a circle meet, and from some point of those on the section a straight line is drawn parallel to one tangent and cutting the section and the other tangent, then, as the squares on the tangents are to each other, so the area contained by the straight lines between the section and the tangent will be to the square cut off at the point of contact.

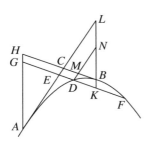

Let there be the section of a cone or circumference of a circle *AB*, and let the straight lines *AC* and *CB*, meeting at *C*, touch it, and let some point *D* be taken on the section *AB*, and through it let *EDF* be drawn parallel to *CB*.

I say that
sq. *BC* : sq. *AC* :: rect. *FE, ED* : sq. *EA*.

For let the diameters *AGH* and *KBL* be drawn through *A* and *B*, and *DMN* through *D* parallel to *AL*; it is at once evident, that
$$DK = KF \text{ (I. 46, 47)},$$
and
$$\text{trgl. } AEG = \text{quadr. } LD \text{ (III. 2)},$$
and
$$\text{trgl. } BLC = \text{trgl. } ACH \text{ (III. 1)}.$$
Since then
$$DK = KF$$
and *DE* is added,
rect. *FE, ED* + sq. *DK* = sq. *KE*.

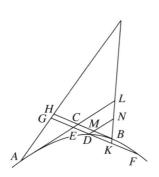

And since triangle *ELK* is similar to triangle *DNK*,
sq. *EK* : sq. *KD* :: trgl. *EKL* : trgl. *DNK*.

And alternately
whole sq. *EK* : whole trgl. *ELK* ::
part subtracted sq. *DK* : part subtracted trgl. *DNK*.
Therefore also
remainder rect. *FE, ED* : remainder quadr. *DL* :: sq. *EK* : trgl. *ELK*.
But

sq. *EK* : trgl. *ELK* :: sq. *CB* : trgl. *BLC*;

therefore also

rect. *FE, ED* : quadr. *LD* :: sq. *CB* : trgl. *LCB*.

But

quadr. *LD* = trgl. *AEG*,

and

trgl. *BLC* = trgl. *ACH*;

therefore also

rect. *FE, ED* : trgl. *AEG* :: sq. *CB* : trgl. *ACH*.

Alternately

rect. *FE, ED* : sq. *CB* :: trgl. *AEG* : trgl. *ACH*.

But

trgl. *AEG* : trgl. *ACH* :: sq. *EA* : sq. *AC*;

therefore also

rect. *FE, ED* : sq. *CB* :: sq. *EA* : sq. *AC*.

And alternately.

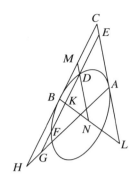

PROPOSITION 17

If two straight lines touching a section of a cone or circumference of a circle meet, and two points are taken at random on the section, and from them in the section are drawn parallel to the tangents straight lines cutting each other and the line of the section, then as the squares on the tangents are to each other, so will the rectangles contained by the straight lines taken similarly.

Let there be the section of a cone or circumference of a circle *AB*; and tangents to *AB*, *AC* and *CB*, meeting at *C*; and let points *D* and *E* be taken at random on the section, and through them at *EFIK* and *DFGH* be drawn parallel to *AC* and *CB*.

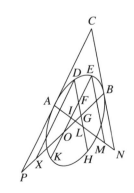

I say that

sq. *CA* : sq. *CB* :: rect. *KF, FE* : rect. *HF, FD*.

For let the diameters *ALMN* and *BOXP* be drawn through *A* and *B*, and let the tangents and parallels be produced to the diameters, and let *DX* and *EM* be

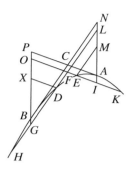

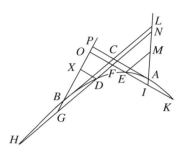

drawn from *D* and *E* parallel to the tangents; then it is evident that

$$KI = IE, \ HG = GD \ \text{(I. 46, 47)}.$$

Since then *KE* has been cut equally at *I* and unequally at *F*,

rect. *KF, FE* + sq. *FI* = sq. *EI* (Eucl. II. 5).

And since the triangles are similar because of the parallels,

whole sq. *EI* : whole trgl. *IME* ::

part subtracted sq. *IF* : part subtracted trgl. *FIL*.

Therefore also

remainder rect. *KF, FE* : remainder quadr. *FM* ::

whole sq. *EI* : whole trgl. *IME*.

But

sq. *EI* : trgl. *IME* :: sq. *CA* : trgl. *CAN*.

Therefore

rect. *KF, FE* : quadr. *FM* :: sq. *CA* : trgl. *CAN*.

But

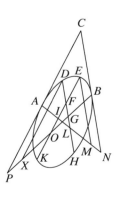

trgl. *CAN* = trgl. *CPB* (III. 1),

   and

      quadr. *FM* = quadr. *FX* (III. 3);

therefore

   rect. *KF, FE* : quadr. *FX* :: sq. *CA* : trgl. *CPB*.

Then likewise it could be shown that

   rect. *HF, FD* : quadr. *FX* :: sq. *CB* : trgl. *CPB*.

Since then

   rect. *KF, FE* : quadr. *FX* :: sq. *CA* : trgl. *CPB*,

and inversely

   quadr. *FX* : rect. *HF, FD* :: trgl. *CPB* : sq. *CB*,

therefore *ex aequali*

   sq. *CA* : sq. *CB* :: rect. *KF, FE* : rect. *HF, FD*.

PROPOSITION 18

If two straight lines touching opposite sections meet, and some point is taken on either one of the sections, and from it some straight line is drawn parallel to one of the tangents cutting the section and the other tangent, then as the squares on the tangents are to each other, so will the rectangle contained by the straight lines between the section and the tangent be to the square on the straight line cut off at the point of contact.

Let there be the opposite sections *AB* and *MN*, and tangents *ACL* and *BCH*, and through the points of contact the diameters *AM* and *BN*, and let some point *D* be taken at random on the section *MN*, and through it let *EDF* be drawn parallel to *BH*.

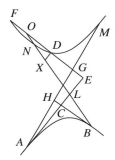

I say that
  sq. *BC* : sq. *CA* :: rect. *FE, ED* : sq. *AE*.

For let *DX* be drawn through *D* parallel to *AE*. Since then *AB* is an hyperbola and *BN* its diameter and *BH* a tangent and *DF* parallel to *BH*, therefore
$$FO = OD \text{ (I. 48)}.$$
And *ED* is added: therefore
$$\text{rect. } FE, ED + \text{sq. } DO = \text{sq. } EO \text{ (Eucl. II. 6)}.$$
And since *EL* is parallel to *DX*, triangle *EOL* is similar to triangle *DXO*. Therefore
$$\text{whole sq. } EO : \text{whole trgl. } EOL ::$$
$$\text{part subtracted sq. } DO : \text{part subtracted trgl. } DXO;$$
therefore also
$$\text{remainder rect. } DE, EF : \text{remainder quadr. } DL :: \text{sq. } EO : \text{trgl. } EOL.$$
But
$$\text{sq. } OE : \text{trgl. } EOL :: \text{sq. } BC : \text{trgl. } BCL;$$
therefore also
$$\text{rect. } FE, ED : \text{quadr. } DL :: \text{sq. } BC : \text{trgl. } BCL.$$
And
$$\text{quadr. } DL = \text{trgl. } AEG \text{ (III. 6, note)},$$
and

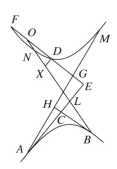

$$\text{trgl. } BCL = \text{trgl. } ACH \text{ (III. 1);}$$

therefore

$$\text{rect. } FE, ED : \text{trgl. } AEG :: \text{sq. } BC : \text{trgl. } ACH.$$

But also

$$\text{trgl. } AEG : \text{sq. } EA :: \text{trgl. } ACH : \text{sq. } AC;$$

therefore *ex aequali*

$$\text{sq. } BC : \text{sq. } AC :: \text{rect. } FE, ED : \text{sq. } EA.^{*}$$

PROPOSITION 19

If two straight lines touching opposite sections meet, and parallels to the tangents are drawn cutting each other and the section, then, as the squares on the tangents are to each other, so will the rectangle contained by the straight lines between the section and the point of meeting of the straight

---

* Eutocius gives an alternative proof of Apollonius', demonstrating another and important case: "For let there be the opposite sections *A* and *B*, and tangents to them *AC* and *CB* meeting at *C*, and let *D* be taken on section *B*, and through it let *EDF* be drawn parallel to *AC*. I say that

$$\text{sq. } AC : \text{sq. } CB :: \text{rect. } EF, FD : \text{sq. } FB.$$

"For let *AHG* be drawn as a diameter through *A*, and through *B* and *G*, *GK* and *BL* parallel to *EF*. Since then *BH* touches the hyperbola at *B*, and *BL* has been drawn ordinatewise,

$$AL : LG :: AH : HG \text{ (I. 36).}$$

But

$$AL : LG :: CB : BK,$$

and

$$AH : HG : AC : KG;$$

therefore also

$$CB : BK :: AC : KG.$$

And alternately

$$AC : CB :: KG : KB,$$

and

$$\text{sq. } AC : \text{sq. } CB :: \text{sq. } GK : \text{sq. } KB.$$

But it was shown

$$\text{sq. } GK : \text{sq. } KB :: \text{rect. } EF, FD : \text{sq. } FB;$$

therefore also

$$\text{sq. } AC : \text{sq. } CB :: \text{rect. } EF, FD : \text{sq. } FB."$$

(Tr.)

lines be to the rectangle contained by the straight lines taken similarly.

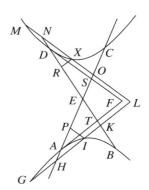

Let there be opposite sections whose diameters are *AC* and *BD* and center in *E,* and let the tangents *AF* and *FD* meet at *F,* and let *GHIKL* and *MNXOL* be drawn from any points parallel to *AF* and *FD.*

I say that

   sq. *AF* : sq. *FD* :: rect. *GL, LI* : rect. *ML, LX.*

Let *IP* and *XR* be drawn through *X* and *I* parallel to *AF* and *FD.*
And since

     sq. *AF* : trgl. *AFS* :: sq. *HL* : trgl. *HLO* :: sq. *HI* : trgl. *HIP,*
therefore

   remainder rect. *GL, LI* : remainder quadr. *IPOL* :: sq. *AF* : trgl. *AFS.*
But

        trgl. *AFS* = trgl. *DTF* (III. 4),
and

       quadr. *IPOL* = quadr. *KRXL* (III. 7);
therefore also

     sq. *AF* : trgl. *DTF* :: rect. *GL, LI* : quadr. *KRXL.*
But

    trgl. *DTF* : sq. *FD* :: quadr. *KRXL* : rect. *ML, LX* (likewise);
and therefore *ex aequali*

     sq. *AF* : sq. *FD* :: rect. *GL, LI* : rect. *ML, LX.*

PROPOSITION 20

If two straight lines touching opposite sections meet, and through the point of meeting some straight line is drawn parallel to the straight line joining the points of contact and meeting each of the sections, and some other straight line is drawn parallel to the same straight line and cutting the sections and the tangents, then, as the rectangle contained by the straight lines drawn from the point of meeting to cut the sections is to the square on the tangent, so is the rectangle contained by the straight lines between the sections and the tangent to the square on the straight line cut off at the point of contact.

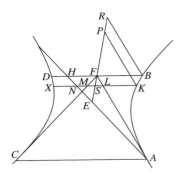

Let there be the opposite sections *AB* and *CD* whose center is *E* and tangents *AF* and *CF,* and let *AC* be joined, and let *EF* and *AE* be joined and produced, and let *BFH* be drawn through *F* parallel to *AC,* and let the point *K* be taken at random, and through it let *KLSMNX* be drawn parallel to *AC.*

I say that

$$\text{rect. } BF, FD : \text{sq. } FA ::$$
$$\text{rect. } KL, LX : \text{sq. } AL.$$

For let *KP* and *BR* be drawn from *K* and *B* parallel to *AF.* Since then

$$\text{sq. } BF : \text{trgl. } BFR :: \text{sq. } KS : \text{trgl. } KSP :: \text{sq. } LS : \text{trgl. } LSF,$$

and

$$\text{sq. } KS : \text{trgl. } KSP ::$$

remainder rect. *KL, LX* (Eucl. II. 5) : remainder quadr. *KLFP* (Eucl. V. 19)

and

$$\text{sq. } BF = \text{rect. } BF, FD \text{ (II. 39, 38)},$$

and

$$\text{trgl. } BRF = \text{trgl. } AFH \text{ (III. 11 and special case)},$$

and

$$\text{quadr. } KLFP = \text{trgl. } ALN \text{ (III. 5)},$$

therefore

$$\text{rect. } BF, FD : \text{trgl. } AFH :: \text{rect. } KL, LX : \text{trgl. } ALN.$$

And

$$\text{trgl. } AFH : \text{sq. } AF :: \text{trgl. } ALN : \text{sq. } AL;$$

then

$$\text{rect. } BF, FD : \text{sq. } FA :: \text{rect. } KL, LX : \text{sq. } AL.$$

PROPOSITION 21

With the same things supposed, if two points are taken on the section, and through them straight lines are drawn, the one parallel to the tangent, the other parallel to the straight line joining the points of contact, and cutting each other and the sections, then, as the rectangle contained by the straight lines drawn from the point of meeting to cut the sections is to the square on the tangent, so will the rectangle contained by the straight lines between the sections and the point of meeting be to the rectangle contained by the straight lines between the section and the point of meeting.

For let there be the same things as before, and let points *G* and *K* be taken, and through them let *NXGOPR* and *KST* be drawn parallel to *AF,* and *GLM* and *KOWIYZQ* parallel to *AC.*

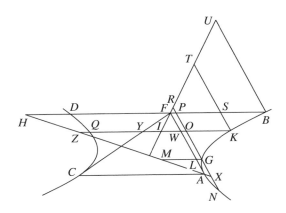

I say that

rect. *BF, FD* : sq. *FA* :: rect. *KO, OQ* : rect. *NO, OG.*

For since

sq. *AF* : trgl. *AFH* :: sq. *AL* : trgl. *ALM* :: sq. *XO* : trgl. *XOZ*

and

sq. *XO* : trgl. *XOZ* :: sq. *XG* : trgl. *XGM,*

therefore

whole sq. *XO* : whole trgl. *XOZ* ::
part subtracted sq. *XG* : part subtracted trgl. *XGM,*

therefore also

remainder rect. *NO, OG* : remainder quadr. *GOZM* :: sq. *AF* : trgl. *AFH.*

But

trgl. *AFH* = trgl. *BUF* (III. 11 porism, special case),

and

quadr. *GOZM* = quadr. *KORT* (III. 12);

therefore

sq. *AF* : trgl. *BFU* :: rect. *NO, OG* : quadr. *KORT.*

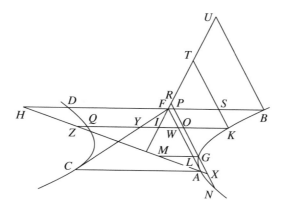

But it was shown (in the course of III. 20)

$$\text{trgl. } BUF : \text{sq. } BF \text{ or rect. } BF, FD \text{ (II. 39, 38) ::}$$
$$\text{quadr. } KORT : \text{rect. } KO, OQ;$$

therefore *ex aequali*

$$\text{sq. } AF : \text{rect. } BF, FD :: \text{rect. } NO, OG : \text{rect. } KO, OQ.$$

And inversely

$$\text{rect. } BF, FD : \text{sq. } FA :: \text{rect. } KO, OQ : \text{rect. } NO, OG.$$

PROPOSITION 22

If two parallel straight lines touch opposite sections, and any straight lines are drawn cutting each other, and the sections, one parallel to the tangent, the other parallel to the straight line joining the points of contact, then as the transverse side of the figure on the straight line joining the points of contact is to the upright, so the rectangle contained by the straight lines between the sections and the point of meeting will be to the rectangle contained by the straight lines between the section and the point of meeting.

Let there be the opposite sections *A* and *B*, and let *AC* and *BD* be parallel and tangent to them, and let *AB* be joined. Then let *EXG* be drawn across parallel to *AB* and *KELM* parallel to *AC*.

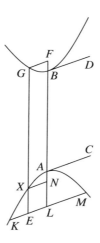

I say that

AB : upright side of the figure :: rect. *GE, EX* : rect. *KE, EM.*

Let *XN* and *GF* be drawn through *G* and *X* parallel to *AC.*

For since *AC* and *BD* are parallels tangent to the sections, *AB* is a diameter (II. 31), and *KL, XN,* and *GF* are ordinates to it (I. 32); then (I. 21)

AB : upright side ::
rect. *BL, LA* : sq. *LK* :: rect. *BN, NA* : sq. *NX* or sq. *LE.*

Therefore

whole rect. *BL, LA* : whole sq. *LK* ::
part subtracted rect. *BN, NA* : part subtracted sq. *LE,*

or

rect. *BL, LA* : sq. *LK* :: rect. *FA, AN* : sq. *LE,*

for

$$NA = BF \text{ (I. 21)};$$

therefore also

remainder rect. *FL, LN* : remainder rect. *KE, EM* :: *AB* : upright.

But

$$\text{rect. } FL, LN = \text{rect. } GE, EX;$$

therefore

AB the transverse side of figure : upright ::
rect. *GE, EX* : rect. *KE, EM.*

PROPOSITION 23

If in conjugate opposite sections two straight lines touching contrary sec-
tions meet in any one section at random, and any straight lines are drawn
parallel to the tangents and cutting each other and the other opposite sec-
tions, then, as the squares on the tangents are to each other, so the rectangle
contained by the straight lines between the sections and the point of meeting
will be to the rectangle contained by the straight lines similarly taken.

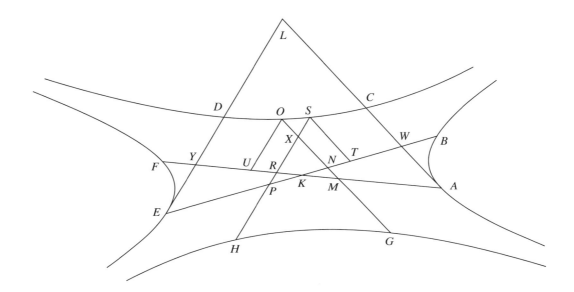

Let there be the conjugate opposite sections *AB, CD, EF,* and *GH,* and their
center *K,* and let *AWCL* and *EYDL,* tangents to the sections *AB* and *EF* meet
at *L,* and let *AK* and *EK* be joined and produced to *B* and *F,* and let *GMNXO*
be drawn from *G* parallel to *AL,* and *HPRXS* from *H* parallel to *EL.*

I say that

$$\text{sq. } EL : \text{sq. } LA :: \text{rect. } HX, XS : \text{rect. } GX, XO.$$

For let *ST* be drawn through *S* parallel to *AL,* and *OU* from *O* parallel to *EL.*
Since then *BE* is a diameter of the conjugate opposite sections *AB, CD, EF*
and *GH,* and *EL* touches the section, and *HS* has been drawn parallel to it,

$$HP = PS \text{ (II. 20; Def. 5),}$$

and for the same reasons

$$GM = MO.$$

And since

$$\text{sq. } EL : \text{trgl. } EWL :: \text{sq. } PS : \text{trgl. } PTS :: \text{sq. } PX : \text{trgl. } PNX,$$

also

remainder rect. HX, XS : remainder quadr. $TNXS$ :: sq. EL : trgl. WLE.

But

$$\text{trgl. } EWL = \text{trgl. } ALY \text{ (III. 4),}$$

and

$$\text{quadr. } TNXS = \text{quadr. } XRUO \text{ (III. 15);}^{*}$$

therefore

$$\text{sq. } EL : \text{trgl. } ALY :: \text{rect. } HX, XS : \text{quadr. } XRUO.$$

But

$$\text{trgl. } AYL : \text{sq. } AL :: \text{quadr. } XRUO : \text{rect. } GX, XO \text{ (same way);}$$

therefore *ex aequali*

$$\text{sq. } EL : \text{sq. } AL :: \text{rect. } HX, XS : \text{rect. } GX, XO.$$

PROPOSITION 24

If in conjugate opposite sections two straight lines are drawn from the center through to the sections, and one of them is taken as the transverse diameter and the other as the upright diameter, and any straight lines are drawn parallel to the two diameters and meeting each other and the sections, and the point of meeting of the straight lines is the place between the four sections, then the rectangle contained by the segments of the parallel to the transverse diameter together with the rectangle to which the rectangle contained by the segments of the parallel to the upright diameter has the ratio which the square on the upright diameter has to the square on the transverse, will be equal to twice the square on the half of the transverse.

---

* This is the case of III. 15 where the tangents are one to each of the opposite sections. Compare with the two cases of III. 12 and III. 18.
For

$$\text{trgl. } TSP - \text{trgl. } KPR = \text{trgl. } AWK \text{ (III. 15),}$$

and

$$\text{trgl. } MOU - \text{trgl. } MNK = \text{trgl. } AWK \text{ (III. 15).}$$

(Tr.)

Let there be the conjugate opposite sections *A*, *B*, *C* and *D* whose center is *E*, and from *E* let the transverse diameter *AEC* and the upright diameter *DEB* be drawn through, and let *FGHIKL* and *MNXOPR* be drawn parallel to *AC* and *DB* and meeting each other at *X*; and first let *X* be within the angle *SEW* or the angle *UET*.

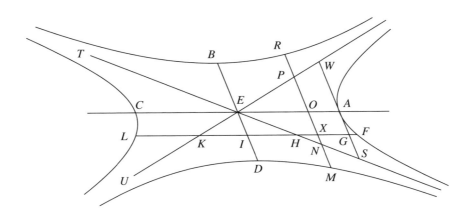

I say that the rectangle *FX*, *XL* together with the rectangle to which the rectangle *RX*, *XM* has the ratio which the square on *DB* has to the square on *AC*, is equal to twice the square on *AE*.

For let the asymptotes of the sections *SET* and *UEW* be drawn, and through *A*, *SGAW* tangent to the section.

Since then

rect. *SA*, *AW* = sq. *DE* (I. 60; II. 1),

therefore

rect. *SA*, *AW* : sq. *EA* :: sq. *DE* : sq. *EA*.

And

rect. *SA*, *AW* : sq. *AE* :: *SA* : *AE* comp. *WA* : *AE*.

But

SA : *AE* :: *NX* : *XH*

and

WA : *AE* :: *PX* : *XK*;

therefore
$$\text{sq. } DE : \text{sq. } AE :: NX : XH \text{ comp. } PX : XK.$$

But
$$\text{rect. } PX,\, XN : \text{rect. } KX,\, XH :: NX : XH \text{ comp. } PX : XK,$$

therefore
$$\text{sq. } DE : \text{sq. } AE :: \text{rect. } PX,\, XN : \text{rect. } KX,\, XH.$$

Therefore also
$$\text{sq. } DE : \text{sq. } AE :: \text{sq. } DE + \text{rect. } PX,\, XN : \text{sq. } AE + \text{rect. } KX,\, XH.$$

And
$$\text{sq. } DE = \text{rect. } PM,\, MN \text{ (II. 11)} = \text{rect. } RN,\, NM \text{ (II. 16)},$$

and
$$\text{sq. } AE = \text{rect. } KF,\, FH \text{ (II. 11)} = \text{rect. } LH,\, HF \text{ (II. 16)};$$

therefore
$$\text{sq. } DE : \text{sq. } AE ::$$
$$\text{rect. } PX,\, XN + \text{rect. } RN,\, NM : \text{rect. } KX,\, XH + \text{rect. } LH,\, HF.$$

And
$$\text{rect. } PX,\, XN + \text{rect. } RN,\, NM = \text{rect. } RX,\, XM;^*$$

therefore
$$\text{sq. } DE : \text{sq. } AE :: \text{rect. } RX,\, XM : \text{rect. } KX,\, XH + \text{rect. } KF,\, FH.$$

---

* For
$$RP = NM \text{ (II. 8)},$$

and
$$RO = OM \text{ (II. 3)},$$

therefore
$$PO = ON.$$

But
$$\text{rect. } PX,\, XN + \text{sq. } OX = \text{sq. } ON \text{ (Eucl. II. 5)},$$

and, for the same reasons,
$$\text{rect. } RN,\, NM + \text{sq. } ON = \text{sq. } OM,$$

and
$$\text{rect. } RX,\, XM + \text{sq. } OX = \text{sq. } OM.$$

Hence
$$\text{rect. } RN,\, NM + \text{sq. } ON = \text{rect. } RX,\, XM + \text{sq. } OX,$$

and, adding equals to equals,
$$\text{rect. } RN,\, NM + \text{sq. } ON + \text{rect. } PX,\, XN + \text{sq. } OX =$$
$$\text{rect. } RX,\, XM + \text{sq. } OX + \text{sq. } ON.$$

Subtracting the common squares,
$$\text{rect } RN,\, NM + \text{rect. } PX,\, XN = \text{rect. } RX,\, XM.$$

(Tr.)

Then it must be shown that

rect. FX, XL + rect. KX, XH + rect. $KF, FH = 2$ sq. AE.

Let the common square AE, that is rectangle KF, FH, be subtracted; therefore it remains to be shown that

rect. FX, XL + rect. KX, XH = sq. AE.

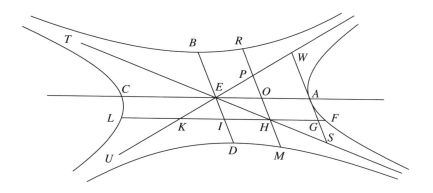

And this is so; for

rect. FX, XL + rect. KX, XH = rect. $LH, HF,$[*]

rect. FX, XL + rect. KX, XH = rect. KF, FH (II. 16),

$\qquad$ = sq. AE (II. 11).

Then let the straight lines FL and MR meet on one of the asymptotes at H. Then

rect. FH, HL = sq. AE,

and

rect. MH, HR = sq. DE (II. 11, 16);

therefore

sq. DE : sq. AE :: rect. MH, HR : rect. FH, HL.

---

[*] By the same manner of proof as in the note above, but using also Euclid II. 6, because of the different position of the point X. (Tr.)

And so we want twice rectangle *FH, HL* to equal twice the square on *AE*.
And it does.

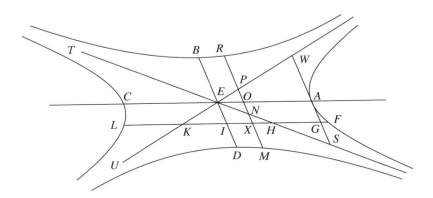

And let the point *X* be within the angle *SEK* or angle *WET*. Then likewise by
the composition of ratios

sq. *DE* : sq. *AE* :: rect. *PX, XN* : rect. *KX, XH*.

And

sq. *DE* = rect. *PM, MN* = rect. *RN, NM*,

and

sq. *AL* = rect. *FH, HL*;

therefore

rect. *RN, NM* : rect. *FH, HL* ::

part subtracted rect. *PX, XN* : part subtracted rect. *KX, XH*.

Therefore also

rect. *RN, NM* : rect. *FH, HL* ::

remainder rect. *RX, XM* : remainder (sq. *AE* − rect. *KX, XH*)

Therefore it must be shown that

rect. *FX, XL* + (sq. *AE* − rect. *KX, XH*) = 2 sq. *AE*.

Let the common square on *AE*, that is rectangle *FH, HL*, be subtracted;
therefore it remains to be shown that

rect. *KX, XH* + (sq. *AE* − rect. *KX, XH*) = sq. *AE*.

And this is so; for

rect. *KX, XH* + sq. *AE* − rect. *KX, XH* = sq. *AE*.

PROPOSITION 25

With the same things supposed, let the point of meeting of the parallels to *AC* and *BD* be within one of the sections *D* and *B*, as set out below, at *X*.

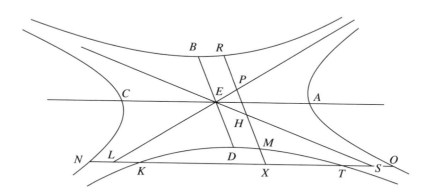

I say that the rectangle contained by the segments of the parallel to the transverse, that is rectangle *OX, XN*, will be greater than the rectangle to which the rectangle contained by the segments of the parallel to the upright diameter, that is rectangle *RX, XM*, has the ratio which the square on the upright diameter has to the square on the transverse by twice the square on the half of the transverse.

For, for the same reasons,

$$\text{sq. } DE : \text{sq. } AE :: \text{rect. } PX, XH : \text{rect. } SX, XL$$

and

$$\text{sq. } DE = \text{rect. } PM, MH,$$

and

$$\text{sq. } AE = \text{rect. } LO, OS \text{ (II. 11)};$$

therefore also

$$\text{sq. } DE : \text{sq. } AE :: \text{rect. } PM, MH : \text{rect. } LO, OS.$$

And since (II. 22)

$$\text{whole rect. } PX, XH : \text{whole rect. } LX, XS ::$$

part subtracted rect. *PM, MH* : part subtracted rect. *LO, OS*, or rect. *ST, TL* therefore also

remainder rect. *RX, XM* : remainder rect. *TX, XK* [*] :: sq. *DE* : sq. *AE*.
Therefore it must be shown that

$$\text{rect. } OX,\, XN = \text{rect. } TX,\, XK + 2 \text{ sq. } AE.$$

Let the common rectangle *TX, XK* be subtracted; therefore it must be shown that

$$\text{rect. } OT,\, TN \text{ (first note to III. 24)} = 2 \text{ sq. } AE.$$

And it is (II. 23).

PROPOSITION 26

And if the point of meeting of the parallels at *X* is within one of the sections *A* and *C*, as set out below, then the rectangle contained by the segments of the parallel to the transverse, that is rectangle *LX, XF*, will be less than the rectangle to which the rectangle contained by the segments of the other parallel, that is rectangle *RX, XG*, has the ratio which the square on the upright diameter has to the square on the transverse by twice the square on half of the transverse.

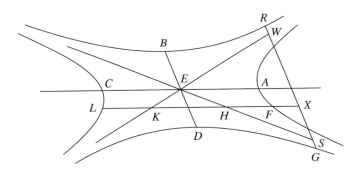

For, since for the same reasons as before

$$\text{sq. } DE : \text{sq. } AE ::$$
$$\text{rect. } WX,\, XS : \text{rect. } KX,\, XH,$$

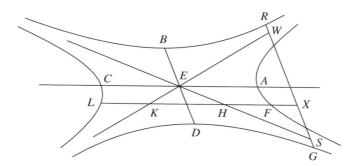

therefore also

whole rect. RX, XG^* : whole rect. KX, XH + sq. AE ::
sq. upright diameter : sq. transverse.

Therefore it must be shown that

rect. LX, XF + 2 sq. AE = rect. KX, XH + sq. AE.

Let the common square on AE be subtracted; therefore it remains to be shown that

rect. LX, XF + sq. AE = rect. KX, XH

or

rect. LX, XF + rect. LH, HF = rect. KX, XH. (II. 16, 11).

And it is, for

rect. LH, HF + rect. LX, XF = rect. KX, XH.[**]

---

[*] For by II. 11

rect. WG, GS = sq. DE;

and

$RW = GS$ (II. 16).

Therefore by the first note to III. 24, and II. 16,

rect. WX, XS + sq. DE = rect. WX, XS + rect. WG, GS = rect. RX, XG.

(Tr.)

[**] This is another case of the first note to III. 24. (Tr.)

PROPOSITION 27

If the conjugate diameters of an ellipse or circumference of a circle are drawn, and one of them is called the upright diameter and the other the transverse, and two straight lines, meeting each other and the line of the section, are drawn parallel to them, then the squares on the straight lines cut off on the straight line drawn parallel to the transverse between the point of meeting of the straight lines and the line of the section plus the figures described on the straight lines cut off on the straight line drawn parallel to the upright diameter between the point of meeting of the straight lines and the line of the section, figures similar and similarly situated to the figure on the upright diameter, will be equal to the square on the transverse diameter.

For let there be the ellipse or circumference of a circle *ABCD*, whose center is *E*, and let two of its conjugate diameters be drawn, the upright *AEC* and the transverse *BED*, and let *NGFH* and *KFLM* be drawn parallel to *AC* and *BD*.

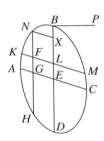 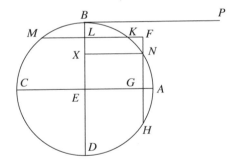

I say that the squares on *NF* and *FH* plus the figures described on *KF* and *FM*, similar and similarly situated to the figure on *AC* will be equal to the square on *BD*.

Let *NX* be drawn from *N* parallel to *AE*; therefore it has been dropped ordinatewise to *BD*. And let *BP* be the upright side. Now since

$$BP : AC :: AC : BD \text{ (I. 15)},$$

therefore also

$$BP : BD :: \text{sq. } AC : \text{sq. } BD.$$

And

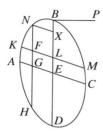

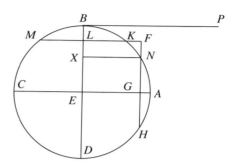

$$\text{sq. } BD = \text{figure on } AC;$$

therefore

$$BP : BD :: \text{sq. } AC : \text{figure on } AC.$$

And

$$\text{sq. } AC : \text{figure on } AC ::$$

sq. NX : figure on NX similar to the figure on AC (Eucl. VI. 22);
therefore also

$$BP : BD :: \text{sq. } NX : \text{figure on } NX \text{ similar to the figure on } AC.$$

And also

$$BP : BD :: \text{sq. } NX : \text{rect. } BX, XD \text{ (I. 21)};$$

therefore

figure on NX or FL similar to the figure on AC = rect. BX, XD.
Then likewise we could show that

figure on KL similar to the figure on AC = rect. BL, LD.
And since the straight line NH has been cut equally at G and unequally at F,

$$\text{sq. } HF + \text{sq. } FN = 2[\text{sq. } HG + \text{sq. } GF] = 2[\text{sq. } NG + \text{sq. } GF]$$
$$\text{(Eucl. VI. 9).}$$

Then for the same reasons also

$$\text{sq. } MF + \text{sq. } FK = 2[\text{sq. } KL + \text{sq. } LF],$$

and the figure on MF and FK similar to the figure on AC are double the
similar figures on KL and LF.

And

figure on KL + figure on FL = rect. BX, XD + rect. BL, LD (above),
and

$$\text{sq. } NG + \text{sq. } GF = \text{sq. } XE + \text{sq. } EL;$$

therefore
   sq. NF + sq. FH + figures on KF and FM similar to the figure on AC =
        2[rect. BX, XD + rect. BL, LD + sq. XE + sq. EL].
And since the straight line BD has been cut equally at E and unequally at X,
        rect. BX, XD + sq. XE = sq. BE (Eucl. II. 5).
And likewise also
        rect. BL, LD + sq. LE = sq. BE;
and so
       rect. BX, XD + rect. BL, LD + sq. XE + sq. LE = 2 sq. BE.
Therefore the squares on NF and FH together with figures on KF and FM
similar to the figure on CA are double the square on BE. But also
        sq. BD = 4 sq. BE;
therefore the squares on NF and FH plus the figures on KF and FM similar
to the figure on AC are equal to the square on BD.

PROPOSITION 28

If in conjugate opposite sections conjugate diameters are drawn, and one of them is called the upright, and the other the transverse, and two straight lines are drawn parallel to them and meeting each other and the sections, then the squares on the straight lines cut off on the straight line drawn parallel to the upright between the point of meeting of the straight lines and the sections have to the squares on the straight lines cut off on the straight line drawn parallel to the transverse between the point of meeting of the straight lines and the sections the ratio which the square on the upright diameter has to the square on the transverse diameter.

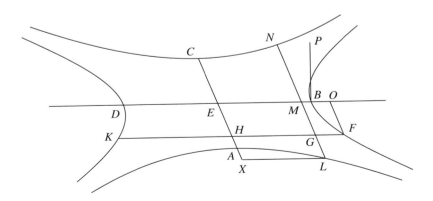

Let there be the conjugate opposite sections *A, B, C,* and *D* and let *AEC* be the upright diameter and *BED* the transverse, and let *FGHK* and *LGMN* be drawn parallel to them and cutting each other and the sections.

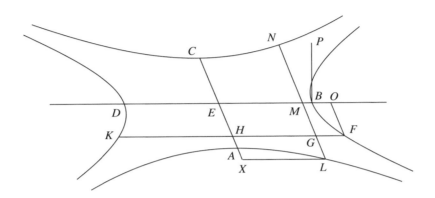

I say that

$$\text{sq. } LG + \text{sq. } GN : \text{sq. } FG + \text{sq. } GK :: \text{sq. } AC : \text{sq. } BD.$$

For let *LX* and *FO* be drawn ordinatewise from *F* and *L*; therefore they are parallel to *AC* and *BD*. And from *B* let the upright side for *BD, BP,* be drawn; then it is evident that

$$PB : BD :: \text{sq. } AC : \text{sq. } BD \text{ (I. 15) } :: \text{sq. } AE : \text{sq. } EB ::$$
$$\text{sq. } FO : \text{rect. } BO, OD \text{ (I. 21) } :: \text{rect. } CX, XA : \text{sq. } LX \text{ (I. 60, 21).}$$

Therefore as one of the antecedents is to one of the consequents, so are all of the antecedents to all of the consequents (Eucl. V. 12); therefore

$$\text{sq. } AC : \text{sq. } BD ::$$
$$\text{rect. } CX, XA + \text{sq. } AE + \text{sq. } OF : \text{rect. } DO, OB + \text{sq. } BE + \text{sq. } LX,$$

or

$$\text{sq. } AC : \text{sq. } BD :: \text{rect. } CX, XA + \text{sq. } AE + \text{sq. } EH :$$
$$\text{rect. } DO, OB + \text{sq. } BE + \text{sq. } ME.$$

But

$$\text{rect. } CX, XA + \text{sq. } AE = \text{sq. } XE,$$

and

$$\text{rect. } DO, OB + \text{sq. } BE = \text{sq. } OE \text{ (Eucl. II. 6);}$$

therefore

$$\text{sq. } AC : \text{sq. } BD :: \text{sq. } XE + \text{sq. } EH : \text{sq. } OE + \text{sq. } EM ::$$
$$\text{sq. } LM + \text{sq. } MG : \text{sq. } FH + \text{sq. } HG.$$

And, as has been shown,
$$\text{sq. } NG + \text{sq. } GL = 2[\text{sq. } LM + \text{sq. } MG],$$
and
$$\text{sq. } FG + \text{sq. } GK = 2[\text{sq. } FH + \text{sq. } HG] \text{ (Eucl. II. 9);}$$
therefore also
$$\text{sq. } AC : \text{sq. } BD :: \text{sq. } NG + \text{sq. } GL : \text{sq. } FG + \text{sq. } GK.$$

PROPOSITION 29

With the same things supposed, if the parallel to the upright diameter cuts the asymptotes, then the squares on the straight lines cut off on the straight line drawn parallel to the upright between the point of meeting of the straight lines and the asymptotes plus the half of the square on the upright diameter has to the squares on the straight lines cut off on the straight line drawn parallel to the transverse between the point of meeting of the straight lines and the sections the ratio which the square on the upright diameter has to the square on the transverse.

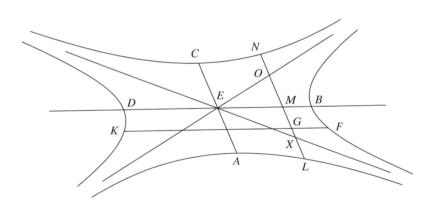

For let there be the same things as before, and let *NL* cut the asymptotes at *X* and *O*.

It is to be shown that
$$\text{sq. } XG + \text{sq. } GO + \text{half sq. } AC : \text{sq. } FG + \text{sq. } GK :: \text{sq. } AC : \text{sq. } BD,$$
or
$$\text{sq. } XG + \text{sq. } GO + 2 \text{ sq. } AE : \text{sq. } FG + \text{sq. } GK :: \text{sq. } AC : \text{sq. } BD.$$

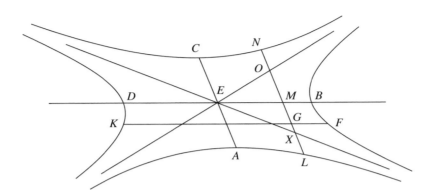

For since
$$LX = ON \text{ (II. 16)},$$
$$\text{sq. } XG + \text{sq. } GO + 2 \text{ rect. } NX, XL = \text{sq. } LG + \text{sq. } GN;^*$$
therefore
$$\text{sq. } XG + \text{sq. } GO + 2 \text{ sq. } AE = \text{sq. } LG + \text{sq. } GN.$$
And
$$\text{sq. } LG + \text{sq. } GN : \text{sq. } FG + \text{sq. } GK :: \text{sq. } AC : \text{sq. } BD \text{ (III. 28)};$$
therefore also
$$\text{sq. } XG + \text{sq. } GO + 2 \text{ sq. } AE : \text{sq. } FG + \text{sq. } GK :: \text{sq. } AC : \text{sq. } BD.$$

---

[*] For
$$OM = MX.$$
Therefore, as in a lemma of Pappus, since
$$2 \text{ rect. } NX, XL + 2 \text{ sq. } MX = 2 \text{ sq. } ML \text{ (Eucl. II. 5)},$$
adding the common square on GM,
$$2 \text{ rect. } NX, XL + 2 \text{ sq. } MX + 2 \text{ sq. } GM = 2 \text{ sq. } ML + 2 \text{ sq. } GM.$$
And
$$2 \text{ sq. } ML + 2 \text{ sq. } GM = \text{sq. } NG + \text{sq. } LG$$
and
$$2 \text{ sq. } MX + 2 \text{ sq. } GM = \text{sq. } OG + \text{sq. } GX \text{ (Eucl. II.9)}.$$
Therefore as above. (Tr.)

PROPOSITION 30

If two straight lines touching an hyperbola meet, and through the points of contact a straight line is produced, and through the point of meeting a straight line is drawn parallel to some one of the asymptotes and cutting both the section and the straight line joining the points of contact, then the straight line between the point of meeting and the straight line joining the points of contact will be bisected by the section.*

Let there be the hyperbola *ABC,* and let *AD* and *DC* be tangents and *EF* and *FG* asymptotes, and let *AC* be joined, and through *D* parallel to *FE* let *DKL* be drawn.

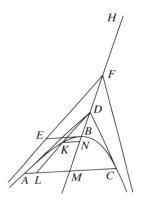

I say that

$$DK = KL.$$

For let *FDBM* be joined and produced both ways, and let *FH* be made equal to *BF,* and through the points *B* and *K* let *BE* and *KN* be drawn parallel to *AC*; therefore they have been dropped ordinatewise (II. 30, 5, 7). And since triangle *BEF* is similar to triangle *DNK,*

* The propositions from 30 to 34 inclusive are one special case, and propositions 35 and 36 are another special case of proposition 37. The first group takes the line drawn through the intersection of the tangents as parallel to an asymptote. The second group takes one of the tangents as an asymptote. Proposition 34, lying between, is special in both these ways.

In proposition 37 we have the line *CF* divided by the section at *D* and *F,* and at *E* by the straight line joining the points of contact, in such a way that

CF : CD :: FE : ED.

This is the same form of the harmonic proportion as we found in I. 34, and *DF* is the harmonic mean between *CF* and *FE.*

If we argue by analogy from this proportion, treating infinity as a definite magnitude, and two such infinities as would occur here as equal and subject to the general laws of magnitudes, we can immediately deduce the special cases of propositions 30 to 36. Thus in the case of the first group, *CF* and *FE* both become infinite, therefore *CD* is equal to *ED.* (Tr.)

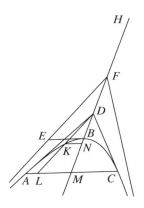

therefore

$$\text{sq. } DN : \text{sq. } NK :: \text{sq. } BF : \text{sq. } BE. \qquad (\alpha)$$

And

$$\text{sq. } BF : \text{sq. } BE :: HB : \text{upright (II. 1);}$$

therefore also

$$\text{sq. } DN : \text{sq. } NK :: HB : \text{upright.}$$

But

$$HB : \text{upright} :: \text{rect. } HN, NB : \text{sq. } NK \text{ (I. 21);}$$

therefore also

$$\text{sq. } DN : \text{sq. } NK :: \text{rect. } HN, NB : \text{sq. } NK. \qquad (\beta)$$

Therefore

$$\text{rect. } HN, NB = \text{sq. } DN.$$

And also

$$\text{rect. } MF, FD = \text{sq. } FB \text{ (I. 37),}$$

because AD touches and AM has been dropped ordinatewise; and so also

$$\text{rect. } HN, NB + \text{sq. } FB = \text{rect. } MF, FD + \text{sq. } DN.$$

But

$$\text{rect. } HN, NB + \text{sq. } FB = \text{sq. } FN \text{ (Eucl. II. 6);}$$

and therefore

$$\text{rect. } MF, FD + \text{sq. } DN = \text{sq. } FN.$$

Therefore DM has been bisected at N with DF added (Eucl. II. 6). And KN and LM are parallel; therefore

$$DK = KL.$$

PROPOSITION 31

If two straight lines touching opposite sections meet, and a straight line is produced through the points of contact, and through the point of meeting a straight line is drawn parallel to the asymptote and cutting both the section and the straight line joining the points of contact, then the straight line between the point of meeting and the straight line joining the points of contact will be bisected by the section.

Let there be the opposite sections A and B, and tangents AC and CB, and let AB be joined and produced, and let FE be an asymptote and through C let CGH be drawn parallel to FE.

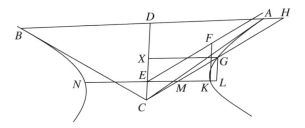

I say that

$$CG = GH.$$

Let CE be joined and produced to D, and through E and G let $NEKM$ and GX be drawn parallel to AB, and through G and K let KF and GL be drawn parallel *to CD*.

Since triangle KFE is similar to triangle MLG,

sq. KE : sq. KF :: sq. ML : sq. LG.

And it has been shown

sq. KE : sq. KF :: rect. NL, LK : sq. LG (α and β of III. 30);

therefore

rect. NL, LK = sq. ML.

Let the square on KE be added to each; therefore

rect. NL, LK + sq. KE = sq. LE = sq. GX = sq. ML + sq. KE.

And

sq. GX : sq. ML + sq. KE :: sq. XC : sq. LG + sq. KF (Eucl. VI. 4; V. 12);

therefore

sq. XC = sq. LG + sq. KF.

And

sq. LG = sq. XE

and

sq. KF = sq. on half of second diameter (II. 1),

= rect. CE, ED (I. 38);

therefore

sq. XC = sq. XE + rect. CE, ED.

Therefore the straight line CD has been cut equally at X and unequally at E (Eucl. II. 5).

And DH is parallel to GX; therefore

$$CG = GH.$$

PROPOSITION 32

*If two straight lines touching an hyperbola meet, and a straight line is pro-
duced through the points of contact, and a straight line is drawn through the
point of meeting of the tangents parallel to the straight line joining the
points of contact, and a straight line is drawn through the midpoint of the
straight line joining the points of contact parallel to one of the asymptotes,
then the straight line cut off between this midpoint and the parallel will be
bisected by the section.*

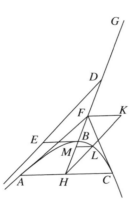

Let there be the hyper-
bola *ABC,* whose center
is *D,* and asymptote *DE,*
and let *AF* and *FC*
touch, and let *CA* and
FD be joined and pro-
duced to *G* and *H*; then
it is evident that

$$AH = HC.$$

Then let *FK* be drawn
through *F* parallel to
AC, and *HLK* through *H*
parallel to *DE.*

I say that

$$KL = HL.$$

Let *LM* and *BE* be drawn through *B* and *L* parallel to *AC*; then, as has been
already shown (III. 30, α, β, and conclusion),

$$\text{sq. } DB : \text{sq. } BE :: \text{sq. } HM : \text{sq. } ML :: \text{rect. } BM, MG : \text{sq. } ML;$$

therefore

$$\text{rect. } GM, MB = \text{sq. } MH.$$

And also

$$\text{rect. } HD, DF = \text{sq. } DB,$$

because *AF* touches, and *AH* has been dropped ordinatewise (I. 37); therefore

rect. *GM, MB* + sq. *DB* = rect. *HD, DF* + sq. *MH* = sq. *DM* (Eucl. II. 6).
Therefore *FH* has been bisected at *M* with *DF* added. And *KF* and *LM* are
parallel; therefore

$$KL = LH.$$

PROPOSITION 33

If two straight lines touching opposite sections meet, and one straight line is produced through the points of contact, and another straight line is drawn through the point of meeting of the tangents parallel to the straight line joining the points of contact, and still another straight line is drawn through the midpoint of the straight line joining the points of contact parallel to one of the asymptotes and meeting the section, and the parallel drawn through the point of meeting, then the straight line between the midpoint and the parallel will be bisected by the section.

Let there be the opposite sections *ABC* and *DEF,* and tangents *AG* and *DG* and center *H,* and asymptote *KH,* and let *HG* be joined and produced, and also let *ALD* be joined; then it is evident that it is bisected at *L* (II. 30). Then let *BHE* and *CGF* be drawn through *G* and *H* parallel to *AD,* and *LMN* through *L* parallel to *HK.*

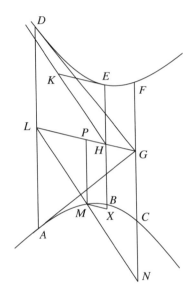

I say that
$$LM = MN.$$

For let *EK* and *MX* be dropped from *E* and *M* parallel to *GH,* and *MP* through *M* parallel to *AD.*

Since then through things already shown (III. 30, α and β)
$$\text{sq. } HE : \text{sq. } EK :: \text{rect. } BX, XE : \text{sq. } XM,$$
therefore
$$\text{sq. } HE : \text{sq. } EK :: \text{rect. } BX, XE + \text{sq. } HE : \text{sq. } KE + \text{sq. } XM. \text{ (Eucl. V. 12)}$$
or
$$\text{sq. } HE : \text{sq. } EK :: \text{sq. } HX : \text{sq. } KE + \text{sq. } XM \text{ (Eucl. II. 6).}$$
But it has been shown (I. 38; II. 1)
$$\text{sq. } EK = \text{rect. } GH, HL$$
and
$$\text{sq. } XM = \text{sq. } HP;$$

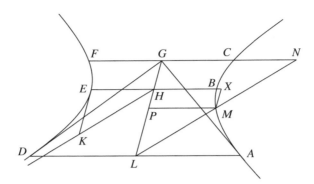

therefore
$$\text{sq. } HE : \text{sq. } EK :: \text{sq. } HX \text{ or sq. } MP : \text{rect. } GH, HL + \text{sq. } HP.$$
And
$$\text{sq. } HE : \text{sq. } EK :: \text{sq. } MP : \text{sq. } PL \text{ (Eucl. VI. 4);}$$
therefore
$$\text{sq. } MP : \text{sq. } PL :: \text{sq. } MP : \text{rect. } GH, HL + \text{sq. } HP.$$
Therefore
$$\text{sq. } PL = \text{rect. } GH, HL + \text{sq. } HP.$$
Therefore the straight line LG has been cut equally at P and unequally at H (Eucl. II. 5).
And MP and GN are parallel; therefore
$$LM = MN.$$

PROPOSITION 34

If some point is taken on one of the asymptotes of an hyperbola, and a straight line from it touches the section, and through the point of contact a parallel to the asymptote is drawn, then the straight line drawn from the point taken parallel to the other asymptote will be bisected by the section.

Let there be the hyperbola AB, and the asymptotes CD and DE, and let a point C be taken at random on CD, and through it let CBE be drawn touching the section, and through *B* let *FBG* be drawn parallel to *CD*, and through *C* let *CAG* be drawn parallel to *DE*.

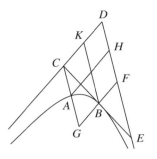

I say that

$$CA = AG.$$

For let *AH* be drawn through *A* parallel to *CD*, and *BK* through *B* parallel to *DE*. Since then

$$CB = BE \text{ (II. 3)},$$

therefore also

$$CK = KD$$

and

$$DF = FE.$$

And since

$$\text{rect. } KB, BF = \text{rect. } CA, AH \text{ (II. 12)},$$

and

$$BF = DK = CK,$$

and

$$AH = DC,$$

therefore

$$\text{rect. } DC, CA = \text{rect. } GC, CK.$$

Therefore

$$DC : CK :: GC : CA.$$

And

$$CD = 2CK;$$

therefore also

$$GC = 2\, CA.$$

Therefore

$$CA = AG.$$

PROPOSITION 35

With the same things being so, if from the point taken some straight line is drawn cutting the section at two points, then as the whole straight line is to the straight line cut off outside, so will the segments of the straight line cut off inside be to each other.

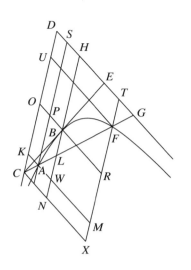

For let there be the hyperbola *AB* and the asymptotes *CD* and *DE,* and *CBE* touching and *HB* parallel, and through *C* let some straight line *CALFG* be drawn across cutting the section at *A* and *F.*

I say that
$$FC : CA :: FL : AL.$$

For let *CNX, KAM, OPBR,* and *FU* be drawn through *C, A, B* and *F* parallel to *DE*; and *APS* and *TFRMX* through *A* and *F* parallel to *CD.*

Since then
$$AC = FG \text{ (II. 8)},$$
therefore also
$$KA = TG \text{ (Eucl. VI. 4)}.$$
But
$$KA = DS;$$
therefore also
$$TG = DS.$$
And so also
$$CK = DU.$$
And since
$$CK = DU,$$
also
$$DK = CU;$$
therefore
$$DK : CK :: CU : CK.$$
And
$$CU : CK :: FC : AC,$$
and
$$FC : AC :: MK : KA,$$
and
$$MK : KA :: \text{pllg. } MD : \text{pllg. } DA \text{ (Eucl. VI. 1)},$$
and

$$DK : CK :: \text{pllg. } HK : \text{pllg. } KN;$$

therefore also

$$\text{pllg. } MD : \text{pllg. } DA :: \text{pllg. } HK : \text{pllg. } KN.$$

But

$$\text{pllg. } DA = \text{pllg. } DB \text{ (II. 12)} = \text{pllg. } ON;$$

for

$$CB = BE \text{ (II. 3)},$$

and

$$DO = OC;$$

therefore

$$\text{pllg. } MD : \text{pllg. } ON :: \text{pllg. } HK : \text{pllg. } KN.$$

And

$$\text{remainder pllg. } MH : \text{remainder pllg. } BK ::$$
$$\text{whole pllg. } MD : \text{whole pllg. } ON.$$

And since

$$\text{pllg. } DA = \text{pllg. } DB,$$

let the common parallelogram DP be subtracted;
therefore

$$\text{pllg. } KP = \text{pllg. } PH.$$

Let the common parallelogram AB be added; therefore

$$\text{whole pllg. } BK = \text{whole pllg. } AH.$$

Therefore

$$\text{pllg. } MD : \text{pllg. } DA :: \text{pllg. } MH : \text{pllg. } AH.$$

But

$$\text{pllg. } MD : \text{pllg. } DA :: MK : KA :: FC : AC,$$

and

$$\text{pllg. } MH : \text{pllg. } AH :: MW : WA :: FL : LA;$$

therefore also

$$FC : AC :: FL : LA.$$

PROPOSITION 36

With the same things being so, if the straight line drawn across from the point neither cuts the section at two points nor is parallel to the asymptote, it will meet the opposite section, and as the whole straight line is to the straight line between the section and the parallel through the point of contact, so will the straight line between the opposite section and the asymptote be to the straight line between the asymptote and the other section.

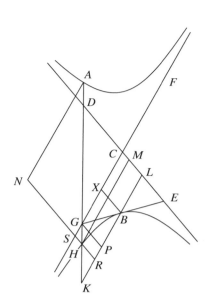

Let there be the opposite sections *A* and *B* whose center is *C* and asymptotes *DE* and *FG*, and let some point *G* be taken on *CG*, and from it let *GBE* be drawn tangent, and *GH* neither parallel to *CE* nor cutting the section in two points (I. 26).

It has been shown that *GH* produced meets *CD* and therefore also section *A*. Let it meet it at *A*, and let *KBL* be drawn through *B* parallel to *CG*.

I say that

$$AK : KH :: AG : GH.$$

For let *HM* and *AN* be drawn from the points *A* and *H* parallel to *CG*, and *BX*, *GP*, and *RHSN* from *B*, *G* and *H* parallel to *DE*. Since then

$$AD = GH \text{ (II. 16)},$$
$$AG : GH :: DH : HG.$$

But

$$AG : GH :: NS : SH,$$

and

$$DH : GH :: CS : SG.$$

And therefore

$$NS : SH :: CS : SG.$$

But

$$NS : SH :: \text{pllg. } NC : \text{pllg. } CH,$$

and

$$CS : SG :: \text{pllg. } RC : \text{pllg. } RG;$$

therefore also

$$\text{pllg. } NC : \text{pllg. } CH :: \text{pllg. } RC : \text{pllg. } RG.$$

And as one is to one so are all to all; therefore

$$\text{pllg. } NC : \text{pllg. } CH :: \text{whole pllg. } NL : \text{whole pllg. } CH + \text{pllg. } RG.$$

And since
$$EB = BG,$$
also
$$LB = BP$$
and
$$\text{pllg. } LX = \text{pllg. } BG.$$
And
$$\text{pllg. } LX = \text{pllg. } CH \text{ (II. 12)};$$
therefore also
$$\text{pllg. } BG = \text{pllg. } CH.$$
Therefore
$$\text{pllg. } NC : \text{pllg. } CH :: \text{whole pllg. } NL : \text{whole pllg. } BG + \text{pllg. } RG$$
or
$$\text{pllg. } NC : \text{pllg. } CH :: \text{pllg. } NL : \text{pllg. } RX.$$
But
$$\text{pllg. } RX = \text{pllg. } LH,$$
since also
$$\text{pllg. } CH = \text{pllg. } BC \text{ (II. 12)},$$
and
$$\text{pllg. } MB = \text{pllg. } XH.$$
Therefore
$$\text{pllg. } NC : \text{pllg. } CH :: \text{pllg. } NL : \text{pllg. } LH.$$
But
$$\text{pllg. } NC : \text{pllg. } CH :: NS : SH :: AG : GH,$$
and
$$\text{pllg. } NL : \text{pllg. } LH :: NR : RH :: AK : KH;$$
therefore also
$$AK : KH :: AG : GH.$$

PROPOSITION 37

If two straight lines touching a section of a cone or circumference of a circle or opposite sections meet, and a straight line is joined to their points of contact, and from the point of meeting of the tangents some straight line is drawn across cutting the line (of the section) at two points, then as the whole straight line is to the straight line cut off outside, so will the segments produced by the straight line joining the points of contact be to each other.

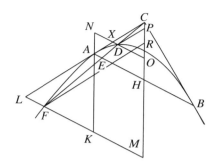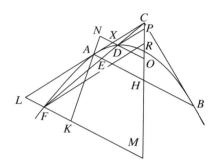

Let there be the section of a cone *AB* and tangents *AC* and *CB* and let *AB* be joined, and let *CDEF* be drawn across.

I say that

$$CF : CD :: FE : ED.$$

Let the diameters *CH* and *AK* be drawn through *C* and *A*, and through *F* and *D*, *DP*, *FR*, *LFM* and *NDO* parallel to *AH* and *LC*. Since then *LFM* is parallel to *XDO*,

$$FC : CD :: LF : XD :: FM : DO :: LM : XO;$$

and therefore

$$\text{sq. } LM : \text{sq. } XO :: \text{sq. } FM : \text{sq. } DO.$$

But

$$\text{sq. } LM : \text{sq. } XO :: \text{trgl. } LMC : \text{trgl. } XCO$$
$$\text{(Eucl. VI. 19)},$$

and

$$\text{sq. } FM : \text{sq. } DO :: \text{trgl. } FRM : \text{trgl. } DPO;$$

therefore also

$$\text{trgl. } LMC : \text{trgl. } XCO ::$$
$$\text{trgl. } FRM : \text{trgl. } DPO ::$$
$$\text{remainder quadr. } LCRF : \text{remainder quadr. } XCPD.$$

But

$$\text{quadr. } LCRF = \text{trgl. } ALK \text{ (III. 2; III. 11)},$$

and

$$\text{quadr. } XCPD = \text{trgl. } ANX \text{ (III. 2; III. 11)};$$

therefore

$$\text{sq. } LM : \text{sq. } XO :: \text{trgl. } ALK : \text{trgl. } ANX.$$

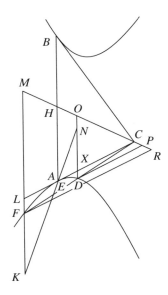
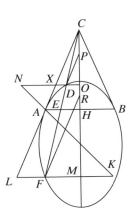

But
$$\text{sq. } LM : \text{sq. } XO :: \text{sq. } FC : \text{sq. } CD,$$
and
$$\text{trgl. } ALK : \text{trgl. } ANX :: \text{sq. } LA : \text{sq. } AX :: \text{sq. } FE : \text{sq. } ED;$$
therefore also
$$\text{sq. } FC : \text{sq. } CD :: \text{sq. } FE : \text{sq. } ED.$$
And therefore
$$FC : CD :: FE : ED.$$

PROPOSITION 38

With the same things being so, if some straight line is drawn through the point of meeting of the tangents parallel to the straight line joining the points of contact, and a straight line drawn through the midpoint of the straight line joining the points of contact cuts the section in two points and the straight line through the point of meeting parallel to the straight line joining the points of contact, then as the whole straight line drawn across is to the straight line cut off outside between the section and the parallel, so will the segments produced by the straight line joined to the points of contact be to each other.

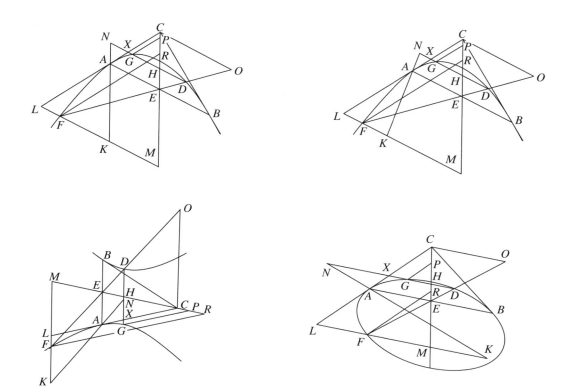

Let there be the section *AB** and tangents *AC* and *BC* and *AB* the straight line joining the points of contact, and *AN* and *CM* diameters; then it is evident that *AB* has been bisected at *E* (II. 30, 39).

Let *CO* be drawn from *C* parallel to *AB*, and let *FEDO* be drawn across through *E*.

I say that

$$FO : OD :: FE : ED.$$

For let *LFKM* and *DHGXN* be drawn through *F* and *D* parallel to *AB*, and through *F* and *G*, *FR* and *GP* parallel to *LC*. Then likewise as before (III. 37)

---

* It could be argued that since Apollonius takes up the case of opposite sections in Props. 39 and 40, Prop. 38 should not include them. However, Heiberg notes that his preferred manuscript shows cases involving the opposite sections. Further, the Arabic version explicitly treats the opposite sections as a special case. So, following those authorities, the diagram for opposite sections is presented

it will be shown that

$$\text{sq. } LM : \text{sq. } XH :: \text{sq. } LA : \text{sq. } AX.$$

And

$$\text{sq. } LM : \text{sq. } XH :: \text{sq. } LC : \text{sq. } CX :: \text{sq. } FO : \text{sq. } OD,$$

and

$$\text{sq. } LA : \text{sq. } AX :: \text{sq. } FE : \text{sq. } ED;$$

therefore

$$\text{sq. } FO : \text{sq. } OD :: \text{sq. } FE : \text{sq. } ED,$$

and

$$FO : OD :: FE : ED.$$

PROPOSITION 39

If two straight lines touching opposite sections meet, and a straight line is produced through the points of contact, and a straight line drawn from the point of meeting of the tangents cuts both of the sections and the straight line

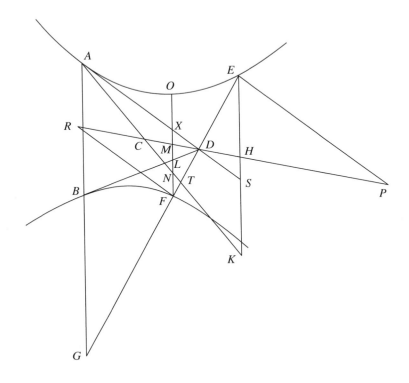

joining the points of contact, then as the whole straight line drawn across is
to the straight line cut off outside between the section and the straight line
joining the points of contact, so will the segments of the straight line pro-
duced by the segments and the point of meeting of the tangents be to each
other.

Let there be the opposite sections *A* and *B* whose center is *C* and tangents
AD and *DB,* and let *AB* and *CD* be joined and produced, and through *D* let
some straight line *EDFG* be drawn across.

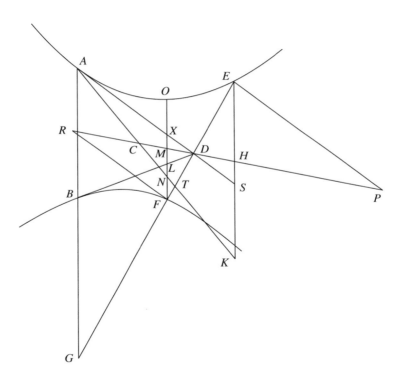

I say that

$$EG : GF :: ED : DF.$$

For let *AC* be joined and produced, and through *E* and *F* let *EHS* and
FLMNXO be drawn parallel to *AB,* and parallel to *AD, EP* and *FR.*

Since then *FX* and *ES* are parallel, and *EF, XS,* and *HM* have been drawn through to them,

$$EH : HS :: FM : MX.$$

And alternately

$$EH : FM :: HS : MX;$$

therefore also

$$\text{sq. } EH : \text{sq. } FM :: \text{sq. } HS : \text{sq. } MX.$$

But

$$\text{sq. } EH : \text{sq. } FM :: \text{trgl. } EHP : \text{trgl. } FRM,$$

and

$$\text{sq. } HS : \text{sq. } MX :: \text{trgl. } DHS : \text{trgl. } XMD;$$

therefore also

$$\text{trgl. } EHP : \text{trgl. } FRM :: \text{trgl. } DHS : \text{trgl. } XMD.$$

And

$$\text{trgl. } EHP = \text{trgl. } ASK + \text{trgl. } DHS \text{ (III. 11)},$$

and

$$\text{trgl. } FRM = \text{trgl. } AXN + \text{trgl. } XMD \text{ (III. 11)};$$

therefore

$$\text{trgl. } DHS : \text{trgl. } XMD :: \text{trgl. } ASK + \text{trgl. } DHS : \text{trgl. } AXN + \text{trgl. } XMD,$$

and

remainder trgl. *ASK* : remainder trgl. *ANX* :: trgl. *DHS* : trgl. *XMD*.

But

$$\text{trgl. } ASK : \text{trgl. } ANX :: \text{sq. } KA : \text{sq. } AN :: \text{sq. } EG : \text{sq. } FG,^{*}$$

and

$$\text{trgl. } DHS : \text{trgl. } XMD :: \text{sq. } HD : \text{sq. } DM :: \text{sq. } ED : \text{sq. } DF.$$

Therefore also

$$EG : FG :: ED : DF.$$

---

* For

$$EG : TG :: KA : TA,$$

and

$$TG : TF :: TA : TN,$$

and

$$[TG - TF] : TG :: [TA - TN] : TA;$$

therefore *ex aequali*

$$EG : FG :: KA : AN.$$

(Tr.)

PROPOSITION 40

With the same things being so, if a straight line is drawn through the point
of meeting of the tangents parallel to the straight line joining the points of
contact, and if a straight line drawn from the midpoint of the straight line
joining the points of contact cuts both of the sections and the straight line
parallel to the straight line joining the points of contact, then as the whole
straight line drawn across is to the straight line cut off outside between the
parallel and the section, so will the straight line's segments produced by the
sections and the straight line joining the points of contact be to each other.

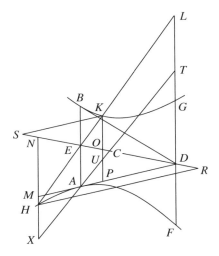

Let there be the opposite sections *A* and *B* whose center is *C*, and tangents
AD and *DB*, and let *AB* and *CDE* be joined; therefore
$$AE = EB \text{ (II. 39)}.$$
And from *D* let *FDG* be drawn parallel to *AB*, and from *E*, *LE* at random.

I say that
$$HL : LK :: HE : EK.$$

From *H* and *K* let *NMHX* and *KOP* be drawn parallel to *AB*, and *HR* and *KS*
parallel to *AD*, and let *XACT* be drawn through.

Since then *XAU* and *MAP* have been drawn across the parallels *XM* and *KP*,

$$XA : AU :: MA : AP.$$

But

$$XA : AU :: HE : EK;$$

and

$$HE : EK :: HN : KO$$

because of the similarity of the triangles *HEN* and *KEO*; therefore

$$HN : KO :: MA : AP;$$

therefore also

$$\text{sq. } HN : \text{sq. } KO :: \text{sq. } MA : \text{sq. } AP.$$

But

$$\text{sq. } HN : \text{sq. } KO :: \text{trgl. } HRN : \text{trgl. } KSO,$$

and

$$\text{sq. } MA : \text{sq. } AP :: \text{trgl. } XMA : \text{trgl. } AUP;$$

therefore also

$$\text{trgl. } HRN : \text{trgl. } KSO :: \text{trgl. } XMA : \text{trgl. } AUP.$$

And

$$\text{trgl. } HNR = \text{trgl. } XMA + \text{trgl. } MND \text{ (III. 11)},$$

and

$$\text{trgl. } KSO = \text{trgl. } AUP + \text{trgl. } DOP \text{ (III. 11)};$$

therefore also

$$\text{trgl. } XMA + \text{trgl. } MND : \text{trgl. } AUP + \text{trgl. } DOP :: \text{trgl. } XMA : \text{trgl. } AUP;$$

therefore also

remainder trgl. *NMD* : remainder trgl. *DOP* :: whole : whole.

But

$$\text{trgl. } XMA : \text{trgl. } AUP :: \text{sq. } XA : \text{sq. } AU,$$

and

$$\text{trgl. } NMD : \text{trgl. } DOP :: \text{sq. } MN : \text{sq. } PO;$$

therefore also

$$\text{sq. } MN : \text{sq. } PO :: \text{sq. } XA : \text{sq. } AU.$$

But

$$\text{sq. } MN : \text{sq. } PO :: \text{sq. } ND : \text{sq. } OD,$$

and

$$\text{sq. } XA : \text{sq. } AU :: \text{sq. } HE : \text{sq. } EK,$$

and

$$\text{sq. } ND : \text{sq. } DO :: \text{sq. } HL : \text{sq. } LK;$$

therefore also

$$\text{sq. } HE : \text{sq. } EK :: \text{sq. } HL : \text{sq. } LK.$$

Therefore

$$HE : EK :: HL : LK.$$

PROPOSITION 41

If three straight lines touching a parabola meet each other, they will be cut in the same ratio.

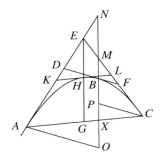

Let there be the parabola *ABC*, and tangents *ADE*, *EFC* and *DBF*.

I say that

$$CF : FE :: ED : DA :: FB : BD.$$

For let *AC* be joined and bisected at *G*.

Then it is evident that the straight line from *E* to *G* is a diameter of the section. (II. 29).

If then it goes through *B*, *DF* is parallel to *AC*, (II. 5) and will be bisected by *EG*, and therefore

$$AD = DE \text{ (I. 35)},$$

and

$$CF = FE \text{ (I. 35)},$$

and what was sought is apparent.

Let it not go through *B*, but through *H*, and let *KHL* be drawn through *H* parallel to *AC*; therefore it will touch the section at *H* (I. 32), and because of things already said (I. 35),

$$AK = KE$$

and

$$LC = LE.$$

Let *MNBX* be drawn through *B* parallel to *EG*, and *AO* and *CP* through *A* and *C* parallel to *DF*. Since then *MB* is parallel to *EH*, *MB* is a diameter (I. 40; I. 51 porism); and *DF* touches at *B*; therefore *AO* and *CP* have been dropped ordinatewise (II. 5; Def. 4). And since *MB* is a diameter, and *CM* a tangent, and *CP* an ordinate,

$$MB = BP \text{ (I. 35)},$$

and so also

$$MF = FC.$$

And since

$$MF = FC$$

and

$$EL = LC,$$

$$MC : CF :: EC : CL;$$

and alternately

$$MC : EC :: CF : CL.$$

But

$$MC : EC :: XC : CG;$$

therefore also

$$CF : CL :: XC : CG.$$

And

$$CL : EC :: CG : CA;$$

therefore *ex aequali*

$$CA : XC :: EC : CF,$$

and *convertendo*

$$EC : FE :: CA : AX;$$

separando

$$CF : FE :: XC : AX.$$

Again since MB is a diameter and AN a tangent and AO an ordinate,

$$NB = BO \text{ (I. 35)},$$

and

$$ND = DA.$$

And also

$$EK = KA;$$

therefore

$$AE : KA :: NA : DA;$$

alternately

$$AE : NA :: KA : DA.$$

But

$$AE : NA :: GA : AX;$$

therefore also

$$KA : DA :: GA : AX.$$

And also

$$AE : KA :: CA : GA;$$

therefore, *ex aequali,*

$$AE : DA :: CA : AX;$$

separando,

$$ED : DA :: XC : AX.$$

And it was also shown

$$XC : AX :: CF : FE;$$

therefore

$$CF : EF :: ED : DA.$$

Again since

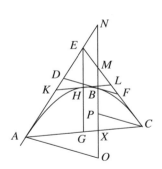

$$XC : AX :: CP : AO,$$
and
$$CP = 2\ BF,$$
and
$$CM = 2\ MF,$$
and
$$AO = 2\ BD,$$
and
$$AN = 2ND,$$
therefore
$$XC : AX :: FB : BD :: CF : FE :: ED : DA.$$

PROPOSITION 42

If in an hyperbola or ellipse or circumference of a circle or opposite sections straight lines are drawn from the vertices of the diameter parallel to an ordinate, and some other straight line at random is drawn tangent, it will cut off from them straight lines containing a rectangle equal to the fourth part of the figure to the same diameter.

For let there be some one of the aforesaid sections, whose diameter is AB, and from A and B let AC and DB be drawn parallel to an ordinate, and let some other straight line CED be tangent at E.

I say that

rect. AC, BD = fourth part of figure to AB.

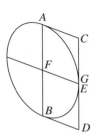

For let its center be F, and through it let FG be drawn parallel to AC and BD. Since then AC and BD are parallel, and FG is also parallel, therefore it is the diameter conjugate to AB (Def. 6); and so

sq. FG = fourth part of figure to AB (Def. 11).

If then FG goes through E in the case of the ellipse and circle,

$$AC = FG = BD \text{ (I. 32 converse, Eucl. I. 33)}$$
and it is immediately evident that

rect. AC, BD = sq. FG or fourth part of figure to AB.

Then let it not go through it, and let *DC* and *BA* produced meet at *K,* and let *EL* be drawn through *E*
parallel to *AC,* and *EM* parallel to *AB.*

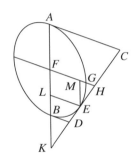

Since then

rect. *KF, FL* = sq. *AF* (I. 37),

KF : AF :: AF : FL,

and

KA : AL :: KF : AF or *FB* (Eucl. V. 18);

inversely

FB : KF :: AL : KA;

componendo or *separando*

BK : KF :: LK : KA.

Therefore also

DB : FH :: EL : CA.

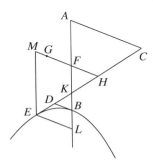

Therefore

rect. *DB, CA* = rect. *FH, EL,*

= rect. *HF, FM.*

But

rect. *HF, FM* = sq. *FG* (I. 38),

= fourth figure to *AB* (Def. 11);

therefore also

rect. *DB, CA* = fourth figure to *AB.*

PROPOSITION 43

If a straight line touch an hyperbola, it will cut off from the asymptotes,
beginning with the center of the section, straight lines containing a rectangle
equal to the rectangle contained by the straight lines cut off by the tangent
at the section's vertex at its axis.

Let there be the hyperbola *AB,* and asymptotes *CD*
and *DE,* and axis *BD,* and let *FBG* be drawn through
B tangent, and some other tangent at random, *CAH.*

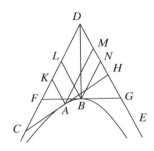

I say that

rect. *FD, DG* = rect. *CD, DH.*

For let *AK* and *BL* be drawn from *A* and *B* parallel to

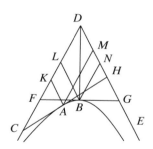

DG, and *AM* and *BN* parallel to *CD.* Since then *CAH* touches,

$$CA = AH \text{ (II. 3)};$$

and so

$$CH = 2AH$$

and

$$CD = 2AM$$

and

$$DH = 2AK.$$

Therefore

$$\text{rect. } CD, DH = 4 \text{ rect. } KA, AM.$$

Then likewise it could be shown

$$\text{rect. } FD, DG = 4 \text{ rect. } LB, BN.$$

But

$$\text{rect. } KA, AM = \text{rect. } LB, BN \text{ (II. 12)}.$$

Therefore also

$$\text{rect. } CD, DH = \text{rect. } FD, DG.$$

Then likewise it could be shown, even if *DB* were some other diameter and not the axis.

PROPOSITION 44

If two straight lines touching an hyperbola or opposite sections meet the asymptotes, then the straight lines drawn to the sections will be parallel to the straight line joining the points of contact.

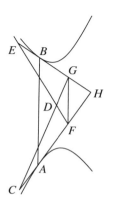

For let there be either the hyperbola or the opposite sections *AB,* and asymptotes *CD* and *DE,* and tangents *CAHF* and *EBHG,* and let *AB, FG,* and *CE* be joined.

I say that they are parallel.

For since

$$\text{rect. } CD, DF = \text{rect. } GD, DE \text{ (III. 43),}$$

therefore

$$CD : DE :: GD : DF;$$

therefore *CE* is parallel to *FG*. And therefore
$$HF : FC :: HG : GE.$$
And
$$FC : AF :: GE : GB;$$
for each is double (II. 3); therefore *ex aequali*
$$HG : GB :: HF : FA.$$
Therefore *FG* is parallel to *AB*.

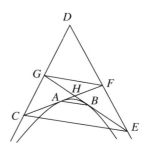

PROPOSITION 45

If in an hyperbola or ellipse or circumference of a circle or opposite sections straight lines are drawn from the vertex of the axis at right angles, and a rectangle equal to the fourth part of the figure is applied to the axis on each side and exceeding by a square figure in the case of the hyperbola and opposite sections, but deficient in the case of the ellipse, and some straight line is drawn tangent to the section and meeting the perpendicular straight lines, then the straight lines drawn from the points of meeting to the points produced by the application make right angles at the aforesaid points. [*]

---

[*] "The points of application" are in modern terminology the foci of the conics. The circle is seen here as an ellipse whose two foci or focal points coincide with the center. This theory is, of course, a special application of Euclid VI. 28 and 29, two theorems on which depends one whole side of Greek geometry.

Apollonius never speaks of the focus of the parabola, but it can be found by analogy with the ellipse.

Thus in the ellipse above
$$\text{rect. } AF, FB = \text{fourth rect. } AB, R$$
where R is the parameter. Or

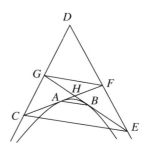

$$\text{rect. } AF, (AB - AF) = \text{fourth rect. } AB, R$$
or
$$AF : \text{fourth } R :: AB : (AB - AF).$$
Then if we consider the ellipse as its axis *AB* gets as large as we please, we can think of it as approaching as near as we please to a parabola with parameter *R*. The ratio *AB* : (*AB* − *AF*) approaches as near as we please to equality and hence also the ratio *AF* : fourth *R*. At the limit we can think of the ellipse as the parabola, its axis *AB* as infinite, and *AB* as equal to *AB* − *AF*. Then *AF* will be equal to a fourth *R*. Thus the focus of a parabola will be defined as the point on its axis at a distance from the

[Footnote continued on next page.]

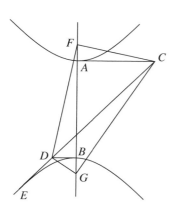

Let there be one of the sections mentioned whose axis is *AB*, and *AC* and *BD* at right angles, and *CED* tangent, and let the rectangle *AF, FB* and the rectangle *AG, GB* equal to the fourth part of the figure be applied on each side (Eucl. VI. 28, 29), as has been said, and let *CF, CG, DF,* and *DG* be joined.

I say that angle *CFD* and angle *CGD* are each a right angle.

For since it has been shown

　　rect. *AC, BD* = fourth figure on *AB* (III. 42),
and since also

　　　　rect. *AF, FB* = fourth figure on *AB*,
therefore

　　　　　rect. *AC, BD* = rect. *AF, FB*.
Therefore

$$AC : AF :: FB : BD.$$
And the angles at points *A* and *B* are right; therefore
　　　　angle *ACF* = angle *BFD* (Eucl. VI. 6),
and

　　　　　angle *AFC* = angle *FDB*.
And since angle *CAF* is right, therefore
　　　　angle *ACF* + angle *AFC* = 1 rt. angle.
And it has also been shown that
　　　　angle *ACF* = angle *DFB*;

therefore

　　angle *AFC* + angle *DFB* = 1 right angle.

Therefore

　　　angle *DFC* = 1 right angle.
Then likewise it could also be shown
　　　angle *CGD* = 1 right angle.

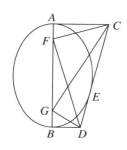

[Footnote continued from previous page.]
vertex equal to one quarter of the parameter. Then many of the properties of the foci of the ellipse can be proved analogously for the parabola. Thus in the case of this proposition, *FD* will become parallel to *CE*. Hence any straight line from the focus of a parabola parallel to a tangent will make a right angle with the straight line drawn from the focus to the intersection of the tangent and the perpendicular to the axis at the vertex. (Tr.)

PROPOSITION 46

With the same things being so, the straight lines joined make equal angles with the tangents.

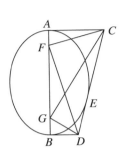
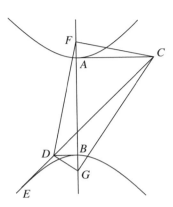

For with the same things supposed, I say that

angle ACF = angle DCG

and

angle CDF = angle BDG.

For since it has been shown that both angle CFD and angle CGD are right angles (III. 45), the circle described about CD as a diameter will pass through points F and G; therefore

angle DCG = angle DFG;

for they are on the same segment of the circle. And it was shown

angle DFG = angle ACF (III. 45);

and so

angle DCG = angle ACF.

And likewise also

angle CDF = angle BDG.

PROPOSITION 47

With the same things being so, the straight line drawn from the point of meeting of the joined straight lines to the point of contact will be perpendicular to the tangent.

For let the same things as before be supposed and let *CG* and *FD* meet each other at *H,* and let *CD* and *BA* produced meet at *K,* and let *EH* be joined.

I say that *EH* is perpendicular to *CD.*

For if not, let *HL* be drawn from *H* perpendicular to *CD.* Since then

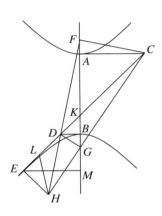 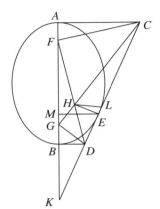

$$\text{angle } CDF = \text{angle } BDG \text{ (III. 46),}$$

and also

$$\text{rt. angle } DBG = \text{rt. angle } DLH,$$

therefore triangle *DGB* is similar to triangle *LHD.* Therefore

$$GD : DH :: BD : DL.$$

But

$$GD : DH :: FC : CH$$

because the angles at *F* and *G* are right angles (III. 45) and the angles at *H* are equal; but

$$FC : CH :: AC : CL$$

because of the similarity of the triangles *AFC* and *LCH* (III. 46); therefore also

$$BD : DL :: AC : CL.$$

Alternately

$$BD : AC :: DL : CL.$$

But

$$BD : AC :: BK : KA;$$

therefore also

$$DL : CL :: BK : KA.$$

Let *EM* be drawn from *E* parallel to *AC*; therefore it will have been dropped ordinatewise to *AB* (II. 7); and

$$BK : KA :: BM : MA \text{ (I.36)}.$$

And

$$BM : MA :: DE : EC;$$

therefore also

$$DL : CL :: DE : EC;$$

and this is absurd. Therefore *HL* is not perpendicular nor is any other straight line except *HE*.[*]

PROPOSITION 48

With the same things being so, it must be shown that the straight lines drawn from the point of contact to the points produced by the application make equal angles with the tangent.

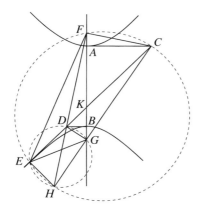

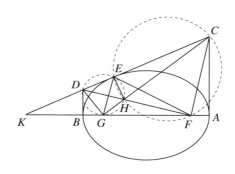

For let the same things be supposed, and let *EF* and *EG* be joined.

I say that

$$\text{angle } CEF = \text{angle } GED.$$

---

[*] There is the analogous theorem for the parabola. *FD* becomes a straight line parallel to *CE* and *CG* a straight line parallel to *AB*. Again *HE* is perpendicular to *CE*, and this can be proved rigorously as well as understood by analogy. (Tr.)

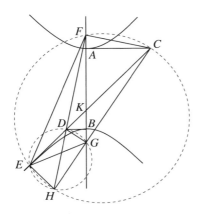

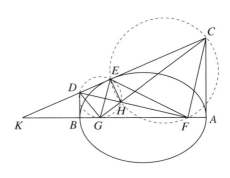

For since angles *DGH* and *DEH* are right angles (III. 45. 47), the circle described about *DH* as a diameter will pass through the points *E* and *G* (Eucl. III. 31); and so

angle *DHG* = angle *DEG* (Eucl. III. 21);

for they are in the same segment. Likewise then also

angle *CEF* = angle *CHF*.

But

angle *CHF* = angle *DHG*;

for they are vertical angles; therefore also

angle *CEF* = angle *DEG*.[*]

PROPOSITION 49

With the same things being so, if from one of the points (of application) a perpendicular is drawn to the tangent, then the straight lines from that point to the ends of the axis make a right angle.

---

[*] Here there is another and important analogous theorem for the parabola. *EG* becomes parallel to *AB,* and

angle *DEG* = angle *CEF.*

(Tr.)

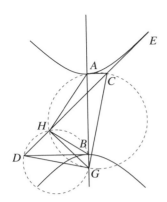 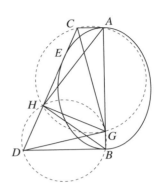

For let the same things be supposed, and let the perpendicular *GH* be drawn from *G* to *CD,* and let *AH* and *BH* be joined.

I say that angle *AHB* is a right angle.

For since angle *DBG* is a right angle and also angle *DHG,* the circle described about *DG* as a diameter will pass through *H* and *B,* and
$$\text{angle } BHG = \text{angle } BDG.$$
But it was shown
$$\text{angle } AGC = \text{angle } BDG \text{ (III. 45)};$$
therefore also
$$\text{angle } BHG = \text{angle } AGC = \text{angle } AHC \text{ (Eucl. III. 21)}.$$
And so also
$$\text{angle } CHG = \text{angle } AHB.$$
But angle *CHG* is a right angle; therefore also angle *AHB* is a right angle.

PROPOSITION 50

With the same things being so, if from the center of the section there falls to the tangent a straight line parallel to the straight line drawn through the point of contact and one of the points (of application), then it will be equal to one half the axis.

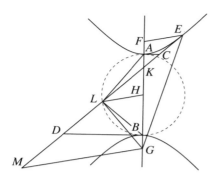
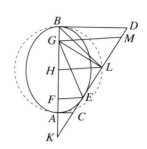

For let there be the same things as before and let H be the center, and let EF be joined, and let DC and BA meet at K, and through H let HL be drawn parallel to EF.

I say that

$$HL = HB.$$

For let EG, AL, LB be joined, and through G let GM be drawn parallel to EF. Since then

$$\text{rect. } AF, FB = \text{rect. } AG, GB \text{ (See III. 45),}$$

therefore

$$AF = GB.$$

But also

$$AH = HB;$$

therefore also

$$FH = HG.$$

And so also

$$EL = LM.$$

And since it was shown (III. 48)

$$\text{angle } CEF = \text{angle } DEG,$$

and

$$\text{angle } CEF = \text{angle } EMG,$$

therefore also

$$\text{angle } EMG = \text{angle } DEG.$$

And therefore

$$EG = GM.$$

But it was also shown

$$EL = LM$$

therefore *GL* is perpendicular to *EM*. And so through what was shown before (III. 49) angle *ALB* is a right angle, and the circle described about *AB* as a diameter will pass through *L*. And
$$HA = HB;$$
therefore also, since *HL* is a radius of the semicircle,
$$HL = HB.$$

PROPOSITION 51

If a rectangle equal to the fourth part of the figure is applied from both sides to the axis of an hyperbola or opposite sections and exceeding by a square figure, and straight lines are deflected from the resulting points of application to either one of the sections, then the greater of the two straight lines exceeds the less by exactly as much as the axis.

For let there be an hyperbola or opposite sections whose axis is *AB* and center *C*, and let each of the rectangles *AD, DB* and *AE, EB* be equal to the fourth part of the figure, and from points *E* and *D* let the straight lines *EF* and *FD* be deflected to the line of the section.

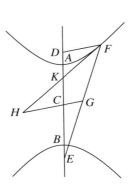

I say that
$$EF = FD + AB.$$

Let *FKH* be drawn tangent through *F,* and *GCH* through *C* parallel to *FD*; therefore
$$\text{angle } KHG = \text{angle } KFD;$$
for they are alternate. And
$$\text{angle } KFD = \text{angle } GFH \text{ (III. 48)};$$
therefore
$$GF = GH.$$
But
$$GF = GE,$$
since also
$$AE = BD$$
and
$$AC = CB$$

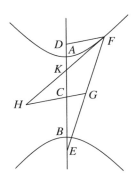

and
$$EC = CD;$$
and therefore
$$GH = EG.$$
And so
$$FE = 2GH.$$
And since it has been shown (III. 50)
$$CH = CB,$$
therefore
$$FE = 2(GC + CB).$$
But
$$FD = 2GC,$$
and
$$AB = 2CB;$$
therefore
$$FE = FD + AB.$$
And so EF is greater than FD by AB.

PROPOSITION 52

If in an ellipse a rectangle equal to the fourth part of the figure is applied from both sides to the major axis and deficient by a square figure, and from the points resulting from the application straight lines are deflected to the line of the section, then they will be equal to the axis.

Let there be an ellipse, whose major axis is AB, and let each of the rectangles AC, CB and AD, DB be equal to the fourth of the figure, and from C and D

let the straight lines *CE* and *ED* have been deflected to the line of the section.

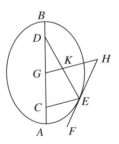

I say that
$$CE + ED = AB.$$

Let *FEH* be drawn tangent, and *G* be center and through it let *GKH* be drawn parallel to *CE*. Since then
$$\text{angle } CEF = \text{angle } HEK \text{ (III. 48)},$$
and
$$\text{angle } CEF = \text{angle } EHK,$$
therefore also
$$\text{angle } EHK = \text{angle } HEK.$$
Therefore also
$$HK = KE.$$
And since
$$AG = GB,$$
and
$$AC = DB,$$
therefore also
$$CG = GD;$$
and so also
$$EK = KD.$$
And for this reason
$$ED = 2HK,$$
and
$$EC = 2KG,$$
and
$$ED + EC = 2GH.$$
But also
$$AB = 2GH \text{ (III. 50)};$$
therefore
$$AB = ED + EC.$$

PROPOSITION 53

If in an hyperbola or ellipse or circumference of a circle or opposite sections straight lines are drawn from the vertex of a diameter parallel to an ordinate, and straight lines drawn from the same ends to the same point on the line of the section cut the parallels, then the rectangle contained by the straight lines cut off is equal to the figure on that same diameter.

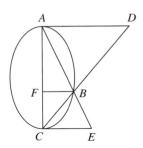

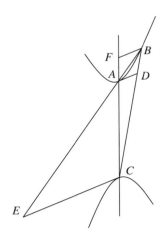

Let there be one of the aforesaid sections *ABC* whose diameter is *AC,* and let *AD* and *CE* be drawn parallel to an ordinate, and let *ABE* and *CBD* be drawn across.

I say that

$$\text{rect. } AD,\ EC = \text{figure on } AC.$$

For let *BF* be drawn from *B* parallel to an ordinate.
Therefore (I. 21)

    rect. *AF, FC* : sq. *FB* :: transverse side : upright side :: sq. *AC* : the figure.
But

$$\text{rect. } AF,\ FC : \text{sq. } FB :: AF : FB \text{ comp. } FC : FB;$$

therefore

$$\text{figure : sq. } AC :: FB : AF \text{ comp. } FB : FC.$$

But

$$AF : FB :: AC : CE,$$

and

$$FC : FB :: AC : AD;$$

therefore

$$\text{figure : sq. } AC :: CE : AC \text{ comp. } AD : AC.$$

And also

$$\text{rect. } AD,\ CE: \text{sq. } AC :: CE : AC \text{ comp. } AD : AC;$$

therefore

$$\text{figure : sq. } AC :: \text{rect. } AD,\ CE : \text{sq. } AC.$$

Therefore

$$\text{rect. } AD,\ CE = \text{figure on } AC.$$

PROPOSITION 54

If two tangents to a section of a cone or to a circumference of a circle meet, and through the points of contact parallels to the tangents are drawn, and from the points of contact, to the same point of the line of the section, straight lines are drawn across cutting the parallels, then the rectangle contained by the straight lines cut off to the square on the straight line joining the points of contact has a ratio compounded of the ratio which the inside segment line joining the point of meeting of the tangents and the midpoint of the straight line joining the points of contact has in square to the remainder, and of the ratio which the rectangle contained by the tangents has to the fourth part of the square on the straight line joining the points of contact.

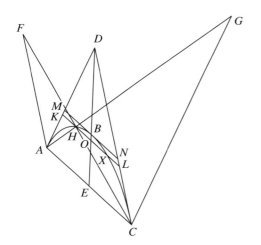
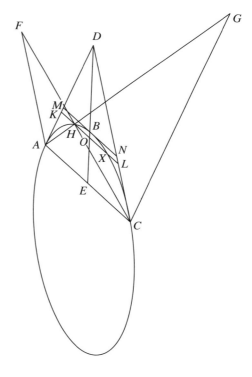

Let there be a section of a cone or circumference of a circle *ABC* and tangents *AD* and *CD,* and let *AC* be joined and bisected at *E,* and let *DBE* be joined, and let *AF* be drawn from *A* parallel to *CD,* and *CG* from *C* parallel to *AD,* and let some point *H* on the section be taken, and let the straight lines *AH* and *CH* be joined and produced to *G* and *F.*

I say that

rect. *AF, CG* : sq. *AC* ::

sq. *EB* : sq. *BD* comp. rect. *AD, DC* : fourth sq. *AC* or rect. *AE, EC.*

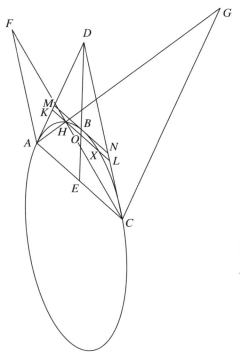

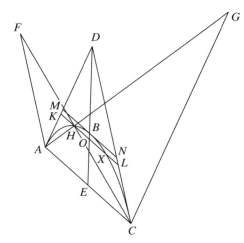

For let *KHOXL* be drawn from *H* parallel to *AC,* and from *B, MBN* parallel to *AC*; then it is evident that *MN* is tangent (II. 29, 5, 6). Since then

$$AE = EC,$$

also

$$MB = BN.$$

and

$$KO = OL$$

and

$$HO = OX \text{ (II. 7)}$$

and

$$KH = XL.$$

Since then *MB* and *MA* are tangents and *KHL* has been drawn parallel to *MB*,

$$\text{sq. } AM : \text{sq. } MB :: \text{sq. } AK : \text{rect. } XK, KH \text{ (III. 16)}$$

or

$$\text{sq. } AM : \text{rect. } MB, BN :: \text{sq. } AK : \text{rect. } LH, HK.$$

And

$$\text{rect. } NC, AM : \text{sq. } AM :: \text{rect. } LC, AK : \text{sq. } AK \text{ (Eucl. VI. 2; V. 18)};$$

therefore *ex aequali*

$$\text{rect. } NC, AM : \text{rect. } MB, BN :: \text{rect. } LC, AK : \text{rect. } LH, HK.$$

But

$$\text{rect. } LC, AK : \text{rect. } LH, HK :: LC : LH \text{ comp. } AK : HK$$

or

$$\text{rect. } LC, AK : \text{rect. } LH, HK :: FA : AC \text{ comp. } GC : CA$$

which is the same as

rect. *GC, FA* : sq. *CA*.

Therefore

rect. *NC, AM* : rect. *MB, BN* :: rect. *GC, FA* : sq. *CA*.

But, with the rectangle *ND, DM* taken as a mean,

rect. *NC, AM* : rect. *MB, BN* ::

rect. *NC, AM* : rect. *ND, DM* comp. rect. *ND, DM* : rect. *MB, BN*;

therefore

rect. *GC, FA* : sq. *CA* ::

rect. *NC, AM* : rect. *ND, DM* comp. rect. *ND, DM* : rect. *MB, BN*.

But

rect. *NC, AM* : rect. *ND, DM* :: sq. *EB* : sq. *BD*,

and

rect. *ND, DM* : rect. *NB, BM* :: rect. *CD, DA* : rect. *CE, EA*;

therefore

rect. *GC, FA* : sq. *CA* :: sq. *BE* : sq. *BD* comp. rect. *CD, DA* : rect. *CE, EA*.

PROPOSITION 55

If two straight lines touching opposite sections meet, and through the point of meeting a straight line is drawn parallel to the straight line joining the points of contact, and from the points of contact parallels to the tangents are drawn across, and straight lines are produced from the points of contact to the same point of one of the sections cutting the parallels, then the rectangle contained by the straight lines cut off will have to the square on the straight line joining the points of contact the ratio which the rectangle contained by the tangents has to the square on the straight line drawn through the point of meeting parallel to the straight line joining the points of contact as far as the section.

Let there be the opposite sections *ABC* and *DEF*, and tangents to them *AG* and *GD*, and let *AD* be joined, and from *G* let *CGE* be drawn parallel to *AD*, and from *A*, *AM* parallel to *DG*, and from *D*, *DM* parallel to *AG*, and let some point *F* be taken on the section *DF*, and let *ANF* and *FDH* be joined.

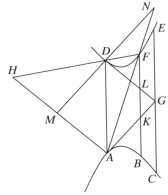

I say that

$$\text{sq. } CG : \text{rect. } AG, GD :: \text{sq. } AD : \text{rect. } HA, DN.$$

For let *FLKB* be drawn through *F* parallel to *AD*.

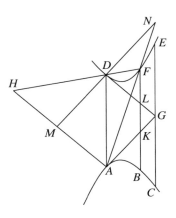

Since then it has been shown that

$$\text{sq. } EG : \text{sq. } GD :: \text{rect. } BL, LF : \text{sq. } DL \text{ (III. 20)},$$

and

$$CG = EG \text{ (II. 38)},$$

and

$$BK = LF \text{ (II. 38)},$$

therefore

$$\text{sq. } CG : \text{sq. } GD :: \text{rect. } KF, FL : \text{sq. } DL.$$

And also

$$\text{sq. } GD : \text{rect. } AG, GD :: \text{sq. } DL : \text{rect. } DL, AK \text{ (Eucl. VI. 2, 1)};$$

therefore *ex aequali*

$$\text{sq. } GC : \text{rect. } AG, GD :: \text{rect. } KF, FL : \text{rect. } DL, AK.$$

But

$$\text{rect. } KF, FL : \text{rect. } DL, AK :: KF : AK \text{ comp. } FL : DL.$$

But

$$KF : AK :: AD : DN,$$

and

$$FL : DL :: AD : HA;$$

therefore

$$\text{sq. } CG : \text{rect. } AG, GD :: AD : DN \text{ comp. } AD : HA.$$

And also

$$\text{sq. } AD : \text{rect. } HA, DN :: AD : DN \text{ comp. } AD : HA;$$

therefore

$$\text{sq. } CG : \text{rect. } AG, GD :: \text{sq. } AD : \text{rect. } HA, DN.$$

PROPOSITION 56

If two straight lines touching one of the opposite sections meet, and parallels to the tangents are drawn through the points of contact, and straight lines cutting the parallels are drawn from the points of contact to the same point of the other section, then the rectangle contained by the straight lines cut off will have to the square on the straight line joining the points of contact the ratio compounded of the ratio which, of the straight line joining the point of meeting and the midpoint, that part between the midpoint and the other section has in square to that part between the same section and the point of meeting, and of the ratio which the rectangle contained by the tangents has to the fourth part of the square on the straight line joining the points of contact.

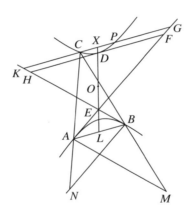

Let there be the opposite sections *AB* and *CD* whose center is *O*, and tangents *AEFG* and *BEHK*, and let *AB* be joined, and bisected at *L*, and let *LE* be joined and drawn across to *D*, and let *AM* be drawn from *A* parallel to *BE*, and *BN* from *B* parallel to *AE*, and let some point *C* be taken on the section *CD*, and let *CBM* and *CAN* be joined.

I say that

rect. *MA, BN* : sq. *AB* ::

sq. *LD* : sq. *DE* comp. rect. *AE, EB* : fourth sq. *AB* or rect. *AL, LB.*

For let *GCK* and *HDF* be drawn from *C* and *D* parallel to *AB*; then it is evident that

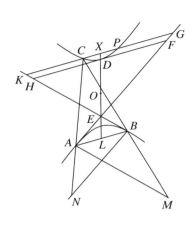

$$HD = DF,$$

and
$$KX = XG,$$

and also
$$XC = XP;$$

and so also
$$CK = GP.$$

And since AB and DC are opposite sections, and BEH and HD are tangents, and KG is parallel to DH, therefore

sq. BH : sq. HD :: sq. BK : rect. PK, KC (III. 18, note).

But
$$\text{sq. } HD = \text{rect. } HD, DF,$$
$$\text{rect. } PK, KC = \text{rect. } KC, CG.$$

Therefore

sq. BH : rect. HD, DF :: sq. BK : rect. KC, CG.

And also

rect. FA, BH : sq. BH :: rect. GA, BK : sq. BK;

therefore *ex aequali*

rect. FA, BH : rect. HD, DF :: rect. GA, BK : rect. KC, CG.

And, with rectangle HE, EF taken as a mean,

rect. FA, BH : rect. HD, DF ::

rect. FA, HB : rect. HE, EF comp. rect. HE, EF : rect. HD, DF;

and

rect. FA, HB : rect. HE, EF :: sq. LD : sq. DE,

and

rect. HE, EF : rect. HD, DF :: rect. AE, EB : rect. AL, LB;

therefore

rect. GA, BK : rect. KC, CG ::

sq. LD : sq. DE comp. rect. AE, EB : rect. AL, LB.

And

rect. GA, BK : rect. KC, CG :: BK : KC comp. GA : CG.

But
$$BK : KC :: MA : AB,$$

and
$$GA : CG :: BN : AB;$$

therefore

rect. MA, BN : sq. AB ::

$MA : AB$ comp. $BN : AB$ ::

sq. LD : sq. DE comp. rect. AE, EB : rect. AL, LB.

Translator's Introduction to Book IV

by Michael N. Fried

Sir Thomas Heath wrote that Apollonius's *Conics* Book IV "is on the whole dull, and need not be noticed at length" (Heath, 1921, II, 157). A great thinker can sometimes write a dull book, and so it is possible that Heath is right in this estimation of Apollonius's fourth book of the *Conics*. But, it is more likely that Heath's dismissal of Book IV comes from deeply rooted prejudices as to what should be the content, interest, and importance of a mathematical text.

Such prejudices come out into the open when, following the remark quoted above, Heath gives his contrasting estimation of Book V: "Book V," he writes, "is of an entirely different order [than Book IV], indeed it is the most remarkable of the Books.... Included in it are a series of propositions which, though worked out by the purest geometric methods, actually lead immediately to the determination of the evolute of each of the three conics; that is to say, the Cartesian equations to the evolutes can be easily deduced from the results obtained by Apollonius" (*ibid.*, pp. 158–159).

For Heath, then, and, unfortunately, not *only* for Heath, what makes a Greek mathematical work impressive or dull is the degree to which the text corresponds, respectively, to advanced or elementary *modern* mathematics when approached and *rewritten* as a modern mathematical text. Indeed, if one does look at the *Conics* as an *algebraic* text disguised in geometric language, where the essence of the conic sections is summarized by their *symptomata* understood as *equations*, then one can understand how Heath, and his like, might see Book IV as truly pedantic and repetitious and, in short, dull.

In reading and translating Book IV, I have tried, on the other hand, to give Apollonius a fair chance, to keep modern algebraic ideas about conics at a distance, and to view the text with eyes trained only on the mathematical and philosophical concerns of Apollonius's contemporaries and on the geometrical character of the previous three books of the *Conics*. Approaching the text this way allows one to see that Book IV, far from being dull, reveals fundamental difficulties in Apollonius's treatment of conic sections. First and foremost of these is, of course, the troubling nature of the opposite sections. But besides that, the book also raises questions as to Apollonius's basic understanding of how conic sections may be present and related to one another in a single plane—and this understanding is crucial in reading the whole of the *Conics*. Let us begin, however, by naively asking, What is Book IV about?

I

In the 1886 classic work, *Die Lehre von den Kegelschnitten im Altertum*, by the Danish mathematician H. G. Zeuthen, whom Heath admittedly followed closely in his own account (see Apollonius (Heath), p. xi), Book IV is seen in the context of the problem of determining a conic through five points. This problem in turn can be related to the problem of the three- and four-line locus, the theme of the final propositions in Book III, and a problem of great interest to Zeuthen, not least because of the ease with which it can be given an algebraic interpretation (see Zeuthen, pp. 126 ff). But is this really what Book IV is about? Is it really directed principally towards this kind of problem-solving enterprise? Apollonius does tell us in the letter introducing Book IV that the book's contents are relevant to problem-solving and analysis, but his reason for saying so seems to be more to fend off criticism that they are of *no* use in this regard than to say that problem-solving and analysis are the main purpose of the book. Moreover, he is not at all specific about what problems he has in mind. All he writes is, "The investigation of these theorems is also of considerable use in the synthesis of problems and limits of possibility (*diorismous*). So, Nicoteles was not speaking truly when, for the sake of his argument with Conon, he said that none of the things discovered by Conon [regarding the number of points at which conic sections can meet one another] were of any use for limits of possibility; but even if the limits of possibility are able to be obtained completely without these things, yet, surely, some matters are more readily perceived by means of them, for example, whether a problem might be done in many ways, and in how many ways, or again, whether it might not be done at all."

As if to distance himself from this kind of argument, Apollonius concludes his letter by saying that, beyond any use it might have, what the fourth book treats is worth studying in its own right. This is not a remark made casually or routinely: Apollonius makes a similar remark in only one other place in the *Conics*—in the prefatory letter to Book V (a somewhat ironic fact in light of Heath's very high estimation of Book V and his very low estimation of Book IV). So, in asking what Book IV is about, we must consider this common remark about the two books at face value and avoid searching for the aim of Book IV in *other* problems, whose importance seems to a modern way of thinking more convincing than those the book actually contains.

The truth is, both in the letter introducing Book I, which also introduces the whole of the *Conics*, and in the letter introducing Book IV itself, Apollonius tells us quite plainly what Book IV is about: "This book treats of the greatest number of points at which sections of a cone can meet one another or meet a circumference of a circle…and, moreover, the greatest number of points at which a section of a cone or a circumference of a circle can meet opposite sections. Besides these questions, there are more than a few others of a similar character." Among these "few others," clearly, are all those concerning the greatest number of points opposite sections can meet opposite sections (that this is one of the principal subjects of Book IV is said explicitly in the introduction to Book I).

Now, Apollonius reports that Conon of Samos treated the case where a conic section or circumference of a circle meets another conic section, but that Conon's demonstrations were incorrect. Apollonius further reports that, in connection with Conon's flawed proofs, Nicoteles remarked that the case in which a conic section meets opposite sections *could* be solved, but, as Apollonius makes sure to say, neither Nicoteles nor anyone else provided a demonstration. As for the greatest number of points in which opposite sections can meet opposite sections, Apollonius says that no one has ever noticed this question, let alone treated it. Thus, as in the introduction to the first book, Apollonius promises his readers that in this book they can expect a fuller and more rigorous treatment (and, therefore, to his mind, a more *correct* treatment) of familiar questions, but also completely new material, which, as in so much of the *Conics*, is precisely that concerning the opposite sections.

The importance of the opposite sections in the *Conics* cannot be overemphasized. The existence of opposite sections may have been known before Apollonius, as the references to Conon and Nicoteles in Apollonius's prefatory letter to Book IV suggest; however, it is highly doubtful that, before the *Conics,* there was anywhere a more than perfunctory treatment of them. The opposite sections are peculiar, and, in the enunciations to propositions in the *Conics*, Apollonius usually separates them from the other conic sections. This peculiarity, in part, has to do with their number, for while there is a sense in which the opposite sections are one curve, as a *visual* object they are clearly two. Thus, for one, like Apollonius, whose work with curves is always governed by a fundamentally geometric outlook,[1] the plural-singular nature of the opposite sections makes them an object of fascination, but it also duly gives rise to a certain uneasiness with them, which one senses already in the first book of the *Conics*.

Consider, for example, the definitions of the "transverse" and "upright diameters" given in Book I. Apollonius writes: "Likewise of any two curved lines lying in one plane I call that straight line the *transverse diameter* which cuts the two curved lines and bisects all the straight lines drawn to either of the curved lines parallel to some straight line…and I call that straight line the *upright diameter* which, lying between the two curved lines, bisects all the straight lines intercepted between the curved lines and drawn parallel to some straight line…." It is clear that, though they refer in completely general terms to two curves, these definitions are directed towards the opposite sections specifically. Apollonius himself is not explicit about this (nor could he be, having not yet defined the opposite sections or any of the other conic sections) and has no diagram hinting at his specific intention. But Eutocius, Apollonius's late 5th century commentator, does have a diagram illustrating the definitions of the "transverse" and "upright diameters" and, although it too aims at generality, it looks suspiciously like a double-branched hyperbola (*i.e.*, opposite sections). In light of the definitions of "transverse" and "upright diameters," then, one might conclude that, for Apollonius, the opposite sections are unambiguously two distinct curves. But the issue appears much less clear when one looks at the way these definitions are used, or rather, not used, in the text.

For example, Apollonius often uses the word "transverse (*plagia*)" to describe the diameter of a *single* curve. Thus, at the end of proposition I.13 he refers to the diameter of the ellipse as the "transverse side,"[2] and in

[1] Apollonius's geometrical outlook and related issues are discussed in much greater depth in *Apollonius of Perga's* Conica: *Text, Context, Subtext*, by myself and Sabetai Unguru.

[2] In this connection, one should note that at least in the Heiberg edition of Eutocius's commentary (pp. 200–201), the text describing the transverse and upright diameters contains not only the diagram suggesting the opposite sections, which I

proposition I.14, the proposition introducing the opposite sections, he refers to the diameter of each of the hyperbolas—and Apollonius does treat the opposite sections here as *two* hyperbolas—as the "transverse side of the figure (*tou eidous hē plagia pleura*)." In neither case, however, does Apollonius refer explicitly to the "transverse *diameter*," thus leaving the whole matter rather ambiguous. Moreover, the "upright diameter," whose definition makes sense only with regard to two curves, never appears in the propositions of Book I and appears only briefly in Book II (e.g., propositions II.37, 38). Instead, where one expects to find the "upright diameter" in Book I, one finds the "second diameter" defined after proposition I.16, or, more often, the "conjugate diameter." Even in the two propositions, II.37, 38, where Apollonius does mention the upright diameter explicitly, he is quick to point out that it is conjugate to the transverse diameter. Apollonius prefers the conjugate diameter to the upright diameter since the former is defined equally for one and two curved lines (*kampulēs grammēs kai duo kampulōn grammōn*) and this allows him, to some degree, to avoid the question of the plural nature of the opposite sections.

In the fourth book, however, Apollonius appears more sure that the opposite sections are, in fact, two curves. And there are numerous places in Book IV where the opposite sections are referred to as two curves explicitly; proposition IV.15, for instance, begins, "If in opposite sections a point is taken *between the two sections* (*metaxu tōn duo tomōn*) [emphasis added]..." This does not mean that the uneasiness with the opposite sections has vanished; it has only turned into meticulous care. It is this that is largely behind the seemingly endless cases that make the book tiring for some readers. For the non-perfunctory reader, however, these cases reveal particularly clearly the manner in which Apollonius attacks the difficulties of the opposite sections. As an example, let us look at the sequence IV.1–8.

Proposition IV.1 says, "If a point be taken outside a section of a cone or circumference of a circle, and from this point two lines are extended towards the section, of which one touches the section and the other cuts the section at two points, and if the line cut off inside the curve be divided in that ratio which the whole line cut off has to the part outside bounded between the point and the curve, so that the homologous lines are at the same

---

mentioned above, but also one suggesting an ellipse. Perhaps Eutocius shared Apollonius's ambivalence regarding the doubleness or singleness of the opposite sections; but whatever was the case, by juxtaposing the diagrams this way Eutocius was certainly reflecting Apollonius's ambiguous usage of *plagia*.

point, then the line drawn from the point of contact to the point of division [*i.e.*, the point dividing the inside segment] will intersect the curve, and the line drawn from the point of intersection to the outside point will touch the curve." Thus, let *D* be a point outside of a conic section *ABG*; let *DB* touch the section at *B* and *DEG* cut the section at *E* and *G*. Then if *DEG* be divided

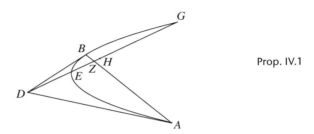

Prop. IV.1

at a point *Z* such that

$$GZ : ZE :: GD : DE,$$

and the line *BZ* be joined and extended, then *BZ* will cut the section again at a point *A,* while *DA* will touch the section at *A*. On the face of it, this is simply the converse of III.37, which states: "If two straight lines touching a section of a cone or circumference of a circle or opposite sections meet, and a straight line is joined to their points of contact, and from the point of meeting of the tangents some straight line is drawn across cutting the line (of the section) at two points, then as the whole straight line is to the straight line cut off outside, so will the segments produced by the straight line joining the points of contact be to each other." But III.37 refers to "a conic section or a circumference of a circle *or opposite sections*" [emphasis added]; IV.1 refers only to "a conic section." And, IV.2, therefore, almost teases in its opening sentence, "The things above were demonstrated for all the sections together," for still nothing has been said about *the opposite sections*. Apollonius might have mentioned the opposite sections in the enunciation to IV.1, and then furnish the proofs related to them later, as he does. Indeed, there are many examples in the *Conics* where Apollonius will state a proposition in general terms, but give a proof only for a special case, leaving the remaining cases for subsequent propositions (as in I.52–53, to take an arbitrary example). Here, however, Apollonius lets the opposite sections enter quietly in IV.4, and the cases concerning them unfold little by little in the propositions following.

There are three possible reasons why Apollonius proceeds in this way. The first is that by their presence in III.37 and absence in IV.1,

Apollonius draws the reader's attention to the case of the opposite sections. The second possibility is that he is, perhaps, mocking predecessors, such as Nicoteles, for thinking that by answering the question of the greatest number of points in which any two *conic sections* can meet they had also answered the question regarding *the opposite sections*. But I think a more compelling reason is that Apollonius saw the less problematic cases of the parabola, ellipse, and, particularly, the hyperbola as a gateway to the understanding of the cases concerning the opposite sections. In this connection, it is important to point out that there is little or no logical dependence among the propositions IV.1–8. Apollonius passes from the three initial conic sections to the opposite sections in a way that seems governed more by analogy than by strict deduction. By analogy I mean "any mode of reasoning in which one object or complex of objects is likened or assimilated to another (of the two particular instances between which a resemblance is apprehended or suggested, one is generally unknown or incompletely known, while the other is, or assumed to be, better known)...," as Lloyd puts it.[3]

What one sees in IV.1–8, then, is Apollonius advancing from the more easily conceived parabola, ellipse, and hyperbola, outward into the realm of the opposite sections. In the first step, in IV.2, Apollonius hardly proves anything; he merely explains that the proof in IV.1 is valid for the hyperbola so long as the point D is inside the angle contained by the asymptotes. Proposition IV.2, however, sets the hyperbola in its asymptotes, and the asymptotes are crucial in comprehending the opposite sections; among other things, the asymptotes, together with the transverse diameter, unify the opposite sections, since, as Apollonius proves in II.15, the asymptotes of opposite sections are common.[4] But the asymptotes turn out to be crucial also in mapping out the opposite sections, in being concrete markers for their geography. Thus, starting with IV.2, the point D has a definite place with respect the asymptotes and, on that account, with respect to the

---

[3] G. E. R. Lloyd, *Polarity and Analogy*, p. 175.

[4] The asymptotes also, in this sense, unify opposite conjugate sections, as Apollonius proves in II.17. unlike the opposite sections, however, the opposite sections and their conjugate sections are never in any way considered as one since the pair are not produced by a single conical surface.

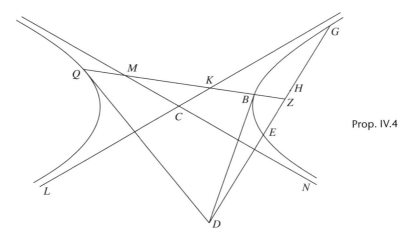

Prop. IV.4

opposite sections as well. So, in IV.4, D is in the angle adjacent to that contained by the asymptotes, and BZ will, accordingly, cut *the opposite section* at a point Q, and DQ will be tangent at Q. In IV.5, D is now on the asymptote, and BZ, in this case, remains parallel to the asymptote and, therefore, will never again cut either of the opposite sections.

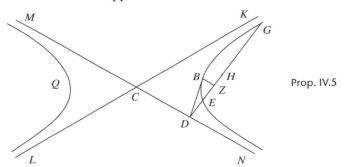

Prop. IV.5

Strictly speaking, IV.5 is not a converse of III.37, but of III.35. Yet, having no new enunciation (*i.e.*, no new *protasis*), it is presented as a continuation of IV.4. Moving from cutting to not cutting, therefore, it should be viewed as a "negative analogy"[5] to propositions IV.1–4. Proposition IV.6, on the other hand, does have a new enunciation: "If a point be taken outside a hyperbola, and from this point two lines are drawn to the section, one of which touches the section while the other is parallel to one of the asymptotes, and if the part of the latter line inside the section be set equal to the part cut off between the section and the point, then the line joined from the point of contact of the former line to the produced point will intersect the

5 Lloyd, *loc. cit.*

section, and the line drawn from the
point of intersection to the point outside will touch the section." Proposition
IV.6 is also the converse of a proposition different from III.37, namely,
III.30. Nevertheless, in light of what
we have just said regarding IV.4–5, it

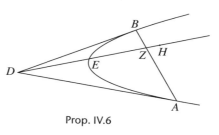

Prop. IV.6

too should be viewed as a "negative analogy" to IV.1, or, rather, to IV.2,
since, like IV.2, it begins with the hyperbola alone. The identical lettering in
the diagram makes it easy to follow the connection: point D is outside the
hyperbola and inside the angle contained by the asymptotes; BD touches the
section at B, while DE, drawn parallel to the asymptote instead of being
drawn cutting the section at two points, cuts the section at a *single point, E*.
Then, with line ZE set equal to DE, and line BZ joined and extended, BZ will
cut the section again at a point A, while DA will touch the section at A.

Propositions IV.7 and IV.8 follow IV.6 according to the same pattern
seen in IV.2–5. Thus, IV.7 has D in the angle adjacent to that contained by
the asymptotes, and BZ
cuts the opposite section
at Q, while, in IV.8, D is
on the asymptote, and
BZ is parallel to the second asymptote. Although
propositions 7 and 8 are
also converses to partic-
ular propositions in
Book III, namely, III.31
and III.34, respectively,
it should be clear from
the review given here
that their place among
the first eight proposi-
tions of Book IV is
determined by their ana-
logical relationship to
IV.1; otherwise (since, as

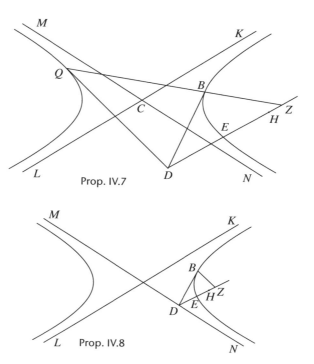

Prop. IV.7

Prop. IV.8

I noted above, deductive structure does not determine the order of propositions IV.1–8), they could have just as easily been arranged in the order of the
propositions in Book III to which they are converses.

Although Heath sees the sequence of propositions beginning with IV.9 only as providing a method of constructing two tangents to a conic section from a given external point,[6] one can easily take these, rather, as a further continuation of the sequence IV.1–8. Proposition IV.9 states, "If from the same point [point *D* in the propositions above] two lines are drawn each cutting a section of a cone or circumference of a circle in two points, and if the segments cut off inside are divided in the same ratio as the wholes are to the segments cut off outside, so that the homologous lines are at the same point, then the line drawn through the points of division will intersect the section in two points, and lines drawn from the points of intersection to the point outside will touch the curve." Thus, instead of a line touching the conic section and a line cutting the section at two points, as in IV.1, the initial configuration in IV.9 consists of two lines, *DQE* and *DHZ each* cutting the

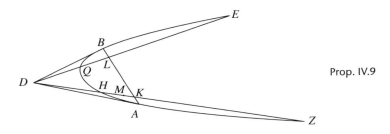

Prop. IV.9

section at two points *Q, E* and *Z, H* respectively; then with *DQE* and *DHZ* divided at points *L* and *K*, such that

$$EL : LQ :: DE : QD$$

and

$$ZK : KH :: DZ : DH,$$

the proposition shows that the line joining *L, K* will cut the section at two points *B, A*, and that the lines *DB* and *DA* will touch the section. Like IV.1, proposition IV.9 refers to a section of a cone but not to the opposite sections. Furthermore, the propositions subsequent to IV.9 follow precisely the same pattern, even presenting, to a great extent, the same wording, as those subsequent to IV.1; in ways just like those in IV.1–8, Apollonius moves from the

---

6 That these propositions *can* be the basis of such a method is not in question here; they can. But this cannot be Apollonius's principal aim in the sequence of propositions starting with IV.9. If it were, why is this aim mentioned neither in the letter opening the book nor in the general description of Book IV in the prefatory letter to Book I? Moreover, since he provides constructions of tangents to conic sections elsewhere, why not present an explicit construction here?

section of a cone (*i.e.* the parabola, ellipse, and hyperbola), to the hyperbola and its asymptotes, to the opposite sections. It is also not hard to discern the pattern set out in IV.1–8 in the overall structure of the book with its main theorems, IV.25, 38, and 55. Thus, IV.25 demonstrates that sections of a cone cannot meet one another or the circumference of a circle at more than four points; IV.38, that sections of a cone cannot meet opposite sections at more than four points; and IV.55, concerning only opposite sections, that opposite sections cannot meet one another at more than four points.

The analogizing steps from sections of a cone, to hyperbola and its asymptotes, to opposite sections, and from tangent to asymptote, from touching to being parallel occur, thus, over and over, and both in the small and in the large. Such repetition, however, should not leave the open-minded reader with a sense of monotony, but of playfulness—and, I think, it is one of the real delights of Book IV. One can see in this playfulness the artful way Apollonius contends with the main challenge of the book, the problem of how the opposite sections, specifically, meet other sections of a cone and other opposite sections and, furthermore, how he gives this problem foundation and context.

II

What has been said so far, then, should convince the reader that Apollonius's handling of the opposite sections in Book IV must be kept in mind both for the sake of understanding and appreciating Apollonius's accomplishments in this book as well as in other parts of the *Conics*. Of course, there are many more details to Apollonius's treatment of his subject than I have presented here, and, as the reader goes through the text, these will undoubtedly provide many opportunities for thoughtful inquiry. That said, the reader must remember that Book IV is not about the opposite sections *per se*; it is about the meeting of conic sections in the plane, and this, right from the start, raises another question, a more general and fundamental one, namely, how does Apollonius understand the *mode* of conic sections' meeting, that is, beyond the specific question regarding the *number* of points at which they meet?

Since conic sections are conceived as meeting somewhere, one ought to begin by asking "Where?" The question is much less innocent than it seems; packed in it is much of the distinction between Apollonius's geometry and Descartes's geometry, analytic geometry. For analytic geometry *begins* with place; it has a single reference plane (I shall speak here only of plane analytic geometry), the Cartesian plane, and any point in it can be given a location from the start—indeed, giving location is all one can do in this plane. The Cartesian plane is an abstraction, an objectless place of places (despite all Aristotle's objections to the very notion of such a thing![7]); in it, things are determined wholly by the location of their points, and the properties of things must be given, therefore, in terms of these locations, or *coordinates*. Such coordinates are conceivable only when one allows a length to correspond abstractly to a number, or, rather, a pair of lengths to a pair of numbers.[8]

In this setting, then, any discussion of objects and their properties is based on the possibility of relating them to numbers, and, further, to variables and equations. Hence, an inquiry into the meeting of curves becomes here an inquiry into the location of meeting points, or rather, into the equations determining such points through the determination of their coordinates. Needless to say, speaking about algebraic equations rather than the objects they determine obviates the theoretical difficulties with the nature of these objects (for example, whether the opposite sections are two or one), since the latter become mere representations of what is contained in the equations.

Now, in Apollonius's geometry, there is also a plane of reference. It is the plane that appears explicitly in the "constructions" at the close of Book I. This plane of reference, or, literally, "the plane set down" (*to hupokeimenon epipedon*), is not an abstract place of conic sections, it is the very plane that produces them—it is the cutting plane. Conic sections are always in this plane, and, unless Apollonius says otherwise, everything said about them must be assumed to be there too. Thus, when in I.24, for example, Apollonius says "If a straight line, meeting a parabola or hyperbola at a point, when

---

7 See the *Physics*, Book IV, on place and the void.

8 For numbers to play their role in that correspondence, they need to be understood in a way radically different from the way they were understood in Euclid's time, namely, as "a multitude composed of units," or, as Jacob Klein puts it, as a "definite number of definite things." This transformation in the understanding of number was the fruit of the 16th and 17th centuries, and it is the reason that analytic geometry had to wait until Descartes' time for its inception. Obviously, a full discussion of these matters would take us too far afield, but the reader can find a profound discussion of them in Jacob Klein's *Greek Mathematical Thought and the Origin of Algebra*, especially chapter 12.

produced both ways, falls outside the section, then it will meet the diameter," he may, with impunity, omit the possibility that the line not be in the cutting plane, in which case, obviously, the proposition would be untrue. Thus, the plane where conic sections occur and where lines and points in connection with them are drawn is that plane where they are *concretely*, namely, in the cutting plane; in this sense alone it is right to call that plane the plane of reference.

It is clear from this that the mode in which conic sections can meet must, for Apollonius, be crucially dependent on how conic sections come to be in a *given* cutting plane. The latter is the subject of propositions I.52–60. In I.52, for example, we are asked "to find in the plane (*heurein en tōi epipedōi*) the section of a cone called parabola." To do this, we start with a plane "set down," a line in the plane given in position (this does not mean that the line has a location specified in advance, as in analytic geometry, but only that it is fixed in place, that it cannot change its place)[9] and bounded at a point, and another line given only in magnitude. The task is accomplished when we are able to *conceive* (*noein*) a cone cut by the plane in such a way that a parabola is produced having the line given in position as its diameter, the bounding point as its vertex, and the line given in magnitude as its upright side. Apollonius's requiring only that a cone "be conceived" suggests that the "construction" in I.52 is not a construction at all; at least it is no more so than I.11 is a construction. Indeed, although proposition I.52 (together with I.53) is presented as a problem, it is, in fact, a true converse to proposition I.11. Propositions I.11–14 show how different planes can cut *a given cone* (strictly speaking, a conic surface) to produce different conic sections, each section having its own characterizing property, its *symptoma*; the sequence of propositions beginning with I.52, on the other hand, shows how different cones can give rise to different conic sections *in a given plane*, each section being prescribed as having a given upright side, diameter, and vertex.

In effect, the sequence of propositions ending Book I is, for conic sections, the counterpart to postulates 1 and 3 in Book I of the *Elements*, which allow one to draw a line and a circle of any size and anywhere in a plane. For Book IV, in particular, the importance of these propositions is

9 See Christian Marinus Taisbak's "Elements of Euclid's *Data*" for an enlightening discussion of what it means to be "given in position."

clear: they provide the theoretical basis for drawing, *in a single plane, any number of conic sections* having any diameter, vertex, and upright side. In this connection, it is telling that the final proposition in Book I involves, for the first time, drawing in a plane *two* sections at once[10]—appropriately enough, a pair of *opposite sections*, the conjugate sections.

This close link between Book IV and the propositions ending Book I is, I think, one of two reasons why Apollonius includes Book IV in that part of the *Conics* he calls "a course in the elements." Indeed, why Book IV *should* be considered among the elements is not obvious at first. Superficially, Book IV seems to have more in common with the books "fuller in treatment," Books V–VIII, rather than with the other "elementary" books, Books I–III: the propositions in Book IV are never used in the subsequent books (at least not in the extant Books V–VII, nor in Halley's reconstruction of Book VIII), and, like the later books, it is restricted to one narrow set of questions. Book VI, which concerns equality and similarity of conic sections and which ends with propositions echoing those at the end of Book I, might seem to have just as good a claim to be called elementary.

Part of the problem is that Apollonius does not say what the criteria are by which a book is termed elementary or not— and, probably, he saw no need to, since a "professional" mathematician like Eudemus who received Books I–III or like Attalus who received Books IV–VII (and, presumably Book VIII as well) would surely find an explication of what it means to be a book of elements somewhat gratuitous. Nevertheless, judging from Aristotle's remarks on "elements" and, especially, Proclus's much later remarks on the elementary character of Euclid's *Elements*, it *is* fairly clear that "elementary" should not be taken to mean "easy" or "for beginners only," but "fundamental."[11]

---

10 This does not include drawing a circle and another conic section; that occurs already in I.55. But drawing a circle in any plane requires no further qualification than Euc. I post. 3.

11 See Aristotle, *Metaphysics* 1014a 26–1014b 16 and Proclus, *Commentary on the First Book of Euclid's Elements*, tr. Glenn R. Morrow, pp. 59–60. In the latter, in particular, Proclus writes, "If we start from the elements, we shall be able to understand the other parts of this science; without the elements we cannot grasp its complexity, and the learning of the rest will be beyond us. The theorems that are simplest and most fundamental and nearest to first principles are assembled here [in the *Elements*] in a suitable order, and the demonstrations of other propositions take them as the most clearly known and proceed from them."

And, in the *Conics*, specifically, being *fundamental* means being close to the *geometrical origin* of the conic sections. That this, indeed, is crucial in the distinction between the first four and second four books can be inferred from the preface to Book V where Apollonius says, "...as for us, we have proven those things [concerning tangents to the sections] in Book I without making use, in our proof of that, of the topic of minimum lines; *for we wanted to make the place where those [things] were put near to our discussion of the derivation of the three sections* [emphasis added], in order to show in this way that in each of the sections there may occur an infinite number of properties and necessities of these things, as is the case with the original diameters."[12] The "derivation" Apollonius speaks of is, of course, the genesis of the conic sections truly *as sections of a cone*, and, in this connection, the reader will recognize the relevance of Apollonius mentioning the *original* diameter too, for it is precisely that diameter which arises *directly* from the sectioning procedure of the cone; it is the closeness to the sectioning procedure, rather than the simplicity of any relation associated with it, that makes the original diameter an *elementary* diameter.

Book IV belongs to the elements, therefore, because its immediate foundation and basic assumption lies in the possibility that different conic sections be placed together anywhere in a given plane, and this relates it intimately to propositions I.52–60, the converses of the derivation of the conic sections and the final climax of Book I.

That Book IV is rooted in the concrete geometrical derivation of the conic sections is, undoubtedly, the main reason it belongs to Apollonius's "course in the elements." But another, secondary, reason should also be mentioned here: it is the close analogy between the theme of Book IV and that of central propositions on circles in Book III of the *Elements*. The analogy between the conic sections and the circle is important throughout the *Conics* and is evident when one compares such propositions as II.26, II.45, II.47, III.16,17 with Euc. III.4, III.1, III.30, III.36, III.35, respectively. Sometimes even the wording of these is the same: for example, II.26 states "If in an ellipse or circumference of a circle two straight lines not through the center cut each other, then they do not bisect each other," while Euc. III.4 states, "If in a circle two straight lines cut one another which are not through the center, they do not bisect each another." Eutocius notes the relationship

---

12 Toomer's translation from the Arabic text of the Bānū Mūsā in *Apollonius* Conics *Books V to VII*. The first and third brackets are mine.

between Euclid's book on circles and Book IV of the *Conics* right at the start of his commentary on Book IV, and although he does not say precisely which propositions from the *Elements* he has in mind, the obvious ones are Euc. III.10, which states, "A circle does not cut a circle at more points than two," and Euc. III.13, which states, "A circle does not touch a circle at more points than one, whether it touch it internally or externally." The former corresponds, of course, to IV.25, 38, and 55, which together say conic sections (including opposite sections) cannot meet one another at more than four points, while the latter corresponds to such propositions as IV.27, 40, 51, 57, which together say that if conic sections (again, including opposite sections) touch at two points they cannot meet at a third point.

Besides these, there are other correspondences, albeit less direct ones. For instance, Euc. III.11,12, which say that the line joining the centers of circles that touch will pass through the point of contact, can be related to IV.34, which says that if ellipses or circles, having the same center, touch at two points, then the line joining the points of contact will pass through the common center. Eutocius specifically relates Apollonius's use of *reductio ad absurdum* in Book IV with Euclid's use of the same in his Book III. This is harder to see, in general, but, even here, one finds a clear connection between the method of proof in, for example, the second part of IV.25 and that of Euc. III.10. The analogy between these propositions in Book III of the *Elements* and the theme of Book IV of the *Conics* signs and seals the character of the latter as a book of elements. For with that analogy, not only is Book IV properly elementary by its propinquity to the geometrical origins of the conic sections, but also by its relationship with circles, whose familiarity, usefulness, and simplicity make them basic objects in any geometrical investigation.

The issues discussed above, namely, Book IV's relationship with the fundamental geometrical derivation of the conic sections in Book I, its preoccupation with the problems of the opposite sections, its link to studies on circles, and, with regard to the latter two, its use of analogical thinking, are, I believe, the right starting points for an informed reading of this book of the *Conics*. They are also issues that are completely obscured by an algebraic interpretation of the *Conics*, such as that of Heath and Zeuthen. For in that interpretation, the cone in I.11–14 is a mere aid to the derivation of the *equations* of conic sections, or, more radically, only a means of *representing* equations somehow already given;[13] the *Conics*, therefore, is built on an

---

13 See Zeuthen, *Die Lehre*, chaps. 2–3, *passim*.

algebraic foundation rather than a geometric one, so that not only can one interpret the results in the *Conics* algebraically (and one who already knows algebra and analytic geometry can do this), but, in a sense, one must interpret them this way. Accordingly, speaking about analogies among the conic sections, opposite sections, and circle, for example, becomes an otiose affair, since the precise relationships among these are laid out almost trivially in the form of their equations—that is, after all, the great strength of algebra. Furthermore, the way curves meet becomes inextricably tied with the solutions of equations. In fact, when one thinks of the third book of Descartes' *La Géométrie* one realizes just how much the nature of curves and their intersections are subordinated to the nature of equations and their roots when geometry is put in a wholly algebraic setting. It is no wonder that Heath and Zeuthen should find Book IV dull and look for problems that it solves beyond its stated goal of investigating the meeting of conic sections. On the contrary, I hope that what has been said above leaves the reader with a sense that Book IV has depth and interest on its own terms.

III

In urging the reader to read Book IV on its own terms, however, I am open to the charge of being inconsistent. After all, what I am presenting here is a *translation* of Book IV, and every translation takes the reader a step away from the text itself. On the other hand, being consistent would mean not translating the *Conics* at all, and that would leave most readers helplessly nodding as interpreters of the text assure them that Apollonius said this or said that. So, right from the start, in taking up the task of translating Book IV I have agreed to live with compromise. The only amends I can make is to state as plainly as possible what principles have guided the translation and where there have been blatant deviations from the text. This, then, I shall now do.

As I said at the start of this introduction, in translating the text the desideratum has always been to keep out all foreign and anachronistic elements as much as possible. Accordingly, there is no modern symbolism in the translation, no matter how seemingly benign, including those symbols and abbreviations that Taliaferro employed in the translation of Books I–III to 'make the reading and the mechanic of study easier' (I do, however, preserve Taliaferro's conventions in the notes to the text). I have also used, with one exception, different words for the different ways in which conic sections are described as meeting. Thus, I translate '*sumballein*' as 'to meet', '*tem-nein*' as 'to cut' and '*sumpiptein*' as 'to intersect'. The exception to this is my translating both '*ephaptesthai*' and '*epipsauein*' as 'to touch'. On the other hand, I have used throughout Latin instead of Greek letters. However, I have employed a single lettering scheme in which (with two exceptions—'theta' and 'psi') the Latin letters correspond to the Greek either in sound or shape, for example, '*G*' corresponds to 'gamma' and '*H*' to 'eta'. (The complete lettering scheme follows this introduction.)

A more serious deviation from the Greek has been the use of the simple passive imperative for the perfect passive imperative. This has not been done without serious reservations. Strictly speaking, where I write "Let some point *D* be taken outside the section," a more literal translation would have, "Let some point *D* have been taken (*eilēphthō*) outside." *eilēphthō* is the perfect passive imperative of *lambanein*. In almost all Greek mathematics this is the form used in that part of a demonstration where constructions are prescribed.[14] I have translated it as an ordinary passive only to avoid an overly awkward English translation, especially where there is a string of such commands. However, the ubiquitous use of this grammatical form may, indeed, point to a significant aspect of Greek mathematical demonstrations. On this subject, Lachterman writes, "With only a few notable exceptions, Euclid chooses to put these verbs [verbs for "operation"] in the perfect passive imperative. Bisecting a line-segment at a point is expressed as 'let it have been cut in two' (*tetmesthō...dicha*). ... The importance of this stylistic trait is twofold: First, Euclid does not give instructions or permission to a reader to carry out a specified operation but casts the operation into impersonal, passive form; second, the perfect tense tells us that the relevant operation has already been executed prior to the reader's encounter with the unfolding proof.... Now we are indeed in unfamiliar territory. As though to strengthen the hand of the Speusippeans, Euclid invites us, not to perform an operation on our own, nor to observe him in performing the operation before our eyes, but rather to consider the operation as already anonymously performed before the 'present moment' in which we are following and taking stock of the movement from enunciation to conclusion. This verbal operator does not so much suppress time as shift it

---

14 Heath, *Euclid*, I, p. 242.

backwards, into an unnoticed past during which neither teacher nor student was necessarily on hand".[15]

Taisbak beautifully puts the case as follows: "It might be appropriate at this stage to introduce that effective constructor, The Helping Hand, the well-known factotum in Greek geometry, who sees that lines be drawn, points be taken, perpendiculars dropped, etc. The perfect imperative passive is its verbal mask. No one who has read Euclid's *Elements* in Greek will have missed it. Never is there any of the commands or exhortations so familiar from our own classrooms: 'Draw the median from vertex *A*,' or 'Cut the circle by that secant,' or 'Let us add those squares together.' The Helping Hand was always there to see that these things were done. I wonder how Europe ever inherited Greek mathematics without the perfect imperative passive."[16]

Another compromise made in the name of readability is the omission of brackets indicating interpolations. Most of these interpolations were made simply to make a Greek sentence an English one. So, having done that, to explain the interpolation of every "the," "if," or "then" here would surely be an act of pedantry in the extreme. Two interpolations, however, are made throughout the text and ought to be pointed out since they have some bearing on whether the text is to be read algebraically or geometrically. In Greek mathematical texts, the phrases "*to hupō tōn AB, CD periechomenon orthogōnion*" and "*to apo tēs AB tetragōnon*," meaning, respectively, "the rectangle contained by *AB, CD*," and "the square on *AB*," are invariably abbreviated to "*to hupō tōn AB, CD*," and "*to apo AB*," "that under *AB, CD*" and "that on *AB*." It is these phrases that appear all too often in translations as "$AB \times CD$" and "AB^2." In the translation, therefore, I have chosen what I consider to be the much lesser of two evils and have written "the rectangle contained by *AB, CD*," and "the square on *AB*" everywhere the phrases "*to hupō tōn AB, CD*," and "*to apo AB*" appear.

As mentioned above, some of the conventions I have adopted deviate from Taliaferro's; however, there are other conventions that I have happily embraced. One of these is repeating diagrams so that a given diagram always appears together with the relevant text. Also, only proper enunciations are

---

15 Lachterman, *The Ethics of Geometry*, p. 65.

16 Taisbak, "Elements of Euclid's *Data*," p. 144.

italicized: where a proposition has no enunciation (for example, IV.19), or where a proposition continues from a previous one (for example, IV.2–5), no italics are used.

As for the Greek text, I have used Heiberg's edition throughout. Heiberg's Latin translation, despite its free use of algebraic notation, has also been of great use, as has been Ver Eecke's French translation.

Acknowledgements

This translation could not have been completed without the advice, support, and encouragement of many people, and they all deserve my thanks. Of these, the first I should thank is my friend and teacher, Sabetai Unguru, who read through the translation and introductory essay, and who, afterwards, was always ready to listen to my questions and to guide me in my deliberations. Next is certainly Joe Cohen who made the initial contact with Green Lion Press and with whom I consulted on many occasions. I am grateful to Bill Donahue for his thoughtful comments on the translation, one of which spared me the embarrassment of a serious error appearing in the manuscript version, and to Hagai Resnik for his expert help in my (often agonizing) dealings with computers. Also, I offer long overdue thanks to Harvey Flaumenhaft who first introduced me to Apollonius a little over twenty years ago; indeed I am sure that our conversations then are still at work in my thoughts about Apollonius now. Finally, I want to thank my wife Yifat and our daughters, Avigail and Inbar, just for making the world a pleasant place and making all hardship, worry, and frustration easy to bear.

Michael N. Fried
Revivim, December 2000

Lettering Scheme for Diagrams in Book IV

α	A	ν	N
β	B	ξ	C
γ	G	o	O
δ	D	π	P
ε	E	ρ	R
ζ	Z	σ	S
η	H	τ	T
θ	Q	υ	U
ι	I	ϕ	F
κ	K	χ	X
λ	L	ψ	V
μ	M	ω	W

A Note on Citations

Unless otherwise noted, references to Heiberg's remarks are to be found on the pages containing the propositions to which they refer in Heiberg's edition of the *Conics*. Book IV of the *Conics* is in Volume II of that edition.

Likewise, references to ver Eecke's translation are by book and proposition number.

Propositions in Euclid's *Elements* are cited as "Eucl." followed by the book number (roman numeral) and the proposition number (arabic numeral).

A citation consisting only of a roman numeral followed by an arabic numeral refers to another proposition in the *Conics*.

Citations included in parentheses in Apollonius's text are not in the original, but are provided by the translator as an aid to the reader.

CONICS

BOOK FOUR

APOLLONIUS to ATTALUS,[1] greeting!

Earlier, I presented the first three books of my eight book treatise on conics to Eudemus of Pergamum, but with his having passed away I decided
to write out the remaining books for you, because of your earnest desire to
have them. To start, then, I am sending you the fourth book. This book
treats of the greatest number of points at which sections of a cone can meet
one another or meet a circumference of a circle, assuming that these do not
completely coincide, and, moreover, the greatest number of points at
which a section of a cone or a circumference of a circle can meet the opposite sections. Besides these questions, there are more than a few others
of a similar character. Conon of Samos presented the first mentioned question[2] to Thrasydaeus without giving a correct demonstration, for which he

---

NOTE: Footnotes are the translator's.

[1] Nothing definite, to my knowledge, is known about Attalus. Toomer, however,
argues convincingly that this Attalus cannot be King Attalus I of Pergamum, who
reigned in Pergamum at the time Apollonius possibly was writing the *Conics*, for, if
King Attalus were intended, surely Apollonius would have referred to him using the
word "King" ($\beta\alpha\sigma\iota\lambda\epsilon\acute{\upsilon}s$) which he does not (see Toomer, "Apollonius of Perga,"
p. 179). Toomer elsewhere has "speculated whether Apollonius's addressee is to be
identified with the mathematician Attalus of Rhodes (known mainly from
Hipparchus' criticism in his commentary on Aratus), but apart from chronological
difficulties, $\H{A}\tau\tau\alpha\lambda os$ is too common a name for any certainty" (*Apollonius* Conics
Books V to VII, p. xii, n. 2).

[2] Following Heath (*Apollonius of Perga*, pp. lxxii–lxxiii), I take the first question to be
that concerning the greatest number of points at which sections of a cone can meet
one another or meet a circumference of a circle, the second, the greatest number of
points at which a section of a cone or a circumference of a circle can meet the opposite sections, and the third, the greatest number of points at which the opposite
sections can meet one another, while the other questions "of a similar character," I
take to be those concerning the number of points conic sections, circles, and opposite sections can touch one another. This seems to me a reasonable division and
one that is completely consistent with the structure of the book.

was rightly attacked by Nicoteles of Cyrene.[3] As for the second question, Nicoteles, in replying to Conon, only mentions that it can be demonstrated, but I have found no demonstration either by him or by anyone else. Regarding the third and similar questions, however, I have not found them even noticed by anyone. And all these things just spoken of, whose demonstrations I have not found anywhere, require many and various striking theorems, of which most happen to be presented in the first three books of my treatise on conics, and the rest in this book. The investigation of these theorems is also of considerable use in the synthesis of problems and limits of possibility ($\delta\iota o\rho\iota\sigma\mu o\acute{\upsilon}\varsigma$). So, Nicoteles was not speaking truly when, for the sake of his argument with Conon, he said that none of the things discovered by Conon were of any use for limits of possibility; but even if the limits of possibility are able to be obtained completely without these things, yet, surely, some matters are more readily perceived by means of them, for example, whether a problem might be done in many ways, and in how many ways, or again, whether it might not be done at all.[4] Moreover, this preliminary knowledge brings with it a solid starting point for investigations, and the theorems are useful for the analysis of

---

[3] Of the three names mentioned here, Conon of Samos, Thrasydaeus, and Nicoteles of Cyrene, only about Conon can we say something definite. We know that he communicated with Archimedes and was greatly respected by him. We also know that he had died before Archimedes published his *Quadrature of a Parabola*, as Archimedes tells us in the letter to Dositheus opening that book. This means that Conon died about the time Apollonius was born, ca. 240 B.C.E., and, therefore, the questions with which Apollonius was engaged in Book IV were already being asked a generation before Apollonius. About Nicoteles, Knorr tentatively mentions the possibility that Nicoteles is none other than Nicomedes. He says, "...Apollonius names one Nicoteles of Cyrene who was especially harsh in his criticisms of Conon's studies of intersecting conics. One is tempted to suggest that a scribal slip in the transmission of the *Conics* has altered an original 'Nicomedes'. Clearly, in both contexts an abrasive individual is encountered, engaged in advanced geometrical studies associated with the work of geometers around the middle of the 3rd century or a bit later. But beyond this, there seems to be no other evidence to argue the identification" (*The Ancient Tradition of Geometric Problems*, p. 282, n. 41).

[4] Literally, "...whether [something] might come to be ($\mathring{\alpha}\nu$ $\gamma\acute{\epsilon}\nu o\iota\tau o$) in many ways, and in how many ways, or again, that it may not come to be." I am following Heiberg here in taking the "something" to be "a problem". Heiberg's translation reads "...velut problema compluribus modis vel tot modis effici posse aut rursus non posse..." Ver Eecke similarly translates the phrase as referring to problems and their solutions: "...soit la possibilité de solutions multiples, ou en nombre déterminé, soit, au contraire, l'impossibilité d'une solution."

limits of possibility.[5] But apart from such usefulness, these things are also worthy of acceptance for the demonstrations themselves: indeed, we accept many things in mathematics for this and no other reason.[6]

PROPOSITION 1

If a point be taken outside a section of a cone or circumference of a circle, and from this same point two lines are extended towards the section, of which one touches the section and the other cuts the section at two points, and if the line cut off inside the curve be divided in that ratio which the whole line cut off has to the part outside bounded between the point and the curve,[7] so that the homologous lines are at the same point,[8] then the line drawn from the point of contact to the point of division [i.e., the point dividing the inside segment] will intersect the curve, and the line

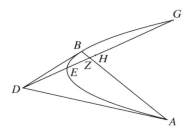

---

[5] P. ver Eecke translates the phrase πρὸς τὰς ἀναλύσεις δὲ τῶν διορισμῶν simply as "...aux développements des discussions." I have chosen, however, to emphasize a more "technical" reading of διορισμός, namely, as "limits of possibility."

[6] Apollonius stresses the intrinsic interest of his subject also in the letter to Attalus that introduces the fifth book. He writes, "As for the propositions in which we speak of the minimum lines, we have separated them out and treated them individually, after much investigation, and have attached the discussion of them to the discussion of the maximum lines which we mentioned above, because of our opinion that students of this science need them for the knowledge of analysis and determination of problems and their synthesis, *not to speak of the fact that they are one of the subjects which deserve investigation in their own right* [emphasis added]," (Toomer's translation from the Arabic in *Apollonius* Conics *Books V to VII*, p. 4).

[7] I am translating γραμμή as "curve" to avoid any confusion with the other *straight* lines (εὐθεῖαι) spoken of in the proposition.

[8] What Apollonius is saying here is that the four lines involved in the proportion are laid out *in the same direction*; thus, referring to the diagram, *GD* and *GZ*, which correspond in the proportion, must both extend from the same point *G*.

drawn from the point of intersection to the outside point will touch the curve.

For let *ABG* be a conic section or circumference of a circle, and let the

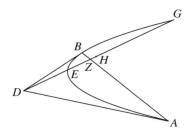

point *D* be taken outside the section; from *D* let *DB* touch the section at point *B* and let *DEG* cut the section at points *E*, *G*, and

     let *GZ* have to *ZE* the ratio that *GD* has to *DE*.

I say that the line drawn from *B* to *Z* will intersect the section, and the line drawn from the point of intersection to *D* will touch the section.[9]

For let *DA* be drawn from *D* touching the section, and let *BA* be joined cutting *EG*, if possible, not at *Z*, but at *H*. Now, since *BD*, *DA* touch the section, *BA* is drawn from the points of contact, and *GD* goes through *AB*, cutting the section at *G*, *E* and intersecting *AB* at *H*,

     as *GD* is to *DE*, so is *GH* to *HE* (III.37)[10]

But this is absurd, for it was assumed that

---

as *GD* is to *DE* so is *GZ* to *ZE*.[11]

Therefore, *BA* does not cut *GE* in a different point from *Z*; therefore, it cuts *GE* at *Z*.

PROPOSITION 2

The things above were demonstrated for all the sections together (κοινῶς). However, regarding the hyperbola only, if *DB* touches the hyperbola while *DG* cuts it at two points *E*, *G*, and if the point of contact, *B*, is contained by *E*, *G*, and *D* is inside the angle contained by the asymptotes, then the proof is carried out similarly. For from *D* it is possible to draw another line, *DA*, touching the hyperbola and the rest of the proof is done similarly.

PROPOSITION 3

The same things being the case,[12] let the points *E*, *G* not contain the point of contact, *B*, between them, and let *D* be inside the angle contained by the asymptotes. Therefore, from *D*, it is possible to draw another line, *DA*, touching to the section, and the rest is proven as before.

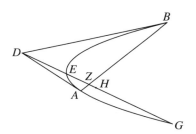

---

is to the straight line cut off outside, so will the segments produced by the straight line joining the points of contact be to each other."

Throughout the text, I shall indicate propositions in the *Conics* only by book and proposition number. Thus, II.49 refers to *Conics*, Book II, proposition 49.

[11] The problem is that then *GZ* : *ZE* :: *GH* : *HE*, so that two *different* points would divide the segment *GE* in the *same* ratio.

[12] Namely, *ABG* is a hyperbola, point *D* is taken outside the hyperbola, *DB* is drawn touching the hyperbola at point *B*, *DEG* is drawn cutting the hyperbola at *E* and *G*, and *EG* is divided at *Z* such that *GZ* : *ZE* :: *GD* : *DE*.

PROPOSITION 4

The same things being the case, if the points of intersection[13] *E, G* contain the point of contact *B,* while the point *D* be in the angle adjacent to that contained by the asymptotes, then the line from the point of contact to the point of division intersects the opposite section, and the line drawn from the point of intersection to point *D* will touch the opposite section.

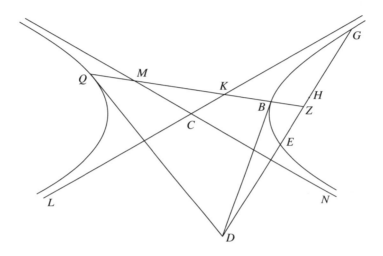

Let *B* and *Q* be opposite sections, let *KL* and *MCN* be asymptotes, and let point *D* be in the angle *LCN.* Furthermore, let *DB* drawn from *D* touch, while *DG* cut one of the sections, let the points of intersection *E, G* contain the point of contact *B,* and

let *GZ* have to *ZE* the same ratio that *GD* has to *DE.*

It is to be shown that the line joined from *B* to *Z* will intersect the section *Q,* and that the line from the point of intersection to *D* will touch the section.

Let *DQ* be drawn from *D* touching the section, and let the line *QB* fall, if possible, not on *Z,* but on *H.* Therefore,

as *GD* is to *DE, GH* is to *HE* (III.37).

But this is absurd, for it was assumed that

as *GD* is to *DE, GZ* is to *ZE.*

---

[13] That is, the points of intersection of *DG* and the hyperbola *ABG.*

PROPOSITION 5

The same things being the case,[14] if the point D be on an asymptote, the line drawn from B to Z will be parallel to that asymptote.

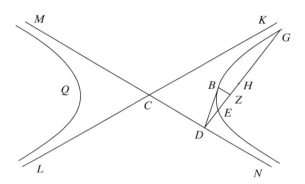

For let these same things be supposed, let point D be on one of the asymptotes MN.

It is to be shown that the line drawn from B parallel to MN will fall on the point Z.[15]

For if not, let the line be BH, if possible. But then

as GD is to DE, GH will be to HE (III.35).[16]

which is impossible.

---

[14] Namely, that B and Q are opposite sections, KL and MCN are the asymptotes, DB touches the section at B, DG cuts the section at E and G, and $GZ : ZE :: GD : DE$.

[15] Showing that the line BZ will be parallel to the asymptote and showing that the line through B parallel to the asymptote passes through Z are the same provided one assumes, what is equivalent to Eucl. I. post. 5, that only one line can be drawn through a given point parallel to a given line. Evidently, Apollonius takes this to be so obvious that he sees no need to point it out.

[16] *Conics* III.35: "With the same things being so [namely, if some point is taken on one of the asymptotes of an hyperbola, and a straight line from it touches the section, and through the point of contact a parallel to the asymptote is drawn], if from the point taken some straight line is drawn cutting the section at two points, then as the whole straight line is to the straight line cut off outside, so will the segments of the straight line cut off inside [by the line drawn parallel to the asymptote] be to each other."

PROPOSITION 6

If a point be taken outside a hyperbola, and from this point two lines are drawn to the section, one of which touches the section while the other is parallel to one of the asymptotes, and if the part of the latter line inside the section be set equal to the part cut off between the section and the point, then the line joined from the point of contact of the former line to the produced point[17] will intersect the section, and the line drawn from the point of intersection to the point outside will touch the section.

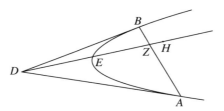

Let *AEB* be the hyperbola and let *D* be some point taken outside, and, to start, let *D* be inside the angle contained by the asymptotes, and from *D* let *BD* touch the hyperbola, let *DEZ* be parallel to one or the other of the asymptotes, and set *EZ* equal to *DE*.

I say that the line joined from *B* to *Z* will intersect the section, and the line from the point of intersection to *D* will touch the section.

For let *DA* be drawn touching the section, and join *BA* cutting *DE*, if possible, not at *Z* but at some other point *H*. Then

$$DE \text{ will be equal to } EH \text{ (III.30).}[18]$$

But this is absurd, for it was assumed that

$$DE \text{ is equal to } EZ.$$

---

[17] That is, the endpoint of the line inside the section that has been set equal to the line cut off between the outside point and the section.

[18] *Conics* III.30: "If two straight lines touching an hyperbola meet, and through the points of contact a straight line is produced, and through the point of meeting a straight line is drawn parallel to some one of the asymptotes and cutting both the section and the straight line joining the points of contact, then the straight line between the point of meeting and the straight line joining the points of contact will be bisected by the section."

PROPOSITION 7

The same things being the case, let D be in the angle adjacent to that contained by the asymptotes.

I say that the same things will come to pass.

For let DQ be drawn touching the hyperbola, and with QB having been joined, let it fall, if possible, not on Z, but on H. Therefore

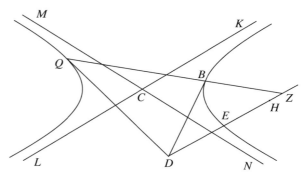

DE is equal to EH (III.31). But this is absurd, for it was assumed that

DE equals EZ.

PROPOSITION 8

The same things being the case, let D be on one of the asymptotes, and let the remaining constructions be the same.[19]

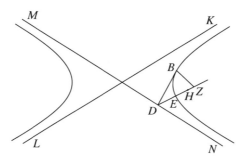

I say that the line drawn from the point of contact to the end of the line cut off[20] will be parallel to the asymptote on which D is situated.

---

[19] I am following ver Eecke here who translates καὶ τὰ λοιπὰ γιγνέσθω τὰ αὐτά as "...et que les autres constructions soient les mêmes." A somewhat more literal translation would be "and let the rest come to be from the same."

[20] The line referred to is the line from D parallel to the other asymptote, and its "end" is the end of the segment inside the hyperbola equal to the segment cut off between D and the hyperbola.

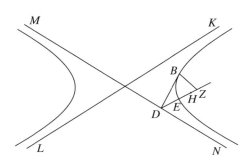

Let there be the things just mentioned, and

set *EZ* equal to *DE*,

and from *B* let line *BH* be drawn, if possible, parallel to *MN*. Therefore,

DE is equal to *EH* (III.34).

But this is absurd, for it was assumed that

DE is equal to *EZ*.

PROPOSITION 9

If from the same point two lines are drawn each cutting a section of a cone or circumference of a circle in two points, and if the segments cut off inside are divided in the same ratio as the wholes are to the segments cut off outside, so that the homologous lines are at the same point, then the line drawn through the points of division will intersect the section in two points, and lines drawn from the points of intersection to the point outside[21] will touch the curve.[22]

Let *AB* be a curve such as we have described,[23] and from a point *D* let *DE* and *DZ* be drawn cutting the curve at *Q*, *E* and at *Z*, *H* respectively. Furthermore, let

the ratio that *EL* has to *LQ* be the same ratio that *DE* has to *QD*,

and

that which *ZK* has to *KH* be the same that *DZ* has to *DH*.

I say that the line joined from *L* to *K* will intersect the section at both ends, and the lines joined from the points of intersection will touch the section.

---

[21] The "point outside" is that from which the two cutting lines extend. Apollonius is not as careful here as he was in Proposition 1 to say that this point is, indeed, outside the curve, although the rest of the enunciation makes that fact abundantly clear.

[22] Heath, among others, has said that IV.9 provides a method for drawing two tangents to a conic section from an exterior point (see *A History of Greek Mathematics*, II, 157–8). While this is true, it is not clear that providing such a method is Apollonius's intention. First, Apollonius does not present proposition IV.9 as a construction, and, second, in II.49, where Apollonius does give a method for drawing a tangent from a point outside a section, he emphasizes that, by the procedure given, two tangents may be drawn.

[23] Namely, a conic section or circumference of a circle.

For since *ED* and *ZD* both cut the section at two points, it is possible to draw a diameter of the section through *D*, and, with that, also lines touching the section on either side.[24] Let lines *DB* and *DA* be drawn touching the section, and join *BA* not passing through *L*, *K*, if

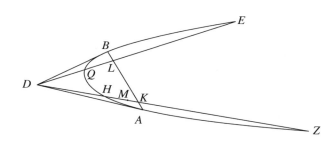

possible, but through only one of the two, or through neither.

First, let it pass through *L* only and let it cut *ZH* at *M*.

As *ZD* is to *DH*, therefore, so is *ZM* to *MH* (III.37).

But this is absurd, for it has been assumed that

as *ZD* is *DH* so is *ZK* to *KH*.

If *BA* passes through neither *L* nor *K*, then the absurdity occurs with regards to each line *DE*, *DZ*.

---

[24] The second part of this statement, namely, that being possible to draw a diameter through *D* means that it is possible to draw tangents from *D*, could plausibly be related to *Conics* II.49 in which Apollonius shows how to draw a tangent from a point outside the section since, with one exception (*viz.,* when the section is a hyperbola and *D* is in the angle adjacent to that contained by the asymptotes, so that one cannot draw a diameter from *D* to the section), Apollonius's construction of the tangent from a point *D* begins by drawing a diameter through *D*; the construction then follows from the fact that with *DA* as the diameter, *DT* the tangent, and *TK* drawn ordinatewise,

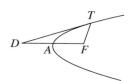

$$DA = AK$$

in the case of the parabola, and

$$BD : DA :: BK : KA$$

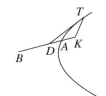

in the case of the hyperbola and ellipse. However, it is not enough just to be able to draw a diameter from point *D*, for if, as was pointed out regarding IV.1, the section is a hyperbola and *D* lies in the angle contained by the extension of the asymptotes beyond the center, then there is no tangent from *D* to the section while there is, indeed, a diameter through *D*. Nor does the mere fact that the two lines both cut the section in two points obviate all difficulties since drawing both a tangent and diameter from *D* demands that *D* be in the angle contained by the asymptotes. But, by *Conics* II.25, this will occur if two lines each meet a hyperbola at two points *and no point of meeting of either is contained by the points of meeting of the other*. For these

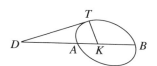

reasons, it seems probable that, like the similar statement in IV.1, this statement is the work of a later and less careful writer than Apollonius (perhaps the author of the remark in IV.1).

PROPOSITION 10

The things above are common to all the sections. However, regarding the hyperbola only, if the other things are assumed, and if the points of intersection of the one line are contained by the points of intersection of the other, and if the point D is inside the angle contained by the asymptotes, the same things said above will happen, as we said above in proposition 2.[25]

PROPOSITION 11

The same things being the case, if the points of intersection of one of the lines do not contain the intersection points of the other line, then point D is in the angle contained by the asymptotes,[26] and the drawing ($\kappa\alpha\tau\alpha\gamma\rho\alpha\phi\acute{\eta}$) and the proof will be the same as in Proposition 9.

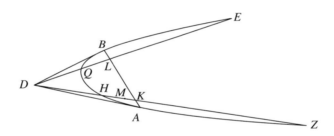

PROPOSITION 12

The same things being the case, if the points of intersection of one of the lines contain those of the other, and if the chosen point is in the angle adjacent to that contained by the asymptotes, then the line drawn through the points of division and extended will intersect the opposite section, and the lines drawn from the points of intersection to point D will touch the opposite sections.

---

25 In fact, since the point D lies in the angle contained by the asymptotes by assumption, one can apply the procedure in IV.2 twice, once for the first line and once for the second.

26 This, as Heiberg notes, is clearly false.

Let *EH* be a hyperbola, *NC*, *OP* the asymptotes, and *R* the center. Furthermore, let point *D* be in the angle *CRP*, let *DE*, *DZ* be drawn cutting the hyperbola each in two points, let *E,Q* be contained by *Z,H*, and let it be that

as ED is to DQ, EK is to KQ,

and that

as ZD is to DH, ZL is to LH.

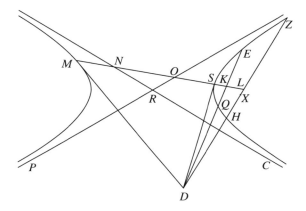

It is to be shown that the [line] through *K* and *L* will intersect both [the section] *EZ* and also the opposite section, and the lines from the points of intersection to *D* will touch the sections.

Let *M* be the opposite section, and from *D* let *DM* and *DS* be drawn touching the section, let *MS* be joined, and, if possible, let it not pass through *K, L* but rather through only one of the two points, or through neither.

First, let it pass through *K* and cut *ZH* at *X*. Therefore,

as ZD is to DH, XZ is to XH (III.37).

But this is absurd, for it has been assumed that

as ZD is to DH, ZL is to LH.

If *MS* passes through neither *K* nor *L*, then the impossibility occurs with regards to each line *ED* or *DZ*.

PROPOSITION 13

The same things being the case, if the point D be on one of the asymptotes, and the remaining things are assumed to be the same, then the line drawn through the points of division will be parallel to the asymptote on which the point is situated, and, extended, will intersect the section. Moreover, the line drawn from the point of intersection to the point situated on the asymptote will touch the section.

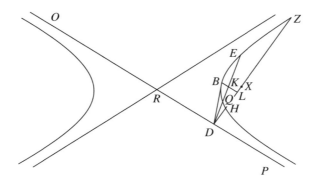

Let there be a hyperbola and its asymptotes, and let *D* be taken on one of the asymptotes. Let lines be drawn and divided as we have said above, and let a line *DB* be drawn from *D* touching the section.

I say that the line drawn from *B* parallel to *PO* passes through *K, L*.

For if not, then surely it will pass through one of these points or through neither.

Let it pass through *K* only. Therefore,

as *ZD* is to *DH*, *ZX* is to *XH* (III.35).

But this is absurd. Therefore the line drawn through *B* parallel to *PO* will not pass through *K* only. Therefore, it will pass through both points.

PROPOSITION 14

The same things being the case, if point D be on one of the asymptotes, and DE cuts the section at two points, while DH, being parallel to the other asymptote, cuts the section at H only, and if it comes to be that as DE is to DQ, EK is to KQ, while HL, equal to DH, is placed in a line with DH,[27] then the line drawn through points K, L will be parallel to the asymptote and will intersect the section, and the line drawn from the point of intersection to D will touch the section.

---

[27] That is, *DH* is extended to *L* such that *DH = HL*.

For, similar to what was said above,[28] drawing *DB* touching the section,
I say that the line drawn from *B* parallel to the asymptote *PO* will pass
through points *K, L*.

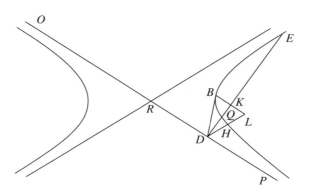

Indeed, if it passes through *K* only, *DH* will not be equal to *HL* (III.34),
which is absurd. And if it passes through *L* only, then it will not be that
as *ED* is to *DQ*, *EK* is to *KQ* (III.35),
and if it passes neither through *K* nor through *L*, the absurdity will occur in
both ways.

Therefore, it will pass through both points.

PROPOSITION 15

*If in opposite sections a point is taken between the two sections,[29] and if a
line from the same touches one of the opposite sections, while another line
cuts each of the opposite sections, and if the ratio that the line between the
point and the one section, which the first line does not touch, has to the line
between the point and the other section is the same as that that a line greater
than the one between the sections has to its excess over the latter, set in a*

---

28 In proposition IV.13.

29 Note that Apollonius is referring to the opposite sections explicitly as *two* sections
here. The importance of this is discussed in the translator's introduction to Book IV.

line with it and with the homologous lines being at the same extremities,[30]
then the line drawn from the extremity of the greater line to the point of
contact will intersect the section, and the line drawn from the point of inter-
section to the original point taken will touch the section.

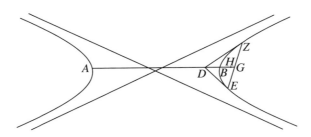

Let there be opposite sections *A*, *B* and let some point *D* be taken between
the sections and in the angle contained by the asymptotes, and from this
point let *DZ* be drawn touching the section while *ADB* be drawn cutting the
section. Furthermore,

                let *AG* have to *GB* the ratio that *AD* has to *DB*.

It is to be shown that the line extended from *Z* to *G* will intersect the section, and the line drawn from the point of intersection to *D* will touch the
section.

For since *D* is in the angle containing the section, it is possible to draw
from *D* another line touching the section (II.49). Let *DE* be drawn, let *ZE* be
drawn and let it pass, if possible, not through *G*, but through *H*. It will be
then that

                as *AD* is to *DB*, *AH* will be to *HB* (III.37),

which is absurd. For it was assumed that

                as *AD* is to *DB*, *AG* is to *GB*.

---

[30] Referring to the diagram, the lines *AD, DB* are said to have the same ratio as *AG,
GB* such that the *corresponding* lines are those with the *same* endpoint—in other
words, that the proportion given is

                AD : *DB* :: *AG* : *GB*

and *not*, say,

                AD : *DB* :: *GB* : *AG*.

PROPOSITION 16

The same things being the case, let *D* be a point in the angle adjacent to that contained by the asymptotes, and let the remaining constructions be the same.

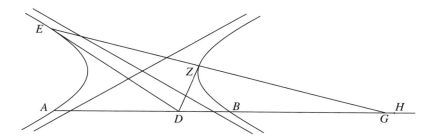

I say that the line joining *Z* to *G* will, when extended, intersect the opposite section, and the line from the point of intersection to *D* will touch the opposite section.

For let the same things be as before, and let point *D* be in the angle adjacent to that contained by the asymptotes, and let *DE* be drawn from *D* touching section *A*, let *EZ* be joined and, when extended, let it not pass through *G*, but through *H*, if that is possible. It will be then that

as *AH* is to *HB*, *AD* will be to *DB* (III.39),

which is absurd. For it was assumed that

as *AD* is to *DB*, *AG* is to *GB*.

PROPOSITION 17

The same things being the case, let point *D* be on an asymptote.

I say that the line drawn from *Z* to *G* will be parallel to the asymptote on which the point *D* is situated.

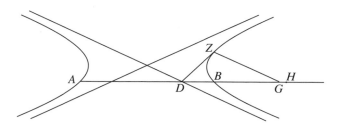

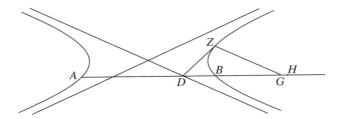

Let there be the same things as before, let point D be on one of the asymptotes, let a line be drawn through Z parallel to the asymptote, and, if possible, let it not fall on G, but on H. It will then be that

as AD is to DB, AH will be to HB (III.36),[31]

which is absurd. Therefore, the line from Z parallel to the asymptote will fall on G.

PROPOSITION 18

If in opposite sections a point is taken between the sections and from that point two lines are drawn cutting each of the sections, and if, as the lines between one of the sections and the point[32] are to those between the other section and the same point, so are lines greater than those cut off between the opposite section[33] to their excess over the latter, then the line drawn through the extremities[34] of the greater lines will intersect the sections, and the lines drawn from the points of intersection to the original point taken will touch the curves.

---

31 *Conics* III.36: "With the same things being so [namely, if some point be taken on one of the asymptotes of an hyperbola, and a straight line from it touch the section, and through the point of contact a parallel to the asymptote is drawn], if the straight line drawn across from the point neither cuts the section at two points nor is parallel to the asymptote it will meet the opposite section, and as the whole straight line is to the straight line between the section and the parallel through the point of contact, so will the straight line between the opposite section and the asymptote be to the straight line between the asymptote and the other section."

32 The point being that taken between the sections, and the lines, those drawn through the point.

33 These lines are the extension of the original two lines. Apollonius does not use the phrase ἐπ' εὐθείας here, as he does in IV.15, probably because he takes this to be obvious after the previous constructions.

34 The extremities are those on the side where these lines exceed the lines between the sections, that is, the extremities located inside the section.

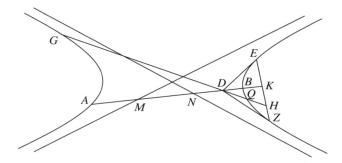

Let there be opposite sections *A*, *B*, and let point *D* be between the sections. Let it be assumed first that *D* be in the angle contained by the asymptotes, and through *D* let *ADB*, *GDQ* be drawn.

AD, therefore, is greater than *DB*,

while

GD is greater than *DQ*,

since

BN is equal to *AM* (II.16).[35]

Furthermore,

let *AK* have to *KB* the ratio that *AD* has to *DB*,

and

let *GH* have to *HQ* the ratio that *GD* has to *DQ*.

I say that the line through *K*, *H* intersects the sections, and the lines from *D* to the points of intersection[36] will touch the section.

For since *D* is inside the angle contained by the asymptotes, it is possible to draw two lines touching the section (II.49). Let *DE*, *DZ* be drawn,[37] and let *EZ* be joined. It will, thus, pass through points *K*, *H*.[38] For if it passes

---

[35] *Conics* II.16: "If in opposite sections some straight line is drawn cutting each of the straight lines containing the angle adjacent to the angles containing the sections, it will meet each of the opposite sections in one point only, and the straight lines cut off on it by the sections from the asymptotes will be equal."

[36] It is not entirely clear why Apollonius changes the order here, that is, from drawing lines from the points of intersection to *D*, as it is stated in the enunciation, to drawing lines from *D* to the points of intersection; mathematically, of course, the two formulations are equivalent.

[37] *DE* and *DZ* are understood to be the two lines touching the section just referred to.

[38] I am following Heiberg here in eliminating the next sentence in the Greek text, which reads "For if not, either it will pass through one of these [points] only, or through neither." It is possible that the interpolator was   *(continued on next page)*

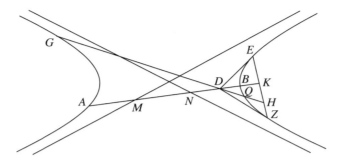

through one of these points only the other line will be cut in the same ratio by another point,[39] which is impossible. If it passes through neither point, the same impossibility will occur in both lines.

PROPOSITION 19

Let point *D* be taken then in the angle adjacent to that contained by the asymptotes, and let lines be drawn cutting the section and divided as said above.

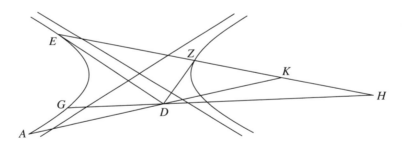

I say that the line produced through *K*, *H* will intersect each of the opposite sections, and the lines from the points of intersection to point *D* will touch the section.

For let *DE*, *DZ* be drawn from *D* touching each of the sections. Therefore the line through *E*, *Z* will pass through *K*, *H*. For if not, it will surely go

---

trying to strengthen the connection between this and propositions IV.9, 12, 13 where this same phrase appears.

[39] That is, two points will cut the same line segment into the same ratio.

through one of the two, or through neither, and again one will similarly infer from this an absurdity.[40]

PROPOSITION 20

If the point be taken on an asymptote, and the remaining things are the same, then the line drawn through the extremities of the lines exceeding[41] *will be parallel to the asymptote on which the point is situated, and the line drawn from the point to the point of intersection of the section and the line drawn through the extremities of the lines exceeding will touch the section.*

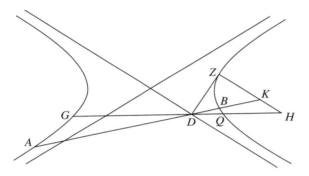

Let there be opposite sections *A*, *B*, and let point *D* be on one of the asymptotes, and let the remaining things be the same.

I say that the line through *K*, *H* intersects the section, and the line from the point of intersection to *D* will touch the section.

Let *DZ* be drawn from *D* touching the section, and a line be drawn from *Z*

---

[40] By *Conics* III.39, which states that "If two straight lines touching opposite sections meet, and a straight line is produced through the points of contact, and a straight line drawn from the point of meeting of the tangents cuts both of the sections and the straight line joining the points of contact, then as the whole straight line drawn across is to the straight line cut off outside between the section and the straight line joining the points of contact, so will the segments of the straight line produced by the segments and the point of meeting of the tangents be to each other."

[41] Recall, according to the construction in IV.18, to which IV.20 still refers, lines have been drawn through *D* and extended into the section; the "lines exceeding" are the parts of the former lines that are inside the section, that is, which *exceed* the part between the opposite sections.

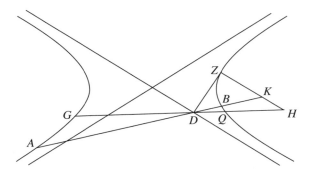

parallel to the asymptote on which D is situated. It will then pass through K, H. For if not, it will either pass through one of the two or neither, and the same absurdities will occur as before (III.36).

PROPOSITION 21

Again let there be opposite sections A, B, and point D on one of the asymptotes, let DBK, being parallel to one of the two asymptotes, meet the section at one point only, B,[42] but let GDQ meet both of the sections. Furthermore,

let GH be to HQ as GD is to DQ,

and

let DB be equal to BK.

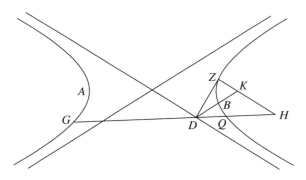

---

[42] Indeed, this must be the case by *Conics* II.13: "If in the place bounded by the asymptotes and the section some straight line is drawn parallel to one of the asymptotes, it will meet the section in one point only."

I say that the line through the points K, H will intersect the section and will be parallel to the asymptote on which point D is situated, and that the line drawn from the point of intersection to D will touch the section.

For let DZ be drawn touching the section, and let a line be drawn from Z parallel to the asymptote on which D is situated. It will thus pass through K, H. For if not, the absurdity spoken of before will occur (III.36).

PROPOSITION 22

Similarly, let there be opposite sections and their asymptotes and let D be similarly taken.[43] Let GDQ be taken cutting the sections, while DB be taken parallel to one of the two asymptotes. Moreover,

as GD is to DQ, let GH be to HQ,

and

let BK be equal to DB.

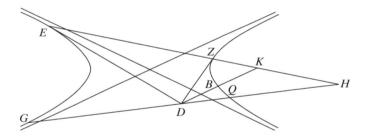

I say that the line through K, H will intersect each of the opposite sections, and the lines from the points of intersection to D will touch the section.

Let DE, DZ be drawn touching the section, let EZ be joined, and, if possible, let it not pass through K, H, but through one of the two points or

---

43 This seems to imply that D be taken on the asymptote; as Heiberg and others have pointed out, this part of the enunciation should read, "...and, similarly, let D be taken in the angle adjacent to that contained by the asymptotes." Halley, in his edition, makes the appropriate correction.

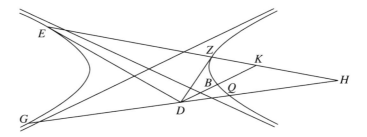

neither. If, on the one hand, it passes through *H* only,

DB will not be equal to *BK*

but to some other line, which is absurd (III.31).[44] If, on the other hand, it passes through *K* only, it will not be that

as *GD* is to *DQ*, *GH* is to *HQ*,

but, some line to some other line (III.36). If yet it passes through neither of the points *K*, *H*, then both impossibilities will occur.

PROPOSITION 23

Again let there be opposite sections *A*, *B*, and let point *D* be in the angle adjacent to that contained by the asymptotes. Let *BD* be drawn cutting section *B* at one point only, and thus parallel to one of the two asymptotes, and let *DA* be drawn similarly to section *A*,[45] and

let *DB* be equal to *BH*,

and

DA to *AK*.

I say that the line through *K*, *H* meets the sections and the lines drawn from the points of intersection to *D* will touch the sections.

Let *DE*, *DZ* be drawn touching the sections, let *EZ* be joined, and, if possible, let it not pass through *K*, *H*. So, either it will pass through one of the two points or through neither of them, and either

DA will not equal *AK*,

---

44 *Conics* III.31: "If two straight lines touching opposite sections meet, and a straight line is produced through the points of contact, and through the point of meeting a straight line is drawn parallel to the asymptote and cutting both the section and the straight line joining the points of contact, then the straight line between the point of meeting and the straight line joining the points of contact will be bisected by the section."

45 That is, *DA* cuts *A* at one point only and is parallel to the second asymptote.

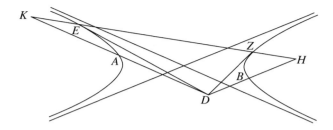

but some other line, which is absurd, or

DB will not equal *BH*,

or neither will equal neither, and again the same absurdity will occur in both situations (III.31).

Therefore, *EZ* will pass through *K, H*.[46]

PROPOSITION 24

A section of a cone will not meet a section of a cone or a circumference of a circle in such a way that part of them will be the same while another part will not be common.[47]

For, if possible, let the section of a cone *DABG* meet the circumference of the circle *EABG*, let the same part of these, *ABG*, be common, and let *AD* and *AE* not be common. Let point *Q* be taken on this part,[48] let *QA* be joined, and through an arbitrary point *E*[49] draw line *DEG* parallel to *AQ*. Moreover, divide *AQ* in two equal parts at *H*, and through *H* draw diameter *BHZ*. Therefore, the line through *B* parallel to *AQ* touches

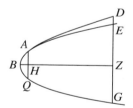

---

[46] With this, the first part of Book IV containing the various converses to *Conics* III.30–39 is finished. Now the book begins the treatment of its proper subject, the meeting of conic sections with one another.

[47] This is restated in *Conics* VI.6 as follows: "If there is a conic section, and part of it coincides with another part of another section so as to fit on it, then the [first] section is equal to the [second] section [*i.e.* they coincide throughout]."

[48] On *ABG*.

[49] *E* is an arbitrary point in the part *AE* or *AD*, that is, in the parts of the sections which do not coincide.

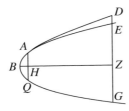

each of the sections, and also will be parallel to

DEG.[50] Also, in the one section

$$DZ \text{ will be equal to } ZG,$$

while in the other

$$EZ \text{ will be equal to } ZG \text{ (I.46–47),}[51]$$

so that, also,

$$DZ \text{ and } ZE \text{ are equal,}$$

which is impossible.

PROPOSITION 25

A section of a cone does not cut a section of a cone or circumference of a circle at more than four points.

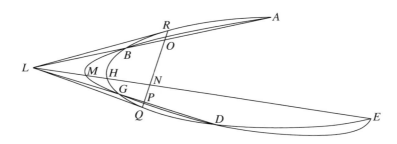

For, if possible, let them cut at five points A, B, G, D, E, and let the points of intersection A, B, G, D, E be taken in succession so that no point of intersection between them is left out, and let AB, GD be joined and produced. So,

---

[50] The first conclusion, that the line through B parallel to AQ touches the section, is by *Conics* I.32, while the second, that the line through B will be parallel to DEG is by Euclid's *Elements* I.30.

[51] *Conics* I.46 states: "If a straight line touching a parabola meets the diameter, then the straight line drawn through the point of contact parallel to the diameter in the direction of the section bisects the straight lines drawn in the section parallel to the tangent," while *Conics* I.47 states the analogous fact for an hyperbola, ellipse, and circumference of a circle, namely, "If a straight line touching an hyperbola or ellipse or circumference of circle meets the diameter, and if through the point of contact and the center a straight line is drawn in the direction of the section, then it bisects the straight lines drawn in the section parallel to the tangent."

these will intersect outside the sections in the cases of the parabola and hyperbola (II.24–25).[52] Let them intersect at L, and

let the ratio that AL has to LB be that which AO has to OB,

while

the ratio that DL has to LG be that which DP has to PG.

Therefore, the line joined from P to O and produced will intersect the section on each side, and the lines joining the points of intersection and L will touch the section (IV.9). So let it intersect at Q, R and let QL, LR be joined. Hence, they touch the section. Therefore, seeing that there is no point of intersection between B, G, the line EL cuts each of the sections.[53] Let it cut at M, H. Therefore, in one of the sections,

EN will be to NH as EL is to LH,

while in the other,

EN will be to NM as EL is to LM.

But this is absurd,[54] so that also what was assumed at the start is absurd.

---

[52] By *Conics* II.24 for the parabola and II.25 for the hyperbola. These state: "If two straight lines meet a parabola/hyperbola each at two points, and if a point of meeting of neither one of them is contained by the points of meeting of the other, then the straight lines will meet each other outside the section (but [in the case of the hyperbola] within the angle containing the section)."

[53] If the two curves met at a point between B and G, Apollonius could not immediately conclude that the line EL will have distinct intersections with the two curves. In this connection, it is also important to note an implicit dependence here on the convexity (to use a slightly anachronistic term) of the conic sections proven, essentially, in *Conics* I.10. In curves that are not convex, one might well have a situation such as that in the figure below, where EL does not pass through the segment BG.

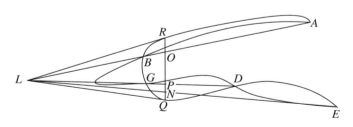

[54] From the two proportions

$EN : NH :: EL : LH$

and

$EN : NM :: EL : LM,$ (continued on next page)

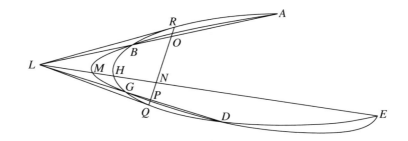

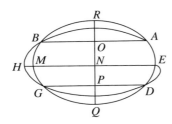

If *AB*, *DG* be parallel, the sections will, of course, be ellipses or a circumference of a circle. Let *AB*, *GD* be cut in two equal parts at *O*, *P*, and let *OP* be joined and produced on each side. It will, then, intersect the sections. So, let it intersect at *Q*, *R*. *QR* will then be a diameter of the sections, and *AB*, *GD* set up ordinatewise (II.28).[55] Let *ENMH* be drawn from *E* parallel to *AB*, *GD*. Therefore, *EMH* cuts *QR* and each of the curves, because there are no other intersections besides *A*, *B*, *G*, *D*.[56] In one of the sections, then,

NM will equal *EN*,

while in the other,

NE will equal *NH* (I def. 4),

so that

NM equals *NH*,

which is absurd.

---

it follows that

LH : NH :: EL : EN :: LM : NM,

or,

LH : HN :: LM : MN.

Hence, two distinct points *H* and *M* divide the segment *LM* in the same ratio. This is the absurdity to which Apollonius most likely refers.

55 "...τεταγμένως δὲ ἐπ' αὐτὴν κατηγμέναι αἱ ΑΒ, ΓΔ."

56 The intention must be that there are no other points of intersection through which a line from *E* might pass.

PROPOSITION 26

If, of the curves mentioned above,[57] *some touch at one point, then they will not meet each other in more than two other points.*

For let two of the above mentioned curves touch at point *A*.

I say that they will not meet in more than two other points.

For if possible let them meet at *B*, *G*, *D*, and let

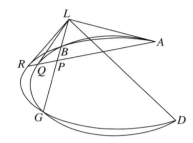

points of intersection be taken in succession with no point of intersection between them being left out. Let *BG* be joined and produced, and, from *A*, let *AL* be drawn touching the section. *AL* will, thus, touch the two sections and intersect *GB*.[58] Let it meet it at *L*, and let it be that

as GL is to LB, GP is to PB.

Let *AP* be joined and produced. It will, thus, intersect the sections and the lines drawn from the points of intersection to L will touch the section (IV.1).

Let it be produced and intersect at *Q, R,* and let *QL, LR* be joined. These will touch the section. Therefore, the line joining *D* to *L* will cut each of the sections, and the earlier mentioned absurdity will occur.[59] Therefore, the sections will not cut one another at more than two points.

If, in an ellipse or circumference of a circle, *GB* be parallel to *AL*, the demonstration will be similar to that given above[60] once *AQ* is shown to be a diameter.

---

[57] A section of a cone or circumference of a circle.

[58] Apollonius assumes that, in the case of the parabola and hyperbola, the point *A* is not contained by the points *B* and *G* (i.e., *A* does not lie on the closed figure contained by the section and line *BG*). Since this is unavoidable in the cases of the ellipse and circle, it is only with regards to those cases that Apollonius considers the possibility of *GB* being parallel to *AL*. Thus his remark at the end of the proof.

[59] That in the first part of IV.25.

[60] Namely, the demonstration in the second part of IV.25.

PROPOSITION 27

If, of the curves mentioned above, some touch one another at two points, they will not meet one another at another point.

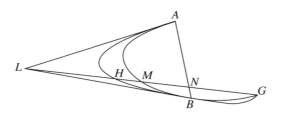

For let two of the curves mentioned above touch one another at two points A, B.

I say that they will not meet one another at another point.

For if possible, let them meet also at G, and, to start, let G be outside the points of contact A, B,[61] and let lines be drawn from A, B touching the sections. Therefore, they will touch both of the curves. Let them touch and be produced to L, as in the first drawing,[62] and let GL be drawn. It cuts, then, each of the sections. Let it cut them at H, M, and let ANB be joined. Therefore, in one of the sections

GN will be to NH as GL is to LH,

while in the other,

GN will be to NM as GL is to LM (III.37),

which is absurd.

PROPOSITION 28

If GH be parallel to the lines touching the section at points A, B, as in the ellipses in the second drawing,[63] then joining AB we conclude that it is a diameter (II.27),[64] so that each of the lines GH and GM are bisected at N

---

[61] Apollonius means, of course, that G is not to be in the segment bounded by A and B.

[62] The first diagram in IV.25.

[63] The second diagram in IV.25.

[64] *Conics* II.27: "If two straight lines touch an ellipse or circumference of a circle, and if the straight line joining the points of contact is through the center of the section, the tangents will be parallel; but if not, they will meet on the same side of the center." Of course, the second part of the enunciation ("but if not…") is equivalent to the converse of the first part, and it is this that Apollonius is applying in IV.28.

(I def. 4), which is absurd. Therefore, the curves do not meet one another at another point, but only at *A, B*.

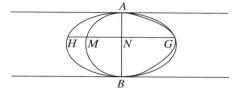

PROPOSITION 29

Let *G* be between the points of contact, as in the third drawing.[65]

It is evident that the curves do not touch one another at *G* since it has been

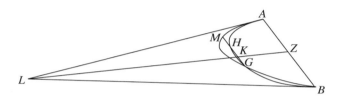

assumed that the curves touch at two points only. Indeed, let them cut one another at *G*. Let lines *AL, LB* be drawn from *A, B* touching the sections, let *AB* be joined and bisected at *Z*. Therefore, the line drawn from *L* to *Z* will be a diameter (II.29).[66]

The diameter will surely not pass through *G*. For if it did pass through, then the line drawn through *G* parallel to *AB* will touch each of the sections (II.5–6),[67] and this is impossible.[68]

---

[65] As is point *B* in relation to points *R* and *A* in IV.26.

[66] *Conics* II.29: "If in a section of a cone or circumference of a circle two tangents meet, the straight line drawn from their point of meeting to the midpoint of the straight line joining the points of contact is a diameter of the section." Therefore, *LZ* is a diameter, and, therefore, it falls on the center of the hyperbola.

[67] By *Conics* II.5 for the parabola or hyperbola and II.6 for the ellipse and circle. These state that "If the diameter of a parabola or hyperbola (ellipse or circumference of a circle) bisects some straight line (not through the center [in the case of the ellipse or circle]), the tangent to the section at the end of the diameter will be parallel to the bisected straight line."

[68] This is impossible because the curves were shown to cut and not touch one another at *G*. As in IV.25, this again depends on the convexity of the curves. For if the curves were not necessarily convex they might have a point where the tangent cuts the curve, or to use an anachronistic term, that they might have an inflection point at *G*, in which case the curves could have a common (continued on next page)

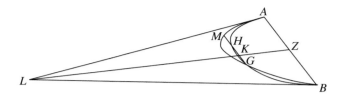

So, from *G* let *GKHM* be drawn parallel to *AB*. In the one section then

GK will be equal to KH,

while in the other

KM will be equal to KG,[69]

so that

KM is equal to KH,

which is impossible.

Similarly, if the lines touching the sections are parallel, the impossibility will be proven in the same way as above.[70]

PROPOSITION 30

A parabola cannot touch a parabola at more points than one.

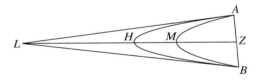

For, if possible, let parabolas *AHB*, *AMB* touch at *A*, *B*, and let lines *AL*, *LB* be drawn touching the parabolas. They will, thus, touch both sections and will intersect at *L*. Let *AB* be joined and bisected at *Z*, and let *AZ* be drawn.

---

tangent at *G*, yet also intersect one another at *G*. I might add that I know of no other curves having inflection points that Apollonius might have known other than some among Nicomedes's conchoids and Perseus's spiric curves. Moreover, it is not at all clear how thoroughly these curves were investigated. In particular, it is not clear that they were shown to have, in some cases, a tangent that cuts the curve.

[69] By the definition of the diameter given in Book I (def.4).

[70] As in IV.28. This statement is unnecessary since point *G* *must*, in this case, be in the segment bounded by *A, B*; the proof would not be similar, but *identical* to IV.28.

Now, since the two curves *AHB*, *AMB* touch one another at *A, B*, they will not meet each other at another point (IV.27–29), so that *LZ* cuts each of the sections. Let it cut at *H, M*. In the one section, then,

LH will be equal to *HZ* (I.35),[71]

while, in the other,

LM will equal *MZ*,

which is impossible.

Therefore, a parabola cannot touch a parabola at more points than one.

PROPOSITION 31

A parabola falling outside a hyperbola will not touch the hyperbola at two points.

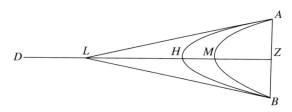

Let there be a parabola *AHB* and a hyperbola *AMB*, and, if possible, let them touch at *A, B*. Let lines be drawn from *A, B* touching each of the sections that touch at *A, B*, and let the lines intersect at *L*. Let *AB* be joined and bisected at *Z*, and let *LZ* be joined.

Now, since the sections *AHB*, *AMB* touch at *A, B*, they will not meet at another point; therefore, *LZ* cuts the sections at one and then another point. Let it cut them at *H, M*, and produce *LZ*. It will fall on the center of the hyperbola (II.29). Let *D* be the center. By the properties of the hyperbola,

as *ZD* is to *DM*, *MD* is to *DL*,

---

[71] *Conics* I.35: "If a straight line touches a parabola, meeting the diameter outside the section, the straight line drawn from the point of contact ordinatewise to the diameter will cut off on the diameter beginning from the vertex of the section a straight line equal to the straight line between the vertex and the tangent, and no straight line will fall into the space between the tangent and the section."

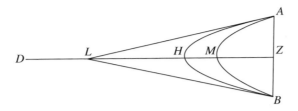

and

the remainders *ZM* to *ML* (I.37)[72].

Therefore,

ZM is greater than *ML*.

But by the properties of the parabola,

ZH is equal to *HL* (I.35),

which is impossible.[73]

PROPOSITION 32

A parabola falling inside an ellipse or circumference of a circle will not touch the ellipse or circumference of the circle at two points.

---

[72] Conics I.37 : "If a straight line touching an hyperbola or ellipse or circumference of a circle meets the diameter, and from the point of contact to the diameter a straight line is dropped ordinatewise, then the straight line cut off by the ordinate from the center of the section with the straight line cut off by the tangent from the center of the section will contain an area equal to the square on the radius of the section [*i.e.* the line from the center to the hyperbola or ellipse], and with the straight line between the ordinate and the tangent will contain an area having the ratio to the square on the ordinate which the transverse has to the upright." Referring to the figure above, the first part of this proposition states that

rect.*ZD,DL* = sq.*DM*,

from which it follows that

ZD : *DM* :: *DM* : *DL*,

and also that

ZD : *DM* :: *DM* : *DL* :: (*ZD* – *DM*) : (*DM* – *DL*) :: *ZM* : *ML*.

[73] Here we are using the fact that the parabola lies outside the hyperbola, for this means that

HZ > *MZ*.

So,

MZ > *ML* > *HL*

implies that

HZ > *HL*,

but

HZ = *HL*.

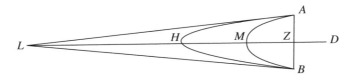

For let there be an ellipse or circumference of a circle *AHB* and a parabola *AMB*, and, if possible, let them touch at two points *A*, *B*, and let lines be drawn from *A*, *B* touching the sections and intersecting at *L*. Let *AB* be joined and bisected at *Z*, and let *LZ* be joined.

LZ will cut each section at one point then another, as we said above.

Let it cut them at *H*, *M*, and let *LZ* be produced to *D*, and let *D* be the center of the ellipse or circle.

Therefore, by the properties of the ellipse and circle,

$$\text{as } LD \text{ is to } DH, DH \text{ is to } DZ,$$

and

$$\text{the remainders } LH \text{ to } HZ \text{ (I.37).}$$

And

$$LD \text{ is greater than } DH.$$

Therefore,

$$LH \text{ is greater than } HZ.$$

But, by the properties of the parabola

$$LM \text{ is equal to } MZ \text{ (I.35),}$$

which is impossible.

PROPOSITION 33

A hyperbola will not touch a hyperbola, having the same center, at two points.

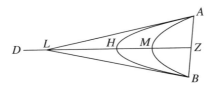

For let hyperbolas *AHB*, *AMB*, having the same center *D*, touch at *A*, *B*, if possible. Let lines *AL*, *LB* be drawn from *A*, *B* touching the hyperbolas and intersecting one another, and let *DL* be joined and produced.

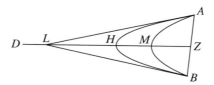

Moreover, let *AB* be joined. Therefore, *DZ* bisects *AB* at *Z*.[74] *DZ* then cuts the sections at *H, M* (IV.29). By the properties of the hyperbola *AHB*,

the rectangle contained by *ZD, DL*

will be equal to the square on *DH*,

while by the properties of the hyperbola *AMB*,

the rectangle contained by *ZD, DL*

will be equal to the square on *DM* (I.37).

Therefore,

the square on *MD* is equal to the square on *DH*,

which is impossible.

PROPOSITION 34

If an ellipse touches an ellipse or circumference of a circle, having the same center, at two points, then the line joining the points of contact falls on the center.

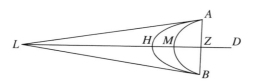

Let the above mentioned curves touch one another at points *A, B*. Let *AB* be joined, and let lines touching the sections be drawn through *A, B*, and, if possible, intersecting at *L*. Let *AB* be bisected at *Z*, and let *LZ* be joined. Therefore, *LZ* is a diameter of the sections (II.29).

If possible, let the center be *D*. Therefore,

the rectangle contained by *LD, DZ*

will be equal to the square on *DH*

---

[74] *DZ* is *DL* produced to *Z*; it bisects *AB* since, by *Conics* II.30, *DL* is a diameter and *AB* is drawn ordinatewise.

by the properties of the one section, but to the square on *MD* by the properties of the other, so that

the square on *HD* is equal to the square on *DM* (I.37),

which is impossible. Therefore, the lines from *A, B* touching the sections do not intersect. Therefore, they are parallel, and for the same reason, *AB* is a diameter (II.27), so that it falls on the center, which was to be proved.

PROPOSITION 35

A section of a cone or circumference of a circle will not intersect a section of a cone or circumference of a circle, not having its convexity in the same direction,[75] at more than two points.

For if possible let a section of a cone or circumference of a circle *ABG* meet a section of a cone or circumference of a circle *ADBEG*, not having its convexity in the same direction, at more points than two, *A, B, G.*

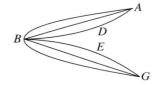

Since three points *A, B, G* have been taken on the curve *ABG*, if *AB, BG* are joined, they will contain an angle having its concavity in the same direction[76] as that of the curve *ABG*. For the same reason *ABG* contain an angle[77] whose concavity is in the same direction as that of curve *ADBEG*. Therefore, the curves we have been speaking of have both their concave and convex parts in the same direction, which is impossible.

PROPOSITION 36

If a section of a cone or circumference of a circle intersects one of the opposite sections at two points, and the curves between the points of

[75] Literally, "not having its convexities towards the same parts (μὴ ἐπὶ τὰ αὐτὰ μέρη τὰ κυρτὰ ἔχουσα)."

[76] "...ἐπὶ τὰ αὐτὰ τοῖς κοίλοις..."

[77] That is, "the lines *AB, BG* contain an angle..."; Halley has "*AB, BΓ*" in his text in place of "*ABΓ*."

intersection have their concavity in the same direction, then the curve, being produced at the points of intersection, will not intersect the other opposite section.

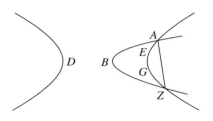

Let there be opposite sections *D*, *AEGZ*, and let there be a section of a cone or circumference of a circle *ABZ* intersecting one of the two opposite sections at two points *A, Z*, and let the sections *ABZ*, *AGZ* have their concavity in the same direction.

I say that *ABZ* produced will not intersect section *D*.

For let *AZ* be joined. Since *D*, *AGZ* are opposite sections and line *AZ* cuts a hyperbola in two points, produced, it will not intersect the opposite section *D* (II.33).[78] Neither, therefore, will the curve *ABZ* intersect section *D*.[79]

PROPOSITION 37

If a section of a cone or circumference of a circle intersects one of the opposite sections, it will not intersect the remaining section at more points than two.

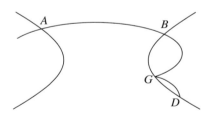

Let there be opposite sections *A*, *B*, and let a section of a cone or circumference of a circle *ABG* meet section *A*, and let *ABG* cut the opposite section *B* at points *B, G*.

I say that it will not intersect *BG* at another point.

---

[78] *Conics* II.33: "If a straight line meeting one of the opposite sections, when produced both ways, falls outside the section, it will not meet the other section, but will fall through the three places of which one is that contained by the angle containing the section, and two are those contained by the angle adjacent to the angle containing the section."

[79] Since, by *Conics* I.10, section *ABZ* will not meet *AZ* at any other point besides *A* and *Z*.

For if possible, let it intersect *BG* at *D*. Therefore, *BGD* meets section *BG*, not having its concavity in the same direction, at more points than two, which is impossible (IV.35).

This will be shown similarly if curve *ABG* touches the opposite section.

PROPOSITION 38

A section of a cone or circumference of a circle will not intersect opposite sections at more points than four.

This is evident from the fact that, intersecting one of the opposite sections, it cannot intersect the remaining section at more than two points (IV.37).

PROPOSITION 39

If a section of a cone or circumference of a circle touches one of the opposite sections in the concave part of the latter,[80] it will not intersect the other opposite section.

Let there be opposite sections *A, B*, and let *GAD* touch section *A*.

I say that *GAD* will not intersect *B*.

Let *EAZ* be drawn from point *A* touching the section. Then it touches each of the sections at point *A*, hence, it will not intersect *B*,[81] so that neither will *GAD* intersect *B*.

---

[80] In other words, so that the tangent is on the same side of both sections.

[81] This follows from *Conics* II.3, which states that "If a straight line touches an hyperbola, it will meet both of the asymptotes…" For if *EAZ* is produced, it will, by II.3, fall in the two angles adjacent to that contained by the asymptotes, and, therefore, it cannot meet the opposite section.

PROPOSITION 40

If a section of a cone or circumference of a circle touches each of the two opposite sections at one point, it will not intersect the opposite sections at another point.

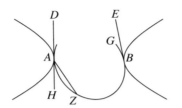

Let there be opposite sections *A, B*, and let a section of a cone or circumference of a circle touch each of the sections *A, B* at points *A, B*.

I say that the curve *ABG* will not intersect the sections *A, B* at another point.

Indeed, since curve *ABG* touches section A and meets (συμπίπτουσα) *B* at one point, it will not, therefore, touch *A* in the direction of its concavity. Similarly, it will be shown that neither will it touch B in the direction of its concavity. Let *AD, BE* be drawn touching sections *A, B*; these, then, will touch the curve *ABG*. For, if possible, let one of them cut the curve, and let it be *AZ*. Therefore, between *AZ*, touching the section *A*, and the section *A*, a line *AH* has been interposed, which is impossible.[82] Therefore, it touches *ABG*, and, because of this, it is evident that *ABG* does not meet opposite sections at another point.

PROPOSITION 41

If a hyperbola intersects one of the opposite sections in two points having its convexity turned oppositely, then the opposite section of the hyperbola will not intersect the other opposite section.

Let there be opposite sections *ABD, Z*, let hyperbola *ABG* meet *ABD* at points *A, B*, the former having its convexity turned oppositely to the concavity of the latter,[83] and let *E* be the opposite section of *ABG*.

---

82 By *Conics* I.36, which, besides giving the main property of tangents to central conic sections, tells us that, "...another straight line will not fall into the space between the tangent and the section of the cone."

83 Literally, "having convex [parts] turned oppositely to concave [parts]."

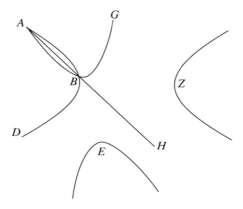

I say that *E* will not intersect *Z*.

Let *AB* be joined and extended to *H*. Since, indeed, the line *ABH* cuts the hyperbola *ABD*, and extended it falls outside each section, it will not intersect section *Z* (II.33). Similarly, because line *ABH* cuts the hyperbola *ABG*, it will not intersect opposite section *E*; neither, therefore, will *E* intersect *Z*.

PROPOSITION 42

If a hyperbola intersects each of two opposite sections, its opposite section will intersect neither of the opposite sections at two points.

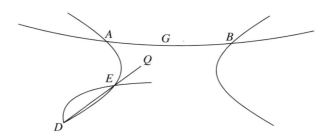

For let there be opposite sections *A, B*, and let hyperbola *AGB* intersect each of the opposite sections *A, B*.

I say that the opposite section of *AGB* will not intersect sections *A, B* at two points.

For, if possible, let it meet one of the opposite sections at *D, E*, and let *DE*

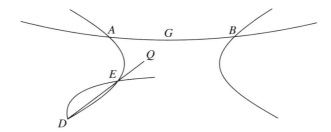

be joined and produced. Because of the section DE,[84] line DE will not intersect section AB (II.33), while, on the other hand, because of section AED,
DE will not intersect section B, since it passes through the three places
(II.33);[85] but this is impossible. Similarly it will be shown that AGB will not
intersect B at two points.

For the same reasons, neither will it touch either of the opposite sections.
For drawing QE touching, it will touch each of the sections, so that, because
of section DE, it will not intersect section AG, whereas, because of section
AE, it will not meet section B, so that neither will AG meet B, which is contrary to what was assumed.

PROPOSITION 43

If a hyperbola cuts each of two opposite sections at two points[86] *having its
convexity turned oppositely to each of them, the opposite section of the hyperbola will intersect neither of the opposite sections.*

Let there be opposite sections A, B, and let hyperbola $GABD$ cut each of
the sections A, B at two points containing convexities turned oppositely.

I say that its opposite section EZ intersects neither of the sections A, B.

---

84 That is, because line DE passes through two points of section DE, the opposite
section of section AGB.

85 These are the three places referred to in *Conics* II.33, namely, that contained by
the angle containing the section, and those contained by the angle adjacent to the
angle containing the section.

86 Here, one can see clearly that Apollonius does not treat the opposite sections as
just another conic section. If it were just another conic section, IV.25 would make
this proposition gratuitous.

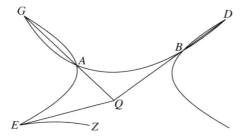

For, if possible, let it intersect section A at point E, and let GA, DB be joined and extended; these, then, will intersect one another (II.25). Let them intersect at Q; Q will then be in the angle contained by the asymptotes of section $GABD$ (II.25). And EZ is the opposite section of $GABD$. Therefore, the line joining E to Q will fall in the angle contained by AQB.[87] Again since GAE is a hyperbola, and GAQ, QE intersect, and the intersection points G, A do not contain E, point Q will be between the asymptotes of section GAE.[88] And BD is the opposite section of GAE. Therefore, the line from B to Q falls inside the angle contained by GQE[89]; which is impossible, for it also fell in the angle contained by AQB. Therefore, EZ will not intersect one of the opposite sections A, B.[90]

---

[87] There is no specific proposition being relied upon here. The argument, however, seems to be this. Since Q is within the angle contained by the asymptotes and E is a point on the opposite section, the line EQ must cut both asymptotes. Accordingly, EQ cuts the section AB at one point only (II.11). Now, if EQ were to fall outside the angle AQB it would either cut the section at two points or not at all. Therefore, it must fall within AQB. In fact, if P and R be points on the section such that QP and QR are parallel to the asymptotes, then the angle AQB will contain the angle PQR (II.14), and EQ will fall within PQR. A similar argument could be made if one or both of the lines AQ, BQ were tangent to the section, as is indeed the case in propositions 47, 48, and 51.

[88] As both Commandino and Halley observed (see Heiberg II p. 65), this conclusion does not follow from what has been stated. The problem is that one does not know what the fate of QE is after it intersects section GAE: if it intersects GAE at a second point or if it is tangent to GAE at E then Q will, indeed, be in the angle contained by the asymptotes of GAE, but if it intersects GAE only at E, and is not tangent, then Q need not be in the angle contained by the asymptotes.

[89] This is for the same reason EQ falls within AQB. See note 87, above.

[90] Since this proposition is problematic, as are some of the others like it, particularly IV.47, I am including the following alternative proof for IV.43 given by Eutocius in his commentary on the *Conics* (in Heiberg II p. 358): *(continued on next page)*

PROPOSITION 44

If a hyperbola cuts one of two opposite sections at four points, the opposite section of the hyperbola will not intersect the other of the two opposite sections.

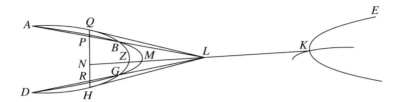

Let there be opposite sections *ABGD*, *E*, and let a hyperbola cut *ABGD* at four points, *A*, *B*, *G*, *D*, and let its opposite section be *K*.

I say that *K* will not intersect *E*.

For if possible, let it intersect at *K*. Let *AB*, *GD* be joined and produced; they will, then, intersect one another. Let them intersect at *L*, and

let *AP* have to *PB* the ratio that *AL* has to *LB*,

and

Let there be opposite sections *A*, *B*, let hyperbola *GABD* cut each of the opposite sections at the points *G*, *A*, *B*, *D*, and let the opposite section of this hyperbola be *EZ*. I say that *EZ* will meet neither of the opposite sections.

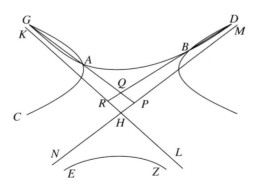

Let *DB*, *GA* be joined and extended, and let them meet at point *Q*. Therefore, *Q* will be (II.25) between the asymptotes of section *GAB* [i.e. in the angle contained by them]. Let the asymptotes of *GABD* be *KHL*, *MHN*. Clearly, then, lines *NH*, *HL* contain section *EZ* (II.15). Moreover, *GA* cuts section *GAC* at two points, *G*, *A*; produced in both directions, therefore, *GA* will not meet *DBO* (II.33), but will be between section *BO* and line *LH*. Similarly, line *DBQ* will not meet section *GAC*, but will be between section *AC* and line *HN*. Since *QP*, *QR*, which do not meet section *A*, *B*, contain the asymptotes *NH*, *HL*, they will, *a fortiori*, contain the section *EZ*, and *EZ* will nowhere meet the opposite sections [*A*, *B*].

let *DR* have to *RG* the ratio that *DL* has to *LG*.

Therefore, the line produced through *P*, *R* will intersect the sections on each side, and the lines from *L* to the points of intersection will touch the sections (IV.9). Let *KL* be joined and produced. It will cut the angle contained by *BLG* and the sections at one and then another point. Let it cut them at *Z*, *M*.

By the properties of the opposite sections *AQZH*, *K*,

as *NK* is to *KL*, *NZ* is to *ZL*,

while by the properties of the opposite sections *ABGD*, *E*,

as *NK* is to *KL*, *NM* is to *ML*,[91]

which is impossible.

Therefore *E*, *K* will not intersect one another.

PROPOSITION 45

If a hyperbola intersects one of two opposite sections at two points, having its concavity in the same direction as that section, while it intersects the other of the two opposite sections at one point, then the opposite section of the hyperbola will intersect neither of the opposite sections.

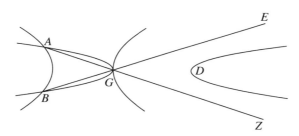

Let there be opposite sections *AB*, *G*, and let the hyperbola *AGB* intersect *AB* at points *A*, *B*, and let it intersect the section *G* at one point *G*, and let *D* be the opposite section of *AGB*.

I say that *D* will intersect neither of the sections *AB*, *G*.

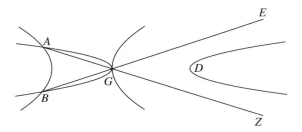

For let *AG*, *BG* be joined and produced. Therefore, *AG*, *BG* will not intersect section *D* (II.33). Neither will they intersect section *G* at another point besides *G*. For if they meet section *G* at another point, they will not intersect the opposite section *AB* (II.33), whereas it is assumed that they do intersect. Therefore, lines *AG*, *BG* meet section *G* at one point *G*, while they do not meet *D* at all. Therefore, *D* will be contained in the angle *EGZ*, so that section *D* will not intersect *AB*, *G*.

PROPOSITION 46

If a hyperbola meets one of two opposite sections at three points, the opposite section of the hyperbola will not intersect the other opposite section at more than one point.

Let there be opposite sections *ABG*, *DEZ*, and let hyperbola *AMBG* meet *ABG* at three points *A*, *B*, *G*, and let *DEK* be the opposite section of *AMG*.[92]

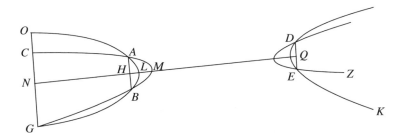

---

92 The text adds here, "while *DEZ* [is the opposite section] of *ABG*." This is an obvious and unnecessary interpolation (and Heiberg marks it as such in his edition), since it has already been said that *ABG* and *DEZ* are opposite sections.

I say that *DEK* will not meet *DEZ* at more points than one.

For if possible let them meet at *D*, *E* and let *AB*, *DE* be joined.

Now, they will either be parallel or not.

To start, let them be parallel, and let *AB*, *DE* be bisected at *H,Q*, and let *HQ* be joined; therefore, *HQ* is a diameter for all the sections (II.36), and *AB*, *DE* are set up ordinatewise. Let *GNCO* be drawn from *G* parallel to *AB*. It will, then, be drawn ordinatewise to the diameter and it will cut the sections, one and then the other. For if it were to cut them at the same point, the sections would no longer meet at three points, but four. In section *AMB*, then,

> *GN* will be equal to *NC*,

while in *ALB*,

> *GN* will be equal to *NO*.

And, therefore,

> *ON* is equal to *NC*,

which is impossible.

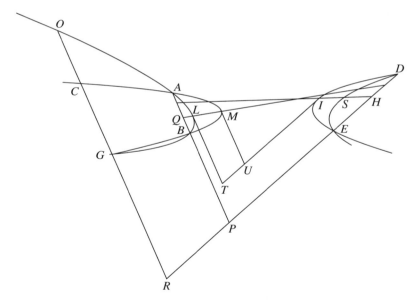

So, let lines *AB*, *DE* not be parallel, but, being produced, let them intersect at point *P*. Let *GO* be drawn parallel to *AP* and let it intersect *DP* produced at point *R*, and let *AB*, *DE* be bisected at *H, Q* [sic], through *H, Q* let diameters *HSI*, *QLM* be drawn, while from *I, L, M* let *IUT*, *MU*, *LT* be drawn touching the sections; *IT*, then, will be parallel to *DP*, while *LT*, *MU* will be parallel to *AP*, *OR* (II.5).

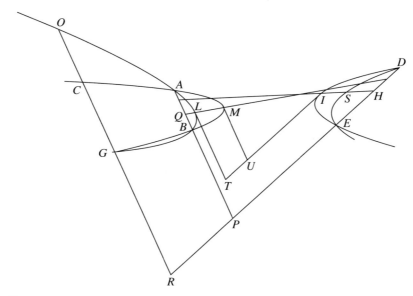

Since,

as the square on *MU* is to the square on *UI*,
the rectangle contained by *AP*, *PB* is to the rectangle contained by *DP*, *PE*
(III.19),[93]

but,

as the rectangle contained by *AP*, *PB*
is to the rectangle contained by *DP*, *PE*,
the square on *LT* is to the square on *TI*,

and, therefore,

as the square on *MU* is to the square on *UI*,
the square on *LT* is to the square on *TI*.

For the same reasons,

as the square on *MU* is to the square on *UI*,
the rectangle contained by *CR*, *RG*
is to the rectangle contained by *DR*, *RE*,

---

93 *Conics* III.19: "If two straight lines touching opposite sections meet, and parallels to the tangents are drawn cutting each other and the section, then, as the squares on the tangents are to each other, so will the rectangle contained by the straight lines between the section and the point of meeting of the straight lines be to the rectangle contained by the straight lines taken similarly." Where I have written "the rectangle contained by *AP, PB*," and "*DP, PB*," the text has "*APB*" and "*DPB*." Similarly, where I have written "the rectangle contained by *CR, RG*" and "*DR, RE*," the text has "*CRG*" and "*DRE*." Unless noted otherwise, I shall continue this convention.

and

as the square on *LT* is to the square on *TI*,
the rectangle contained by *OR, RG*
is to the rectangle contained by *DR, RE*.

Therefore,

the rectangle contained by *OR, RG* is equal to
the rectangle contained by *CR, RG*,

which is impossible.[94]

PROPOSITION 47

If a hyperbola touches one of two opposite sections, while it cuts the other at two points, then the opposite section of the hyperbola will intersect neither of the opposite sections.

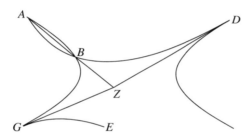

For let there be opposite sections *ABG, D,* and let some hyperbola *ABD* cut *ABG* at *A, B,* and touch section *D* at point *D,* and let *GE* be the opposite section of *ABD.*

I say that *GE* intersects neither of the opposite sections *ABG, D.*

For, if possible, let it intersect *AB* at *G,* and let *AB* be joined, and let a line be drawn through *D* touching section *ABD* and intersecting line *AB* at *Z.*

---

[94] For since

$$\text{sq.}MU : \text{sq.}UI :: \text{sq.}LT : \text{sq.}TI,$$

we have

$$\text{rect. } CR, RG : \text{rect. } DR, RE :: \text{rect. } OR, RG : \text{rect. } DR, RE,$$

and, therefore, by Euclid V.9,

$$\text{rect. } CR, RG = \text{rect. } OR, RG,$$

so that, by Euclid VI.1,

$$CR = OR.$$

This is impossible because *O* and *C* are different points, as Apollonius emphasizes at the start of the proof.

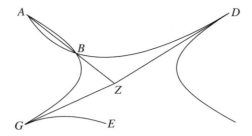

Therefore, point Z will be inside the angle contained by the asymptotes of section ABD (II.25 porism). And GE is the opposite section of ABD. Therefore, the line from G to Z falls inside the angle contained by BZ, ZD.[95] Again, since ABG is a hyperbola, and AB, GZ intersect, and the points of intersection A, B, do not contain G, point Z is between the asymptotes of section ABG.[96] And D is the opposite section of ABG. Therefore, the line from D to Z falls in the angle contained by AZ, ZG, which is absurd.[97] For, it fell in the angle contained by BZ, ZD.

Therefore, GE does not intersect one of the opposite sections ABG, D.

PROPOSITION 48

If a hyperbola touches one of two opposite sections at one point, while it intersects it at two, then the opposite section of the hyperbola will not intersect the other opposite section.

Let there be opposite sections ABG, D, and let some hyperbola AHG touch ABG at point A, and let it intersect ABG at points B,G, and let E be the opposite section of AHG.

I say that E will not intersect D.

For, if possible, let it intersect at point D, let BG be joined and extended to Z, and let line AZ be drawn from A touching the section.

As in the earlier demonstration, it will be shown that point Z is inside that angle contained by the asymptotes (II.25 porism). Moreover, AZ will touch

---

[95] The reasoning here is similar to that in IV.43: see note 87 to that proposition.

[96] As in IV.43, and for the same reasons, this is incorrect.

[97] The result does not follow.

both sections, and *DZ* produced will cut the sections at points *H, K* between points *A, B*.[98]

Let *GL* have to *LB* that ratio which

GZ has to ZB,

and let *AL* be joined and produced; it will cut the sections, one and then the other (IV.1). Let it cut them at *N, M*. The lines, therefore, from *Z* to *N, M* will touch the sections (IV.1), and, as in the earlier demonstration, by the properties of the one section,

CK is to KZ as CD is to DZ,

while, by the properties of the other section,

CH is to HZ as CD is to DZ (III.37),[99]

which is impossible.

Therefore, it does not intersect the opposite section.

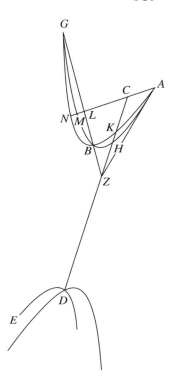

PROPOSITION 49

If a hyperbola, touching one of two opposite sections, intersects the same at another point, then the opposite section of the hyperbola will not intersect the other opposite section at more points than one.

Let there be opposite sections *ABG, EZH*, and let some hyperbola *DAG* touch *ABG* at point *A*, and let it cut *ABG* at *G*, and let *EZQ* be the opposite section of *DAG*.

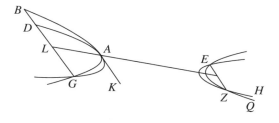

---

[98] The reasoning here is similar to that in IV.43: see note 87 to that proposition.

[99] See note 91, above.

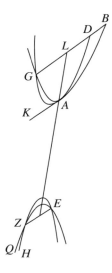

I say that it will not intersect the other opposite section at more points than one.

For if possible let it meet at two points *E, Z,* and let *EZ* be joined, and through *A* let *AK* be drawn touching the sections.

Now, *EZ* and *AK* will be parallel or not parallel.

To start, let them be parallel, and let the diameter bisecting *EZ* be drawn; therefore, it will pass through *A,* and it will be the diameter of the two conjugates (II.34).[100] Let line *GLDB* be drawn through *G* parallel to *AK, EZ.* Therefore, it will cut the sections at one point and then another. In the one section then

GL will be equal to *LD,*

whereas in the remaining section

GL will be equal to *LB.*

This is impossible.

So, let *AK, EZ* not be parallel, let them intersect at *K,* and let *GD,* drawn parallel to *AK,* intersect *EZ* at *N.* Let *AM,* bisecting *EZ,* cut the sections at *C, O,* and let *CP, OR* be drawn from *C, O,* touching the sections.

Therefore,

as the square on *AP* is to the square on *PC,*

so is the square on *AR* to the square on *RO,*[101]

and, for this reason,

as the rectangle contained by *DN, NG*

is to the rectangle contained by *EN, NZ,*

the rectangle contained by *BN, NG*

is to the rectangle contained by *EN, NZ.*[102]

---

100 *Conics* II.34: "If some straight line touch one of the opposite sections and a parallel to it be drawn in the other section, then the straight line drawn from the point of contact to the midpoint of the parallel will be a diameter of the opposite sections." Referring to the common diameter of the two opposite sections as the "diameter of *the two conjugates* (διάμετρος τῶν δύο συζυγῶν)," is unusual and somewhat misleading. Nevertheless, the phrase occurs again in IV.50, 53.

101 Since *PC* and *OR* are parallel.

102 By *Conics* III.19,

sq. *AP* : sq. *PC* :: rect. *DN, NG* : rect. *EN, NZ*

and

sq. *AR* : sq. *RO* :: rect. *BN, NG* : rect. *EN, NZ,*

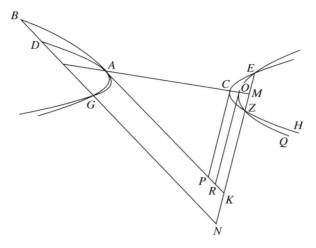

Therefore,

the rectangle contained by *DN, NG*

is equal to the rectangle contained by *BN, NG,*

which is impossible.

PROPOSITION 50

If a hyperbola touches one of two opposite sections at one point, the op-posite section of the hyperbola will not intersect the other opposite section at more points than two.

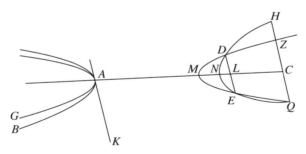

Let there be opposite sections *AB, EDH,* and let a hyperbola *AG* touch *AB* at *A*, then let *EDZ* be the opposite section of *AG*.

---

so that, since

sq. *AP* : sq. *PC* :: sq. *AR* : sq. *RO,*

we have

rect. *DN, NG* : rect. *EN, NZ* :: rect. *BN, NG* : rect. *EN, NZ.*

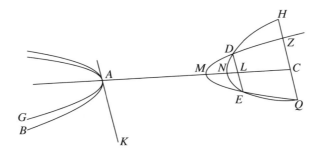

I say that *EDZ* will not intersect *EDH* at more points than two.

For, if possible, let *EDZ* meet *EDH* at three points *D, E, Q*, let *AK* be drawn touching sections *AB*, *AG*, let *DE* be joined and extended, and, to start, let *AK*, *DE* be parallel. Let *DE* be bisected at *L*, and let *AL* be joined. *AL* will then be a diameter for the two conjugates (II.34) and will cut the sections between *D, E*, at points *M, N*.[103] Let *QZH* be drawn from *Q* parallel to *DE*.

In the one section, then,

QC will equal *CZ*,

while in the other,

QC will equal *CH*,

so that also

CZ equals *CH*,

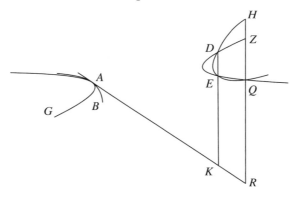

which is impossible.

So let *AK*, *DE* not be parallel, but let them intersect at [point] *K*, and let the remaining constructions be the same. Let *AK* be produced and let it meet *ZQ* at point *R*.

---

[103] Here, there is an interpolation which adds, "so that *DLE* is bisected at [point] *L*," but since this has *already* been said, the interpolation in fact adds nothing.

As before, we will show that, in section *ZDE*,

> the rectangle contained by *ZR*, *RQ* is to the square on *RA*

> as the rectangle contained *DK*, *KE* is to the square on *AK*,

whereas in section *HDE*,

> the rectangle contained by *HR*, *RQ* is to the square on *RA*

> as the rectangle contained *DK*, *KE* is to the square on *AK* (III.19).

Therefore,

> the rectangle contained by *HR*, *RQ*

> is equal to the rectangle contained by *ZR*, *RQ*,

which is impossible.[104]

Therefore, *EDZ* does not meet *EDH* at more points than two.

PROPOSITION 51

*If a hyperbola touches each of two opposite sections, the opposite section
of the hyperbola will intersect neither of the opposite sections.*

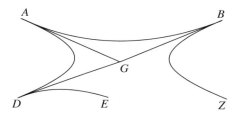

Let there be opposite sections *A*, *B*, and let hyperbola *AB* touch each of
these at points *A*, *B*, and let the opposite section of *AB* be *E*.

I say that *E* will intersect neither of the sections *A*, *B*.

For, if possible, let it intersect *A* at *D*, and let lines be drawn from points
A, *B*, touching the sections; they will intersect one another in the angle con-
tained by the asymptotes of section *AB* (II.25 porism). Let them intersect at
G, and let *GD* be joined.

---

[104] Since

$$\text{rect. } ZR, RQ : \text{sq. } RA :: \text{rect. } HR, RQ : \text{sq. } RA,$$
$$\text{rect. } ZR, RQ = \text{rect. } HR, RQ,$$

and hence,

$$ZR = HR,$$

which is impossible.

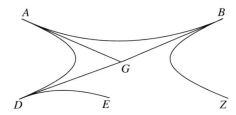

The line *GD*, therefore, will be in the place between *AG, GB*.[105] But it is between *BG, GZ*,[106] which is absurd.

Therefore, *E* does not intersect *A, B*.

PROPOSITION 52

If each of two opposite sections touch each of two opposite sections at one point, each having its concavity in the same direction, then they will not intersect at another point.

For let opposite sections touch one another at points *A, D*.

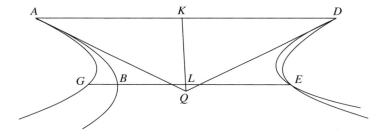

I say that they will not meet at another point.

For, if possible, let them meet at *E*. Since, indeed, a hyperbola touching one of the opposite sections at *D* has intersected it at *E*, therefore, section *AB* will not meet section *AG* at more points than one (IV.49).[107] Let *AQ, QD* be drawn from *A, D* touching the sections, let *AD* be joined, let *EBG* be drawn

---

[105] The reasoning here is similar to that in IV.43: see note 87 to that proposition.

[106] Since *BG* touches section *BZ*, it does not touch section *AD*, and, therefore, *DG* must fall within the angle contained by *GB* and the extension beyond *G* of *AG*.

[107] *AB* and *AG* are, of course, the opposite sections, respectively, of the hyperbola and of one branch of the opposite sections that the hyperbola meets.

through E parallel to AD, and let the second diameter QKL of the opposite sections be drawn from Q (II.38).

It will then bisect AD at K. And, therefore, EB, EG will be bisected at L (II.39).[108] Therefore,

$$BL \text{ is equal to } LG,$$

which is impossible.

Therefore, the sections will not intersect at another point.

PROPOSITION 53

If a hyperbola touches one of two opposite sections at two points, the opposite section of the hyperbola will not intersect the other opposite section.

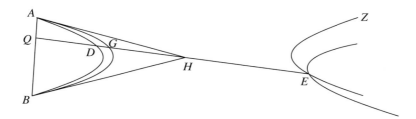

Let there be opposite sections ADB, E, and let hyperbola AG touch ADB at two points A, B, and let Z be the opposite section of AG.

I say that Z will not intersect E.

For, if possible, let it intersect at point E, and let AH, HB be drawn from A, B, touching the sections, let AB and EH be joined, and let EH be produced. It will cut the sections at one point and then another;[109] let it be as $EHGDQ$.

---

[108] *Conics* II.39: "If two straight lines meeting touch opposite sections, the straight line drawn through the center and the point of meeting of the tangents bisects the straight line joining the points of contact." Since QLK has been drawn as a second diameter, it must pass through the center of the sections, and, therefore, by II.39, bisect AD. This means, also, that AD is drawn ordinatewise with respect to diameter QLK (since, by *Conics* II.41, no other line through K can be bisected by QLK), so that EBG, being parallel to AD, are also drawn ordinatewise and, therefore, bisected by QLK.

[109] IV.27 shows that there can be no intersection point between A and B through which $EHGDQ$ might pass.

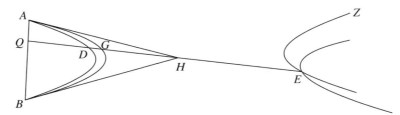

Since *AH, HB* indeed touch the sections, and *AB* joins the points of contact, in one of the conjugate sections

QD will be to *DH* as *QE* is to *EH*,[110]

while in the other,

QG will be to *GH* as *QE* is to *EH*,

which is impossible.

Therefore, section *Z* does not meet section *E*.

PROPOSITION 54

If a hyperbola touches one of two opposite sections, having its convexity turned oppositely, then the opposite section of the hyperbola will not intersect the other opposite section.

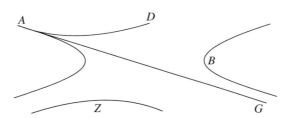

For let there be opposite sections *A, B*, and let some hyperbola *AD* touch sections *A* at point *A*, and let the opposite section of *AD* be *Z*.

I say that *Z* will not intersect *B*.

Let line *AG* be drawn from point *A* touching the sections; therefore, because of the properties of section *AD*, *AG* will not intersect *Z*, while, because of the properties of section *A*, it will not intersect *B* (II.33), so that *AG* falls between sections *B, Z*. It is evident then that *B* will not intersect *Z*.

PROPOSITION 55

Opposite sections will not intersect opposite sections at more points than four.

For let there be one pair of opposite sections *AB, GD*, and another pair of opposite sections *ABGD, EZ*, and, to start, let *ABGD* cut each of *AB, GD* at four points *A, B, G, D* containing convexities turned oppositely, as in the first drawing. Therefore, the opposite section of *ABGD*, that is, *EZ*, will not intersect *AB, GD* (IV.43).

But let *ABGD* cut *AB* at *A,B* and *G* at one point *G*, as in the second drawing. Therefore *EZ* does not intersect section *G* (IV.41). If *EZ* meets *AB*, it will meet at one point only; for if it meets it at two points, its opposite section, *ABG*, will not meet the other opposite section *G* (IV.43). But it has been assumed that it meets it at one point *G*.

If, as in the third drawing, *ABG* cuts *ABE* at two points *A, B*, while *EZ* meets *ABE* at one point, *EZ* will not intersect section *D* (IV.41), whereas, intersecting *ABE*, it will not intersect *ABE* at more points than two.[111]

If, as in the fourth drawing, *ABGD* cuts each of two opposite sections at one point, *EZ* will intersect neither at two points (IV.42). [So that by the things already said, and their converse,

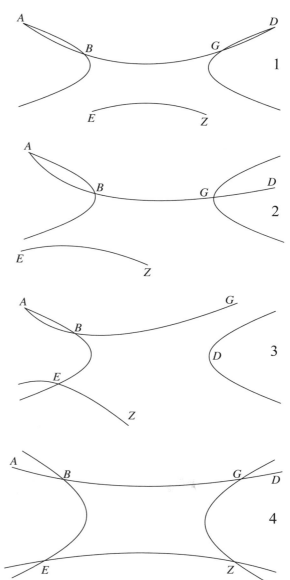

---

[111] Since *ABE* meets *ABG*, it cannot, by IV.37, meet the opposite section of *ABG*, i.e., *EZ*, at more points than two.

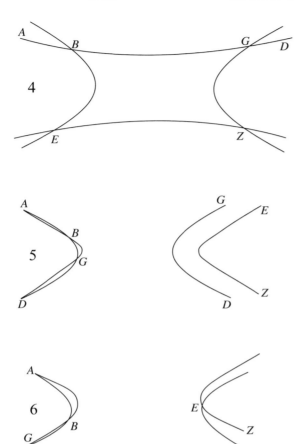

ABGD, GZ will not meet the opposite sections *BE, EZ* at more points than four.][112]

If the sections have their concavities in the same direction, and one cuts the other at four points *A, B, G, D*, as in the fifth drawing, *EZ* will not intersect the other opposite section (IV.44). Of course, *EZ* will not intersect *AB*; for, again, *AB* will not intersect the opposite sections *ABGD, EZ* at more points than four (IV.38); neither will *GD* intersect *EZ*.

If, as in the sixth drawing, *ABGD* meets the other section at three points, *EZ* will intersect the other at one point only (IV.46).

And we will say the same things as before for the remaining cases.

So, since what was proposed is clear in all possible configurations (διαστολὰς), opposite sections will not meet opposite sections at more points than four.

PROPOSITION 56

If opposite sections touch opposite sections at one point, they will not intersect at more than two other points.

Let there be opposite sections *AB, BG* and others, *D, EZ*,[113] let *BGD* touch *AB* at *B*, let them have their convexity turned oppositely, and,

112 With Heiberg, one must see the bracketed sentence as useless and suspect.

113 In this formulation, we are to understand that *D* is the opposite section of *AB*, and *EZ* is the opposite section of *BG*. Since exactly the same wording is used in IV.57 but with the usual order of the sections (*i.e.*, the section and its opposite being paired together), I suspect the inversion here is a scribal error.

first, let *BGD* intersect *GD* at two points *G, D*, as in the first figure (σχήματος).

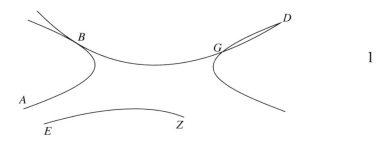

1

Indeed, since *BGD* cuts *GD* at two points having its convexity turned oppositely, *EZ* will not intersect *AB* (IV.41). Again, since *BGD* touches *AB* at *B*, having its convexity turned oppositely, *EZ* will not intersect *GD* (IV.54). Therefore, *EZ* will not intersect either of the sections *AB, GD*; therefore, the sections will meet at the two points *G, D*, only.

But let *BG* cut *GD* at one point *G*, as in the second figure. Therefore, *EZ* will not intersect *GD* (IV.54), whereas it will intersect *AB* at one point only.[114] For if *EZ* meets *AB* at two points, *BG* will not intersect *GD* (IV.41). But it was assumed that they meet at one point.

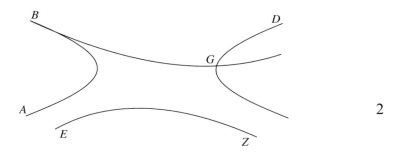

2

---

[114] The argument which follows shows only that *EZ* will intersect *AB* at one point *at most*. This, however, is sufficient for what Apollonius wants to show, namely, that when opposite sections touch at one point they can intersect again at two other points *at most*.

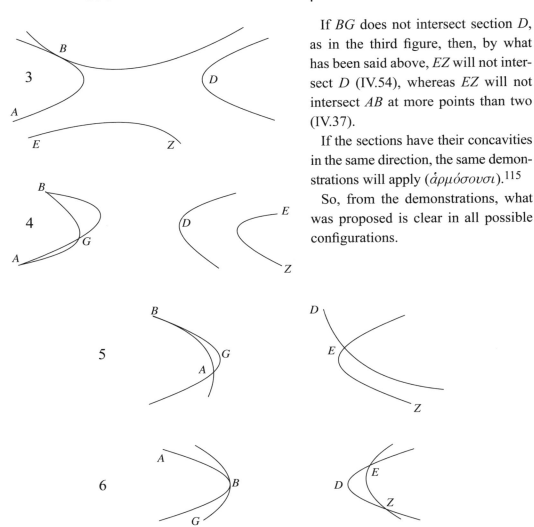

If *BG* does not intersect section *D*, as in the third figure, then, by what has been said above, *EZ* will not intersect *D* (IV.54), whereas *EZ* will not intersect *AB* at more points than two (IV.37).

If the sections have their concavities in the same direction, the same demonstrations will apply (ἁρμόσουσι).[115]

So, from the demonstrations, what was proposed is clear in all possible configurations.

PROPOSITION 57

If opposite sections touch opposite sections at two points, they will not intersect at another point.

Let there be opposite sections *AB, GD*, and others, *AG, EZ*, and, first, let them touch at *A, G*, as in the first figure.

---

[115] The cases presented in diagrams 4–6 are proven by IV.48, 49, 50, respectively.

Indeed, since *AG* touches each of the sections *AB*, *GD* at points *A*, *G*, therefore *EZ* will intersect neither of the sections *AB*, *GD* (IV.51).

So, let them touch as in the second figure. It will be proved similarly that *GD* will not intersect *EZ* (IV.53).

So, let *GA* touch *AB* at *A* and let *D* touch *EZ* at *Z*, as in the third figure.[116] Indeed, since *AG* touches *AB* having its convexities turned oppositely, *EZ* will not intersect *AB*. Again, since *ZD* touches *EZ*, *GA* will not intersect *DZ*.

If *AG* touches *AB* at *A*, while *EG* touches *GD* at *G*, having their concavities in the same direction, as in the fourth figure, they will not intersect at another point (IV.52). *EZ* will not even intersect *AB*.

So, from the demonstrations, what was proposed is clear in all possible configurations.

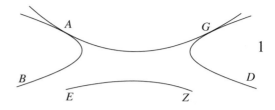

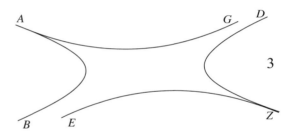

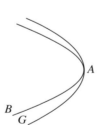
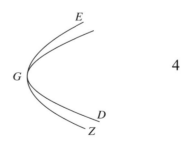

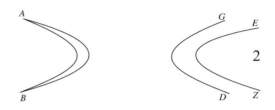

---

116 In fact, this case cannot occur by IV.54.

APPENDIX A:

On Three and Four Line Loci

The three-line locus property of conics is easily deduced for the ellipse, hyperbola, parabola and circle from III. 54; and for the opposite sections from III. 55 and 56. The three-line locus property of conics can be stated thus:

> Any conic section or circle or pair of opposite sections can be considered as the locus of points whose distances from three given fixed straight lines (the distances being either perpendicular or at a given constant angle to each of the given straight lines, although the constant angle may be different for each of the three straight lines) are such that the square of one of the distances is always in a constant ratio to the rectangle contained by the other two distances.

It is shown in III. 54 that in the case of conic sections and circles
rect. *AF, CG* : sq. *AC* ::
sq. *EB* : sq. *BD* comp. rect. *AD, DC* : fourth sq. *AC*.

Now if we consider the straight lines *AD, DC,* and *AC* as fixed and given and therefore straight line *DE* fixed and given as bisecting *AC,* then it is evident that the straight lines *AC, EB, BD, AD, DC,* and therefore the squares on them and the rectangles contained by them, are also fixed and given. Then although as the point *H* is taken at different points along the conic, the straight lines *AF* and *CG* change in magnitude, nevertheless the magnitude of the rectangle *AF, CG,* because of the above proportion remains constant.

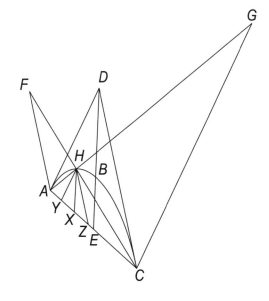

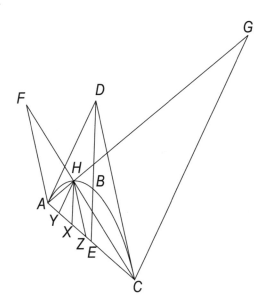

For let *HX* be drawn parallel to *BE*, and *HY* to *AD*, and *HZ* to *DC*. Then *HX* is the distance from *H* to *AC* at a given angle, and *AY* because of parallels represents the distance from *H* to *AD* at another given angle, and *ZC* represents the distance from *H* to *DC* at another given angle. Then by similar triangles

$$CZ : ZH :: AC : AF,$$
$$AY : YH :: AC : CG;$$

therefore compounding

$$\text{rect. } CZ, AY : \text{rect. } ZH, YH ::$$
$$\text{sq. } AC : \text{rect. } AF, CG.$$

Now we have seen that the rectangle *AF, CG* is a constant magnitude as the point *H* changes, and the square on *AC* is constant; therefore their ratio is constant. Therefore

$$\text{rect. } CZ, AY : \text{rect. } ZH, YH \text{ is a constant ratio.} \tag{1}$$

Again by similar triangles

$$ZH : HX :: CD : DE,$$
$$YH : HX :: AD : DE;$$

therefore compounding

$$\text{rect. } ZH, YH : \text{sq. } HX :: \text{rect. } CD, AD : \text{sq. } DE.$$

But rectangle *CD, AD* and the square on *DE* are constant magnitudes as the point *H* changes; therefore their ratio is constant. Therefore

$$\text{rect. } ZH, YH : \text{sq. } HX \text{ is a constant ratio.} \tag{2}$$

Compounding (1) and (2), we get a constant ratio, that is

$$\text{rect. } CZ, A Y : \text{sq. } HX \text{ is a constant ratio.}$$

In other words, as the point *H* changes, the rectangle contained by the distances from *H* to two of the given straight lines (at given angles to those straight lines) has a constant ratio to the square on the distance to the third straight line (at a given angle to that straight line). And it can easily be proved by means of similar triangles that if any other three angles are chosen for the distances, than those chosen here for the demonstration, then the corresponding ratio will be constant, although not equal.

The four-line locus property can be easily deduced from the three-line. If to any conic section we construct four tangents *AG, BE,*

AI, and *EC,* and the straight lines *FG, GI, ID,* and *DF,* joining the points of contact; and draw the distances from any point *H* on the conic to

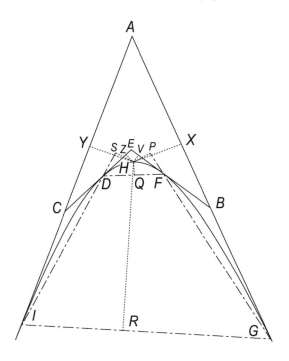

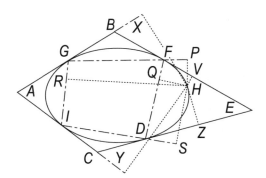

these straight lines at any given angles (perpendiculars are convenient), then by the three-line locus property with respect to triangle *FBG* for any point *H* on the conic

$$\text{rect. } HX, HV : \text{sq. } HP \text{ is constant;} \qquad (\alpha)$$

with respect to triangle *AIG*

$$\text{rect. } HX, HY : \text{sq. } HR \text{ is constant;} \qquad (\beta)$$

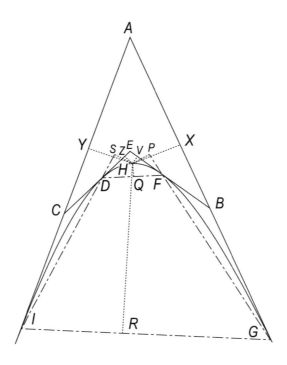

with respect to triangle *DCI*

      rect. *HY, HZ* : sq. *HS* is constant;   (γ)

with respect to triangle *EFD*

      rect. *HZ, HV* : sq. *HQ* is constant.   (δ)

It will be noticed that we have taken in succession a pair of adjacent tangents and the straight line joining their points of contact. It will also be noticed that the rectangles in the four ratios present a cyclical arrangement, so that if the inverse of (α) is compounded with (β), and the inverse of (γ) with (δ), we would have two constant ratios

      pllpd. *HY, HP, HP* : pllpd. *HV, HR, HR,*   (ε)

      pllpd. *HV, HS, HS* : pllpd. *HY, HQ, HQ.*   (ζ)

Again compounding the first of these with the second, we would have finally

      rect. *HP, HS* : rect. *HQ, HR,* a constant ratio.

And this is the property of the four-line locus, namely the locus of points *H* such that the rectangle contained by the distances from points *H* to any two given fixed straight lines *FG* and *ID* has to the

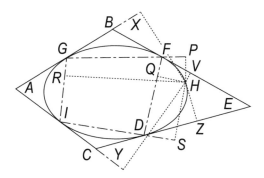

rectangle contained by the distances from *H* to two other fixed straight lines, *IG, FD,* a constant ratio.

The rigorous method of effecting these compoundings is as follows. For inverting (α), by Eucl. XI. 32 we have the constant ratios

sq. *HP* : rect. *HX, HV* :: pllpd. *HP, HP, HY* : pllpd. *HX, HV, HY,*

rect. *HX, HY* : sq. *HR* :: pllpd. *HX, HY, HV* : pllpd. *HR, HR, HV.*

Hence, by definition, the ratio (a constant one) compounded of these two is

pllpd. *H Y, HP, HP* : pllpd. *HV, HR, HR.*

And in the same way we find the constant ratio compounded of the inverse of (γ) and (δ). Now

pllpd. *HY, HP, HP* : pllpd. *HV, HR, HR* ::

HY : HV comp. sq. *HP* : sq. *HR,*

pllpd. *HV, HS, HS* : pllpd. *HY, HQ, HQ* ::

HV : HY comp. sq. *HS* : sq. *HQ.*

If then we take two lines *M* and *N* such that

$$HP : HR :: HR : M, \qquad (\eta)$$
$$HS : HQ :: HQ : N, \qquad (\theta)$$

then

sq. *HP* : sq. *HR* :: *HP* : *M,*

sq. *HS* : sq. *HQ* :: *HS* : *N.*

Hence

ratio *HY : HV* comp. sq. *HP* : sq. *HR* = ratio *HY : H* comp. *HP : M*

ratio *HV : HY* comp. sq. *HS* : sq. *HQ* = ratio *HV : HY* comp. *HS : N.*

But

rect. *HY, HP* : rect. *HV, M* :: *HY : HV* comp. *HP : M,*

rect. *HV, HS* : rect. *HY, N* :: *HV : HY* comp. *HS : N;*

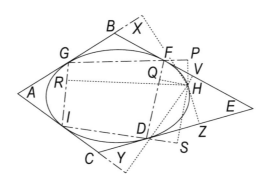

and
> pllpd. *HY, HP, HS* : pllpd. *HV, M, HS* :: rect. *HY, HP* : rect. *HV, M,*
> pllpd. *HV, HS, M* : pllpd. *HY, N, M* :: rect. *HV, HS* : rect. *HY, N;*

and these are constant ratios. Hence compounding, we get the constant ratio
> pllpd. *HY, HP, HS* : pllpd. *HY, N, M,*

which is the same as the constant ratio
> rect. *HP, HS* : rect. *N, M.*

Now, taking *L* and *O* as some constants,
> rect. *HP, HS* : rect. *N, M* :: *L* : *O*

and
> rect. *HP, HS* : rect. *HR, HQ* :: rect. *HR, HQ* : rect. *M, N*

by compounding (η) and (θ). But equal ratios have equal duplicate ratios (Heath's note to Euclid, VI. 22) and hence
> rect. *HP, HS* : rect. *HR, HQ* is constant.

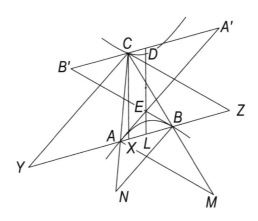

In the case of opposite sections, it is shown in III. 56

$$\text{rect. } MA, BN : \text{sq. } AB ::$$
$$\text{sq. } LD : \text{sq. } DE \text{ comp. rect. } AE, EB : \text{fourth sq. } AB.$$

Then it is evident for the same reasons as before that for different points C the magnitudes MA and BN may change, but the rectangle MA, BN is a constant magnitude.

For as before, let CX be drawn parallel to DE, CY to EA, CZ to EB. By similar triangles

$$AY : YC :: AB : BN,$$
$$BZ : ZC :: AB : MA;$$

therefore compounding

$$\text{rect. } AY, BZ : \text{rect. } YC, ZC :: \text{sq. } AB : \text{rect. } MA, BN.$$

Since rectangle MA, BN is constant as C changes, and also the square on AB is constant, therefore

$$\text{rect. } AY, BZ : \text{rect. } YC, ZC \text{ is a constant ratio.} \qquad (1)$$

Again by similar triangles

$$ZC : CX :: EB : EL,$$
$$YC : CX :: EA : EL,$$

therefore compounding

$$\text{rect. } YC, ZC : \text{sq. } CX :: \text{rect. } EB, EA : \text{sq. } EL.$$

Hence

$$\text{rect. } YC, ZC : \text{sq. } CX \text{ is a constant ratio.} \qquad (2)$$

Compounding (1) and (2), we have a constant ratio

$$\text{rect. } AY, BZ : \text{sq. } CX.$$

But AY and BZ are equal to CA' and CB', the distances from C. This is the property of the three-line locus of section C with respect to the straight lines EA and EB tangents to the other section, and EB the straight line joining their points of contact. And so one opposite section is a three line-locus to the tangents to the other of the opposite sections. That it is also a four-line locus could be shown in the same way as before.

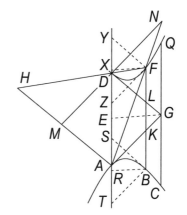

Again from III. 55 we can conclude that both of the opposite sections together are a three-line locus to the triangle formed by a tangent to each of the sections and the straight line joining their points of contact. For by III. 55

$$\text{rect. } HA, DN : \text{sq. } AD :: \text{rect. } AG, GD : \text{sq. } CG.$$

Now since the three last terms of this proportion

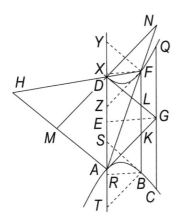

are evidently constants as the point *F* changes, therefore also, although *HA* and *DN* change with *F*, yet rectangle *HA*, *DN* remains constant in magnitude. Then reproducing the figure of III. 55, we drop *YF* parallel to *DL*, and *FZ* to *KA*, and *FX* to *GE*, where *E* is the midpoint of *AD*. Then by similar triangles

$$YD : FY :: AD : HA,$$
$$AZ : FZ :: AD : DN;$$

therefore compounding,

rect. *YD, AZ* : rect. *FY, FZ* :: sq. *AD* : rect. *HA, DN*.

But the last two terms are constant, therefore rect. *YD, AZ* : rect. *FY, FZ* is a constant ratio.   (1)
Again by similar triangles

$$FY : FX :: DG : EG,$$
$$FZ : FX :: AG : EG;$$

therefore compounding,

rect. *FY, FZ* : sq. *FX* :: rect. *DG, AG* : rect. *ED, EG*.

But the last two terms are constant, therefore

rect. *FY, FZ* : sq. *FX* is a constant ratio.   (2)

Compounding (1) and (2), we see that

rect. *YD, AZ* : sq. *FX* is a constant ratio.

But this is the definition of a three-line locus—that the rectangle contained by the distances from any point on the locus to two fixed straight lines have to the square on the distance to a third fixed straight line a constant ratio. But

$$DY = LF,$$
$$AZ = KF,$$

and *FX* is the distance from *F* to *AD*. And so the ratio fulfills the definition.

Furthermore, if we consider *B* the point of intersection of the straight line *KF*, drawn parallel to *AD*, with the other opposite section, and draw *BS* parallel to *FY*, *BR* to *FX*, and *BT* to *FZ*, since they are parallels between parallels,

$$BR = FX,$$
$$KF = AZ,$$
$$TA = BK.$$

But it was shown in the course of III. 55 that

$$BK = LF,$$

$$BL = KF.$$

Hence

$$TA = BK = YD = LF,$$
$$AZ = KF = BL.$$

Therefore

$$\text{rect. } LF, KF : \text{sq. } FX :: \text{rect. } BK, BL : \text{sq. } BR.$$

Hence any point B on one opposite section fulfills the same constant ratio with respect to its distances from the three fixed lines AD, GD, and AG as any point F on the other opposite section.

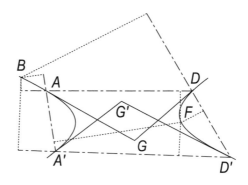

It can be similarly deduced that the opposite sections are together a four-line locus with respect to any four fixed straight lines joining points, two on each section (the points being four points of contact of four tangents, and the straight lines, the straight lines joining them).

To sum up, a parabola, ellipse, circle, and hyperbola are three-line loci with respect to any two tangents to them and a straight line joining the points of contact. One opposite section is also a three-line locus with respect to any two tangents to the other section together with the straight line joining their points of contact. The two opposite sections together are a three line-locus with respect to two tangents, each to one of the sections, together with the straight line joining their points of contact.

The parabola, ellipse, circle, and hyperbola are four-line loci with respect to any inscribed quadrilateral. One opposite section is also a four-line locus with respect to any quadrilateral inscribed in the other section. The two opposite sections together are a four-line locus to any four straight lines joining four points, two lying on each opposite section.

R. Catesby Taliaferro

APPENDIX B:

Corollary to Proposition I.38

The following text was appended to Proposition 38 of Book I as an extrapolation or corollary (Heiberg labels it a corollary in his Latin translation, although it is not so labelled in the Greek text). Left in line in the text, it creates great confusion both because it is not part of the proof of the proposition as Apollonius stated it and because the theorem or corollary is itself incorrectly stated and incorrectly proved. It seems likely that the extrapolation is not originally from Apollonius. For this edition we have removed the extrapolation from the text of I. 38 to this Appendix, where it is reproduced with the diagrams we provided for I. 38. (The mistranslation is corrected in this text so that what appears here is what the Greek text actually says.) We follow it with a correct statement and valid proof for the proportion asserted. (Ed.)

COROLLARY TO I. 38

With the same things supposed, it remains to be shown that, as the straight line between the tangent and the end of the (second) diameter on the same side with the dropped straight line is to the straight line between the tangent and the second diameter, so is the straight line between the other end and the dropped straight line to the straight line between the first end and the dropped straight line.

For since

$$\text{rect. } FG, GH = \text{sq. } GC = \text{rect. } CG, GD$$
$$\text{(in proof of I. 38),}$$

for

$$CG = GD,$$

therefore

$$\text{rect. } FG, GH = \text{rect. } CG, GD;$$

therefore

$$FG : GD :: CG : GH.$$

And *convertendo*

$$GF : FD :: GC : CH.$$

And let the doubles of the antecedents be taken; but

$$2GF = CF + FD$$

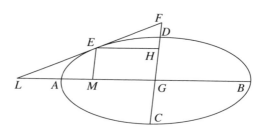

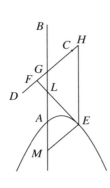

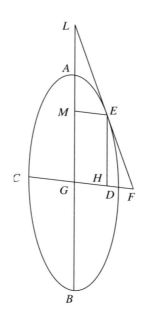

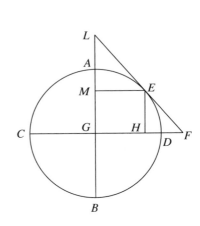

because

$$CG = GD,$$

and

$$2GC = CD;$$

therefore

$$CF + FD : FD :: DC : CH.$$

And *separando*

$$CF : FD :: DH : HC;$$

and this was to be shown.

Then it is clear from what has been said that the straight line *EF* touches the section, either if

$$\text{rect. } FG, \ GH = \text{sq. } GC,$$

or if

$$\text{rect. } FH, \ HG : \text{sq. } HE$$

in the ratio we said [i.e., the upright : the transverse]; for it could be shown conversely.

REVISION OF COROLLARY TO I.38

With the same things supposed, it remains to be shown that the intersections made with the second diameter by the tangent and by the dropped straight line divide the second diameter in the same ratio, in the case of the ellipse, taking the segments in the same order (that is, from the same vertices), and in the case of the hyperbola, taking them in reciprocal order (that is, from opposite vertices).

In the figure given in the proposition,
for the ellipse,
$$CF : FD :: CH : HD;$$
and for the hyperbola,
$$CF : FD :: DH : HC.$$

For since
$$\text{rect. } FG, \ GH = \text{sq. } GC = \text{rect. } CG, \ GD \text{ (as proved in I. 38),}$$
$$FG : GD :: CG : GH. \tag{1}$$

ELLIPSE

By (1) *convertendo*
$$FG : FD :: CG : DH.$$
Doubling the antecedents
$$2FG : FD :: 2CG : DH.$$
But
$$2FG = 2GD + 2DF = CF + FD$$
$$2CG = CD.$$
Therefore
$$CF + FD : FD :: CD : HD$$
and *separando*
$$CF : FD :: CH : HD.$$

$$\text{Q.E.D.}$$

HYPERBOLA

By (1) *invertendo*
$$GD : FG :: GH : CG;$$
and *separando*
$$FD : FG :: CH : CG;$$

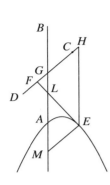

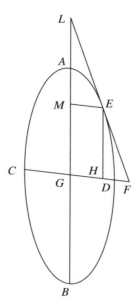

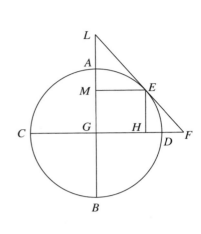

and *invertendo* again

$$FG : FD :: CG : CH.$$

Doubling the antecedents

$$2FG : FD :: 2CG : CH.$$

But[*]

$$2FG = 2(GD - FD) = 2GD - 2DF$$
$$= CD - 2DF = CF - DF;$$

and

$$2CG = CD.$$

Therefore

$$CF - DF : DF :: CD : HC$$

and *componendo*

$$CF : DF :: DH : HC.$$

Q.E.D.

---

[*] The following step assumes that F lies between G and D. If it doesn't, then you must add GD and FD to get FG. (Ed.)

Bibliography

Apollonii Pergaei conicorum libri quatuor.... *Quae omnia nuper Federicus Commandinus...e Graeco convertit & commentariis illustravit,* F. Commandino, editor. Bologna: Alexander Benatius, 1566.

Apollonii Pergaei Conicorum libri octo, Edmund H. Halley, editor. Oxford: Theatrum Sheldonianum, 1710. (Greek text and Latin translation of *Conics* Books I–VII, "restoration" of Book VIII.)

Apollonii Pergaei Quae Graece Exstant cum Commentariis Antiquis, J. L. Heiberg, editor. 2 Vols. Stuttgart: B. Tuebner, 1974. (This is the Greek text of Books I–IV and commentaries by Pappus and Eutocius, edited by Heiberg and originally published in 1891 and 1893 with his Latin translation.)

Apollonius (Heath), *Treatise on Conic Sections, Edited in Modern Notation* by T. L. Heath. Cambridge: Cambridge University Press, 1896. (This is Heath's retelling of Conics with his commentary, not a translation.)

Apollonius of Perga, *Conics*, translation by R. Catesby Taliaferro. Annapolis: St. John's College, *The Classics of the St. John's Program* series, 1939.

Apollonius of Perga, *Conics*, translation by R. Catesby Taliaferro. Chicago: University of Chicago and Encyclopædia Britannica Inc., 1952 (formerly in Volume 11, Great Books of the Western World).

Apollonius (ver Eecke), *Les Coniques d'Apollonius de Perge*. French translation by Paul ver Eecke. Paris: Libraire Scientifique et Technique, Albert Blanchard, 1963.

Apollonius (Toomer), *Conics Books V to VII, The Arabic Translation of the Lost Greek Original in the Version of the Bānū Mūsā,* edited with translation and commentary by G. J. Toomer. 2 vols. New York: Springer-Verlag, 1990.

Apollonius (Taliaferro), *Conics, Books I–III*, tr. R. Catesby Taliaferro. Santa Fe: Green Lion Press, 1997.

Apollonius (Fried), *Conics, Book IV,* tr. Michael N. Fried. Santa Fe: Green Lion Press, 2002

Apollonius de Perge, *Coniques*, edited by Roshdi Rashed, Micheline Decorps-Foulquier, and Michel Federspiel. 4 vols. Berlin/New York: Walter de Gruyter, 2008–2010. (Greek and Arabic texts of Books I–IV with French translations and mathematical and philological commentaries.)

Archimedes (Heiberg and Stamatis), *Opera Omnia*, ed. by J. L. Heiberg and E. S. Stamatis. 3 vols. Stuttgart, 1972.

Dictionary of Scientific Biography (*DSB*), Charles Coulston Gillispie, Editor. New York: Charles Scribner's sons, 1970–1980.

Euclid (Heath), *The Thirteen Books of Euclid's Elements. Translated from the text of Heiberg with introduction and commentary by Sir Thomas L. Heath.* 3 vols. Cambridge: Cambridge University Press, 1926 (repr., New York: Dover Publications, Inc., 1956).

Euclid, *Euclid's Elements, all thirteen books complete in one volume. The Thomas L. Heath translation.* Santa Fe, New Mexico: Green Lion Press, 2013).

Euclid (Heiberg), *Elementa post I. L. Heiberg*, ed. E. Stamatis. 5 vols., Stuttgart: B. G. Teubner, 1969–77.

Fried, Michael N., and Sabetai Unguru, *Apollonius of Perga's* Conica: *Text, Context, Subtext* (Leiden: Brill, 2001).

Heath, Sir Thomas, *A History of Greek Mathematics*, 2 vols. Oxford: At the Clarendon Press, 1921 (repr. New York: Dover Publications, Inc., 1981).

Klein, J., *Greek Mathematical Thought and the Origin of Algebra*, tr. by Eva Brann. Cambridge, Mass.: The M.I.T. Press, 1968.

Knorr, W. R., *The Ancient Tradition of Geometric Problems*. Boston: Birkhauser, 1986 (repr. New York: Dover Publications, Inc., 1993).

Liddell, H. G. and Scott, R. (revised by Jones, H. S.), *A Greek-English Lexicon* (*LSJ*). Oxford: At the Clarendon Press, 1940.

Lloyd, G. E. R, *Polarity and Analogy: Two Types of Argumentation in Early Greek Thought*. Cambridge: At the University Press, 1971.

Mugler, Charles, *Dictionaire historique de la terminologie géometrique des Grecs*. Paris, 1959.

Proclus, *Proclus: A Commentary on the First Book of Euclid's Elements*. Glenn R. Morrow, tr. Princeton: Princeton University Press, 1970.

Sarton, George, *A History of Science*. Cambridge: Harvard University Press, 1959.

Sarton, George, *Introduction to the History of Science*. Huntington: Robert E. Krieger Publishing Co., 1975.

Taisbak, C. M., "Elements of Euclid's *Data*," in *Peri Ton Mathematon*, special issue of *Apeiron*, vol. XXIV, no. 4 (1991), pp. 135–171.

Toomer, G. J., "Apollonius of Perga," in *DSB*, I (1970), pp. 179–193.

Unguru, S., "On the Need to Rewrite the History of Greek Mathematics," *Archive for History of Exact Sciences*, 1975, 15, pp. 67–114.

Unguru, S., "History of Ancient Mathematics: Some Reflections of the State of the Art," *Isis* 70 (1979), pp. 555–565.

Zeuthen, H. G., *Die Lehre von den Kegelschnitten im Altertum.* Kopenhagen, 1886 (repr. Hildesheim: Georg Olms Verlagsbuchhandlung, 1966).

Index

abscissa, defined, 38*n*. (I.20)

analysis and synthesis, method used,
154–157, 159–169 (II.44, II.46, II.47,
II.49–51)

angle containing the section,
notes on 307, 315, 326, 330
see also asymptotes

Apollonius the younger, 117

Archimedes, 290*n*.

asymptotes, 1, 118 (II.2)
of opposite sections, 129–130 (II.15)
of conjugate opposite sections, 131
(II.17)
notes on 292, 295–297, 299–300,
306–307, 310–312, 327, 331–332
see also angle containing the section

Attalus, 289

axial triangle, first constructed, 7 (I.4)
named, 8 (I.4)

axis
of cone, defined, 3 (Def. 2)
of conic section,
defined, 4 (Defs. 7–8)
to find, 155-157 (II. 46–47)
of conic surface, defined, 1 (Def. 1)

base of the cone, defined, 3 (Def. 2)

center
of conic sections,
notes on 292, 399, 314,
318–319, 322 , 345

of hyperbola and ellipse,
defined, 36 (Def. 9)
of opposite sections,
defined, 36 (Def. 10)
to find, 155 (II.45)

circle (circumference of a circle), 289
notes on 289, 292, 298, 314, 317–319,
322
as base of cone,
defined, 3 (Def. 2)
as conic section,
first constructed, 7–8 (I.4),
9 (I.5)
used in generation of conic surface,
3 (Def. 1)

Commandino, F., 331*n*.

concavity,
as an indication of the orientation
of a conic section, 328*n*.

cone, defined, 3 (Def. 2)

cone, oblique, defined, 3 (Def. 3)

cone, right, defined, 3 (Def. 3)

conic sections,
in general, 289
notes on 289, 292, 298, 313–315, 317,
319, 328
coincident, 289, 313*n*.
cutting,
notes on 319–320
intersecting,
notes on 290, 316, 319–320, 331,
345, 349
meeting, 289
notes on 289, 313, 315, 332,
344, 347

conic sections, *cont.*
   touching (tangent),
      notes on 319, 349

conic surface, defined, 3 (Def. 1)

conjugate axes, defined, 4 (Def. 8)
   constructed, 114 (I.60)

conjugate diameter (second diameter),
   345*n.*
   defined, 4 (Def. 6)
   constructed, 33 (I.16)

conjugate opposite sections, 114–115 (I.60),
   340*n.*
   have same property as other conic
      sections, 197 (III.15)

Conon of Samos, 289–290

convexity,
   as a property of conic sections,
      315*n*, 319*n.*
   as an indication of the orientation of a
      conic section,
      notes on 325, 328

diagram,
   figure (*schēma*), 348–351
   drawing (*katagraphē*), 300, 318–320,
      347–348

diameter, defined, 3 (Defs. 4 and 5),
      4 (Defs. 6 and 8)
   first constructed, 12–15 (I.7)
   first named, 15 (I.7 porism)
   to find, 154 (II.44)
   notes on 292, 299, 314, 319–322, 324,
      340, 345

diameter, transverse, defined, 3 (Def. 5)
   of opposite sections, 26 (I.14)

diameter, upright, defined, 3 (Def. 5)

diorismos (limits of possibility), 290–291

dropped straight line,
   explained, 67 (I.38)

ekthesis (setting out), 333*n.*

ellipse, defined, 24, 26 (I.13)
   constructed, 24–26 (I.13)
   to find center of, 155 (II.45)
   to find axis of, 156 (II.47)
   notes on 292, 299, 314, 317–319, 322

Euclid, 314*n.*

Eudemus of Pergamum, 1, 117, 289.

Eutocius of Ascalon, 41, 60, 61, 80, 88, 91,
      100, 104, 111, 188, 204
   note on 331–332

figure, the (εἶδος),
   first introduced, 21 (I.12)

first definitions, 1

focus, *see* points of application

Halley, E., notes on 311, 325, 331

Heath, T. L., notes on 289, 298

Heiberg, J. L.,
   notes on 290, 292, 300, 307–308, 311,
      331, 333–334, 348

hyperbola, defined, 21–22 (I.12)
   constructed, 22–23 (I.12)
   to find center of, 155 (II.45)
   to find axis of, 156 (II.47)
   notes on 293, 294, 295, 296, 297,
      299, 306, 314–315, 317,
      319, 322, 327, 332, 344

Knorr, W. R., 290*n.*

latus rectum (ὀρθία),
   defined, 21*n.* (I.12)

locus properties, 267–275

Nicomedes, notes on 290, 320

Nicoteles of Cyrene, 290

opposite sections (see also hyperbola),
    defined, 26 (I.14)
    constructed, 26–29 (I.14)
    diameter of, 33 (I.16)
    center of, 36 (Def. 10)
    constructing as problems, 113 (I.59),
        114 (I.60)
    in general, 289
        notes on 289, 292, 295, 303,
            306–309, 312, 326–327, 330–334,
            336, 340, 344–345, 347–349
    as two hyperbolas,
        notes on 303, 330

ordinate (see also dropped straight line),
    defined, 3*n*. (Def. 4)

ordinatewise, defined, 3 (Def. 4)

parabola,
    defined, 19, 21 (I.11)
    constructed, 19–21 (I.11)
    to find axis of, 155 (II.46)
    notes on 299, 314–315, 317, 319,
        321–322

parameter,
    defined, 21*n*. (I.11)

points of application, 249n. (III.45)

point of contact,
    notes on 292, 295–296, 306, 309, 312,
        314, 318–319, 321–322, 340

principal diameter,
    defined, existence established, 15
        (I.7 porism)
    named, 95 (I.51 porism)

problems, 290, 291*n*.

protasis (enunciation),
    notes on 298, 307, 311, 318

radius of section, defined, 36 (Def. 9)
    used in construction, 77 (I.43)

second definitions, 36

second diameter, defined, 36 (Def. 11)

subcontrariwise, named, 9 (I.5)

subcontrary section, defined, 9 (I.5)

tangent,
    first appears, first named, 36 (I.17)

tangent to a conic section (line touching
        a conic section),
    first appears, first named, 36 (I.17)
    notes on 292–293, 295–296, 298–299,
        306–307, 309, 312, 314, 318–319,
        321–322, 327–328, 331, 333, 336,
        340, 344–345

Toomer, G. J., notes on 289, 291

triangle as conic section,
    constructed, 6–7 (I.3)

transverse side, 322*n*.

upright diameter, defined, 3 (Def. 5)

upright side (*latus rectum*),
    21 (I.11), 322*n*.

ver Eecke, P.,
    notes on 290–291, 297

vertex
    of cone, defined, 3 (Def. 2)
    of conic section,
        defined, 3 (Def. 4), 321*n*.
    of conic surface,
        defined, 3 (Def. 1)

About the Author

Apollonius of Perga was born about 262 BCE in Perga, on the southern coast of what is now Turkey. He spent some time in Alexandria, and is said to have died about 190 BCE; however, little is known about his life with any degree of certainty. He wrote many mathematical works, chiefly on geometry but also on number theory and calculation, almost none of which have survived. His fame chiefly rests on his *Conics,* as well as his correspondence with Archimedes. Through the astronomer Claudius Ptolemy we learn that Apollonius also contributed to the development of the epicycle/deferent model of planetary motion.

About the Translators

R. Catesby Taliaferro (1907–1989), American mathematician and philosopher, was appointed by St. John's College, Annapolis, to translate Ptolemy's *Almagest,* Apollonius's *Conics,* and Augustine's *De Musica* for the college's fledgling Great Books program. He subsequently taught at the University of Notre Dame. His translations of Ptolemy and Apollonius were included in Volume 16 of the Britannica *Great Books of the Western World,* in 1952. His Augustine translation was published in Vol. 4 of *Fathers of the Church* (Cima Publishing Co., 1947). He also wrote *The Concept of Matter in Descartes and Leibniz* (University of Notre Dame Press, 1964).

Michael N. Fried (1960–) is associate professor in the Program for Science and Technology Education at Ben Gurion University of the Negev. His undergraduate degree in the liberal arts is from St. John's College in Annapolis, Maryland where he read the "great books," books he still reads and loves. He received his M.Sc. in applied mathematics from SUNY at Stony Brook and his Ph.D. in the history of mathematics from the Cohn Institute at Tel Aviv University. His books include *Apollonius of Perga's Conica: Text, Context, Subtext* (Brill, 2001), and *Edmond Halley's Reconstruction of the Lost Book of Apollonius's Conics: Translation and Commentary* (Springer 2011), as well as his translation of Book IV of the Conics (originally published as a separate volume by Green Lion Press in 2002).